Ex Libris
SARGE
HANSON
BPP-03689 © Antioch Pub.

ICE ON THE EQUATOR

Quaternary Geology of Mount Kenya, East Africa

ICE ON THE EQUATOR

Quaternary Geology of Mount Kenya, East Africa

by

William C. Mahaney
York University
Toronto, Canada

Wm Caxton Ltd
Sister Bay, Wisconsin

Published by

Wm Caxton Ltd
Box 709 – Smith Drive & Hwy 57
Sister Bay, Wisconsin 54234

Telephone: (414) 854-2955

Printed in the United States of America

10 9 8 7 6 5 4 3 2 1

Library of Congress Cataloging in Publication Data

Mahaney, W. C.
Ice on the equator : quaternary geology of Mount Kenya, East Africa / William C. Mahaney.
p. cm.
Includes bibliographical references.
ISBN 0-940473-19-4 : $65.00
1. Geology, Stratigraphic--Quaternary. 2. Glacial epoch--Kenya--Kenya, Mount, Region. 3. Geology--Kenya--Kenya, Mount, Region. I. Title.
QE696.M33 1990
551.7'9'096762--dc20 90-41212
CIP

ISBN# 0-940473-19-4

Cover Graphic: View of Mount Kenya from the south (compare Fig 1.3, page 4).

THIS BOOK IS DEDICATED TO LINDA M. MAHANEY WHO SPENT TWELVE MONTHS ON MOUNT KENYA IN 1976 AND 1983/84. WITHOUT HER RESOLVE, UNFAILING ENCOURAGEMENT, AND DRIVE TO LEARN MORE ABOUT THE AFROALPINE, THIS BOOK AND THE RESEARCH UPON WHICH IT IS BASED WOULD HAVE TAKEN MUCH LONGER TO PRODUCE.

TABLE OF CONTENTS

Chapter 5

Naro Moru and Teleki Glaciations

Chapter 6

The Liki Glaciation

Chapter 7

Holocene (pre-Neoglaciation) and Late Glacial Record

Chapter 8

Neoglaciation

Plates

FOREWORD

This book is based on fifteen years of environmental research on Mount Kenya, one of the principal high mountains of East Africa. The Mount Kenya Quaternary History Project started in 1970, when I began to gather published and unpublished information on all the East African mountains, and developed what turned out to be a lengthy correspondence and collaboration with other East African specialists. I selected Mount Kenya for a detailed glacial geological/environmental survey because it is one of the highest mountains in East Africa with an extensive Afroalpine zone, and a lengthy geologic history, spanning the whole of the Quaternary Period. What I originally intended to be a three to five year project slowly lengthened to a decade, and now shows no sign of terminating. The Mount Kenya Quaternary History Project which I have directed since 1970 is fortunate to have numerous collaborators in different parts of the world, many of whom have never seen the mountain. Yet these associates continue to produce new and exciting information in many different disciplines.

This book surveys a long glacial and periglacial sequence on Mount Kenya, its geological and environmenal settings, sequences of paleosols and their significance in understanding the multiplicity of glaciations. In addition, it provides a discussion of relative and absolute age-dating methods used to refine the stratigraphy and interpret the paleoclimate under which deposits evolved over the last two million years. At this time new information on Mount Kenya becomes available almost daily, as samples are processed in laboratories in Africa, Europe and North America, and as old data are viewed in novel ways within the light of this knowledge. The next decade will see the development of a refined time-stratigraphy for Mount Kenya that will provide even more precise information on the sequence and timing of geologic events that continue to shape one of the most imposing and beautiful places on earth.

PREFACE

Over the last fifteen years, since the inception of the Mount Kenya Quaternary History Project, I have been fortunate to have had support from many individuals and institutions in several African, European, and North American countries. In particular, I owe an immense debt of gratitude to my wife, Linda Mahaney, who not only encouraged me to finish this project after many frustrations and delays, but worked in the field over six-month periods in 1976-77, and again in 1983-84. Without her support and attention to organizational details, I would not have completed all the research reported in this volume.

My interest in Mount Kenya grew out of my thesis research at the Institute of Arctic and Alpine Research at Boulder, Colorado, in the 1960s. After completing my thesis research in the mountains of Colorado and Wyoming, I began to look at the possibility of working on one of the high mountains in East Africa. I chose Mount Kenya because it was as old as the Quaternary itself—the volcanic plug dating from beyond the Pliocene/Pleistocene boundary, and it had a magnificent moraine record stretching beyond the last glaciation. In addition, it was accessible by four-wheel drive vehicle which made it relatively easy to reach the lower limit of glaciation at 3000 m. Perhaps most importantly, Mount Kenya is a National Park, and as such it has seen relatively little human impact, and it is inhabited by large herds of buffalo, elephant, and other animals that are becoming increasingly threatened with extinction by illegal poaching. It is my hope that, after reconstructing the paleoenvironment, and establishing a precise time-stratigraphy for Mount Kenya, it will be possible also to reconstruct the evolutionary process, which led to the development of the Afroalpine and subalpine ecosystems, following the last glaciation.

I am indebted to a great number of individuals for assistance in many areas beyond by own discipline of geomorphology. The late Paul Temple, formerly of University College, Dar es Salaam, and John Walsh (former Commissioner of Mines, Nairobi), offered encouragement during the planning stages in 1970-76. Professor Eduard van Zinderen Bakker of the University of the Orange Free State, Bloemfontein, South Africa, offered many helpful suggestions, and provided, along with Dr. J.A. Coetzee, of the same institution, an important corpus of palynological data. In fact, the enormous amount of information in the first few volumes of *Palaeoecology of Africa* (edited by E.M. van Zinderen Bakker) provided immensely useful background data of immeasurable importance in planning the first and second expeditions to Mount Kenya. Since 1987, Brigitte Cooremans and Louis Scott of the Institute of Environmental Sciences, University of The Orange Free State have been analyzing pollen samples from the Meteorological Station Catena.

Of the many individuals and institutions involved in the Mount Kenya project, I am grateful for the financial support received (grants 1576.76, 2306.81, 2672.83, and 3332.86) from the National Geographic Society, which funded the project throughout an eleven-year period. The Natural Sciences and Engineering Research Council of Canada (grant A-9021), and York University provided additional support in the form of operating and minor research grants. Walter Vortisch and Klaus Fecher of Marburg University, West Germany, collaborated on the mineralogical and chemical analysis of several hundred samples from Mount Kenya. René W. Barendregt of Lethbridge University, Alberta, assisted with some section descriptions, artifact analysis, and paleomagnetic investigations of some of the older tills and aeolian sediments between 1984 and 1988. John Spence of Canterbury University, Christchurch, New Zealand, spent two months on Mount Kenya, in 1984, collecting vegetation data at several sites of interest to my own Quaternary investigations. Bruno Messerli (Bern University, Switzerland) discussed various matters relating to the Quaternary of Mount Kenya in the field. Tim Charsley of the U.K. Geological Survey provided invaluable assistance, overcoming bureaucratic hurdles in Nairobi, as well as conducting studies of sediments on the northern flanks of Mount Kenya that have an important bearing on and correlation with the glacial sequence on the Massif itself. Nat Rutter, Gaileen Lyons and Ingrid Moffat of the University of Alberta gave freely of their time and resources to provide the amino acid analyses that added so much to developing a time stratigraphy for the mountain. Dave Huntley and Glenn Berger of Simon Fraser University, British Columbia, carried out the laborious measurements that provided the first four thermoluminescence dates ever obtained anywhere in the tropical mountains. Don Halvorson of the University of North Dakota conducted a lengthy series of laboratory investigations involving scanning electron microscopy and electron microprobe analyses of paleosol chronosequences on the eastern flank of the mountain, in Gorges Valley, and in the Lake Ellis areas. Emile Roche of the Museum for Middle Africa, Tervuren, Belgium, analyzed the pollen in several Mount Kenya bog and lake cores, and provided very important information on paleoclimate over the Holocene period. Ed Derbyshire of Leicester University analyzed a number of oriented diamicton and old till samples for sand and granule orientation. Jerry Stipp and Murray Tamers of Beta Analytic, Coral Gables, Florida and Jaan Terasmae and Howard Melville of Brock University, St. Catharines, Ontario provided invaluable advice on radioisotope dating methods. Mike Boyer of York University and his thesis students – Mike Belanger and Steven de Belle – provided an invaluable mass of data on bacteria and fungi counts (see Chapter 12) with depth in several paleosols, which established the basis for my interpretations of microflora effects on the validity of radiocarbon dates. Kandiah Sanmugadas analyzed hundreds of samples for extractable iron and aluminum in the Geomorphology and Pedology Laboratory of York University. Mary Leakey and S. Wandibba (National Museums of Kenya) kindly identified pottery and stone artifacts from Mount Kenya. Rufus Churcher (ROM) identified various bone samples collected near Naro Maru.

For the successful completion of the mud polygon freezing experiment (Chapter 12), I am indebted to Barry Fahey, Forest Research Institute, Christchurch, N.Z., and J. Brian Reynolds (Guelph University) for advice on carrying out the experimental work. Harry Lamb (Atmospheric Environment Service, North York, Ontario) loaned the CR21X datalogger and thermistors used in the experimental work. Joe Megyes (Atmospheric Environment Service, North York, Ontario) assisted with the computer analysis of the freeze-thaw data.

Students in my mountain geomorphology field schools (1976 and 1983) contributed significantly to sample collection and mapping projects. I am also indebted to two former students – Larry Gowland and Lina Radziunas. Larry Gowland assisted with field work for two months in 1976 and 1983, in addition to normal course requirements. Lina Radziunas completed the first amino acid analysis of buried paleosols on Mount Kenya, as part of her thesis research at York University, and proved that amino acid geochronometry was worth pursuing. Simon Kamakia, Joseph Camillo, Frank Wachira and Simon Murithii assisted with the field excavations in 1976/77, 1983/84, 1986 and 1987. Several dozen colleagues, too numerous to name in full, have assisted with ideas and criticisms they may not recall, but which have contributed immeasurably to the ideas and discussions which evolved into *Ice on the Equator*. Any omissions or misinterpretations are, however, my own responsibility.

Janet Allin drafted most of the illustrations. Heather Spurll computer-formatted various drafts and the final copy of the volume. John Dawson and Jeff Nolte produced all the photographic work, and in some cases they improved the original photography, producing high-quality half-tones in the process. René Barendregt (Lethbridge University), Pete Birkeland (University of Colorado), Mike Boyer (York University), Tim Charsley (U.K. Geological Survey), Alan Hamilton (The Queen's University, Belfast), Ron Hancock (University of Toronto), Rudolf Harmsen (Queen's University), Stefan Hastenrath (University of Wisconsin), Klaus Heine (Regensburg University, F.R.G.), Bill Melhorn (Purdue University), Carlos Schubert (I.V.I.C., Caracas, Venezuela), John Spence (Canterbury University, Christchurch, New Zealand), Jim Teller (University of Manitoba), Walter Vortisch (Philipps University, F.R.G.), Chalmers Clapperton (University of Aberdeen), Joey Coetzee, Eduard van Zinderen Bakker (University of the Orange Free State, S. Africa), Fred Jansen, Netherlands Institute of Sea Research (Den Burg, Holland), and Sid White (Ohio State University) offered constructive criticisms of various chapters in the volume. Their helpful suggestions allowed me to improve my interpretations. To all these individuals I owe and immense debt of gratitude.

During the 1976/77 expedition, F.W. Woodley (former Warden) Mountain National Parks, and P.M. Snyder (Assistant Warden) Mt. Kenya National Park, and their rangers assisted with the field work, particularly when we had to work in areas frequented by poachers. At other times, Fred Pertet (Ministry of Tourism), Perez Olindo (Wildlife Ministry), and L. Butaki (Mines Dept.) provided permits and encouragement to continue the research. Willie and Didi Currey (formerly Naro Moru River Lodge), Eddie Hoarau (Mount Kenya Safari Club, Nanyuki), and Diamond and Babu Mapara (Aquarius Ltd., Nairobi) assisted from time to time with logistical support.

I acknowledge the following copyright holders for permission to use copyrighted material:

Palaeogeography, Palaeoclimatology, Palaeoecology for permission to reprint Fig. 3, p. 50, V. 60, 1987 as Fig. 7.4.

Reidel Publishing Co., Dordrecht, Holland, for permission to reprint a modified version of map 3.3.2 (p. 57) in S. Hastenrath, 1984, *The Glaciers of Equatorial East Africa* as Fig. 8.12, and Figure 4 in J. Oerlemans, editor, **Glacier Fluctuations and Climatic Change,** 1989.

Mountain Research and Development for permission to reprint Fig. 6.21.

Geolgica et Paleontologica for permission to reprint Figure 5.5.

Dr. W. Junk Publishers (reproduced by permission of Kluwer Academic Publishers, Dordrecht, Holland) for permission to reprint Figs. 2.7, 2.8, 2.10, 2.23A, 2.23B, 2.24, 2.25, and 12.21 from M. Coe, **The Ecology of the Alpine Zone of Mount Kenya.**

Palaeoecology of Africa – and the surrounding islands, V.3, 1966 – annual publication, c. 250 pp. Hfl. 125/U.S. $60.00 per volume. A.A. Balkema, P.O. Box 1675, Rotterdam, Netherlands/A.A. Balkema, Old Post Road, Brookfield, Vermont 05036, U.S.A. For permission to reprint Table 2.1 and Figures 2.21, 2.25, and 2.26 from Vol. 3, 1966; Figures 2A, 2B, 5, and 7 from Volume 19, in press, 1990; and Figure 4 from Cooremans and Mahaney, V. 20, 1990, in press.

Balkema Publishers, P.O. Box 1675, Rotterdam, Netherlands/A.A. Balkema, Old Post Road, Brookfield, Vermont 05036, U.S.A. for permission to reprint Figure 4.3 from W.C. Mahaney, editor, **Quaternary and Environmental Research on East African Mountains**, 483 pp. Hfl 135/U.S. $70.00 per volume.

American Journal of Science for permission to reprint Figure 4D from V. 288, 1988, p. 817.

Pergamon Press for permission to reprint Figures 2b, 4, and 5b (as Figures 4.28, 4.32, and 4.33) from **Journal of African Earth Sciences**, 1988, V.7 No.1, p. 219-225, and Figures 12.26 and 12.27 from 1986, V.5, No.4, p.321-327

Les Presses de l'Université de Montréal for permission to reprint Figures 10.1 and

10.2 from **Géographie physique et Quaternaire,** V. 40, No. 2, 1986, p. 171-184, and Figures 3, 5, 6, 7, and 8, V. 42, No.1, 1988, pp 89-96.

L'Anthropologie and Inst. of Paléontologie Humaine, Paris, H. de Lumley, editor, for permission to reprint Figure 11.1.

Acta Geocriogenica, A. Corte, editor, for permission to reprint Figures 12.5, 12.6, 12.8, 12.9, 12.10, and 12.11.

Catena, for permission to reprint Figures 10.9 and 10.10 from V. 16, No. 3, 1989, p. 205-214.

Pergamon Press, for permission to reprint Figures 7.1, 7.5 A and B, and 7.7 A, B, and C from **Quaternary Science Reviews**, 1988, V. 7, No. 2: 211-225.

Elsevier Science Publishers, for permission to reprint Figures 7.5 A and B from **Palaeogeography, Palaeoclimatology and Palaeoecology**, V. 60: 50.

CHAPTER 1

INTRODUCTION

The first European to see Mount Kenya (Figure 1.1) was John Ludwig Krapf who had heard various stories about snow-capped mountains in the interior of Kenya. On December 3, 1849, he observed, and later described, "two large horns or pillars, as if it were rising over an enormous mountain to the northwest of Kilimanjaro, covered with a white substance" (Krapf, 1860). His observations from a summit near Kitui have been cited by many researchers (Dutton, 1929; Coe, 1967) as the first written description of the Mount Kenya Massif, even though local people – especially the Kikuyu, Wanderobo, and Masai – had, along with Arab traders, known that snow (kirira) covered the summit. Indeed, Coe (1967) cited Krapf's encounter with an Uembu tribesman. Rumu Wa Kikandi was said to have described the snow as rolling "down the mountain with a great noise, and that the water flowed into a great lake that fed the Dana (Tana) River." From this, Krapf surmised that glaciers existed on the mountain, although the lake described by Rumu is somewhat of a mystery since only lakes Michaelson and Ellis (Figure 1.2) are known above the montane forest limit. As Coe (1967) points out, the Wanderobo at this time were known to frequent lakes above the timberline. In addition, there is evidence for recent human occupation in the timberline ecotone-ericaceous belt areas (Mackinder, 1900; Zeuner, 1948; Mahaney, Barendregt, Churcher and Spence, 1986). Krapf's important sighting of Mount Kenya was dismissed by numerous geographers, most notably by Cooley (1852), as the product of imagination. In the meantime, other scientists heard of, or related, tales of a snow-capped summit in the general area of the equator in Central Kenya (Rebmann who travelled with Krapf in 1849, cited by Coe, 1967; and Hildebrant, 1878), but no actual observations were reported.

Joseph Thomson approached Mount Kenya from Longonot Crater (*i.e.* from the west) and at the same time he named the Aberdare Mountains (Figure 1.1) after the President of the Royal Geographical Society (Thomson, 1885). From the Aberdares he explored Thomson's Falls, reached the western fringe of the Mount Kenya forest, but did not climb the mountain. Two years later, in 1885, Count Samuel Teleki managed to traverse the western forest along the Naro Moru River and entered the valley that today bears his name (Figure 1.1). From Lieutenant von Höhnel's account of Teleki's expedition, it appears they were the first Europeans to see the magnificent *Lobelias* and *Senecios* in the Afroalpine zone. According to von Höhnel (1894) Teleki's party reached the 15,000-ft. level where the valley forms what appears to be a "bowl-shaped cwm" (Coe, 1967). From this landform shape, Teleki believed the high tarn at the head of the valley (Teleki Tarn, Figure 1.3) to be the remains of a volcanic crater. Both von Höhnel and Teleki returned to the eastern flank in 1892, but were not successful in attempting to penetrate the dense and expansive montane forests there.

An attempt to reach the summit in 1891 on the part of Captain Dundas and C.W. Hobley (Baker, 1967) failed owing to the difficulty of penetrating the forest on the southern side. Hobley compiled notes on some of the volcanic ash and basalt exposures, but no publications resulted from this early work.

Later on, in 1893, the geologist and naturalist J.W. Gregory led the first scientific collecting expedition into East Africa. He was the first European to notice striae on rocks in the timberline area (Gregory, 1894a) from which he deduced that the mountain had been glaciated some time in the past. In the course of his first expedition to the mountain, Gregory made significant contributions to the biogeography of the Afroalpine zone. In particular, he noted that large populations of rodents – especially hyrax (*Procavia Johnstoni Mackinderi*) and groove-toothed rats (*Otomys orestes orestes*) were present in large numbers in Teleki Valley (Figure 1.4), and that the flora contained a large number of northern genera. His conclusion that the plants represent a remnant rather than a dispersive flora (as interpreted by Coe, 1967) is significant

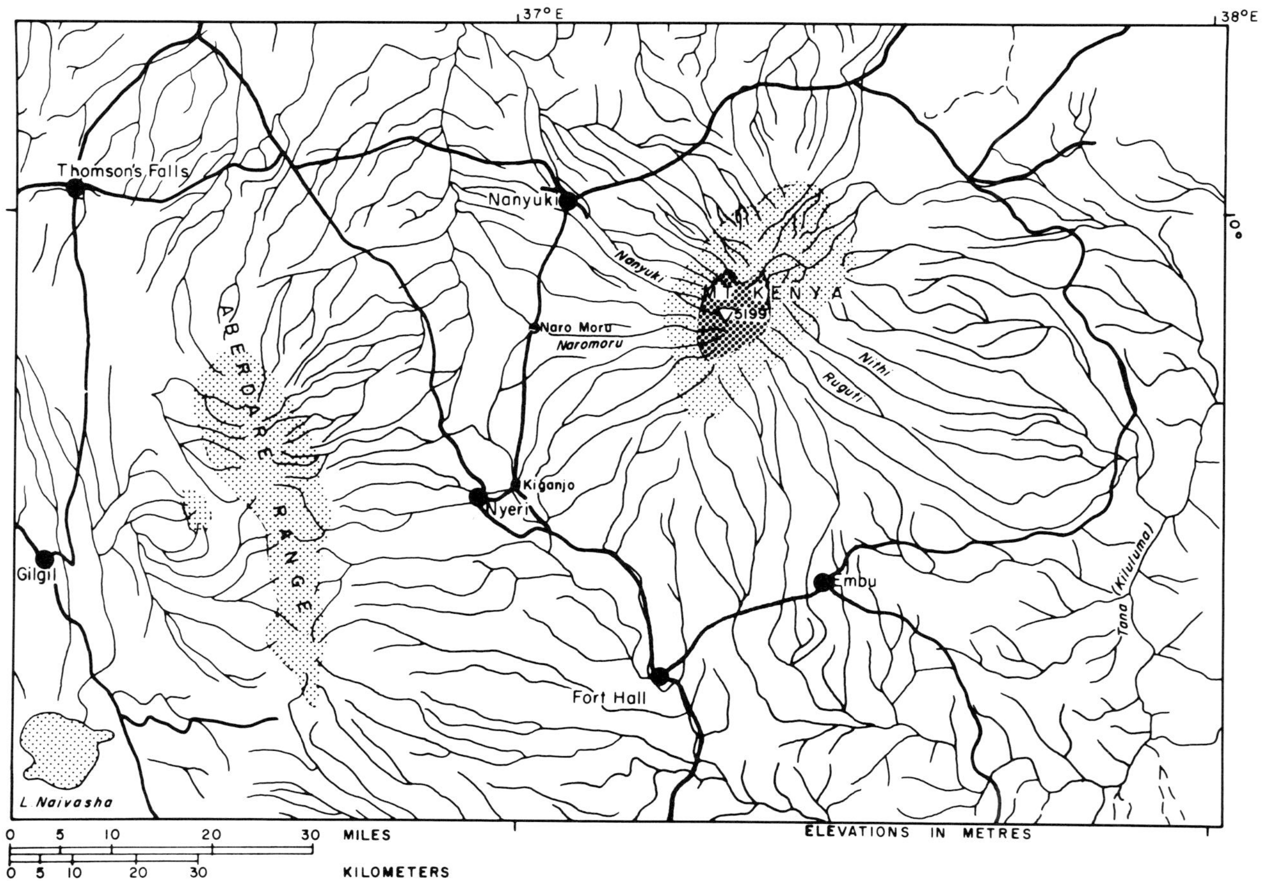

Figure 1.1: Mount Kenya Massif and the Aberdare Range. The large stippled area is the approximate delimitation of the lower Afroalpine Vegetation Belt; the smaller stippled area delimits the upper Afroalpine and nival belts.

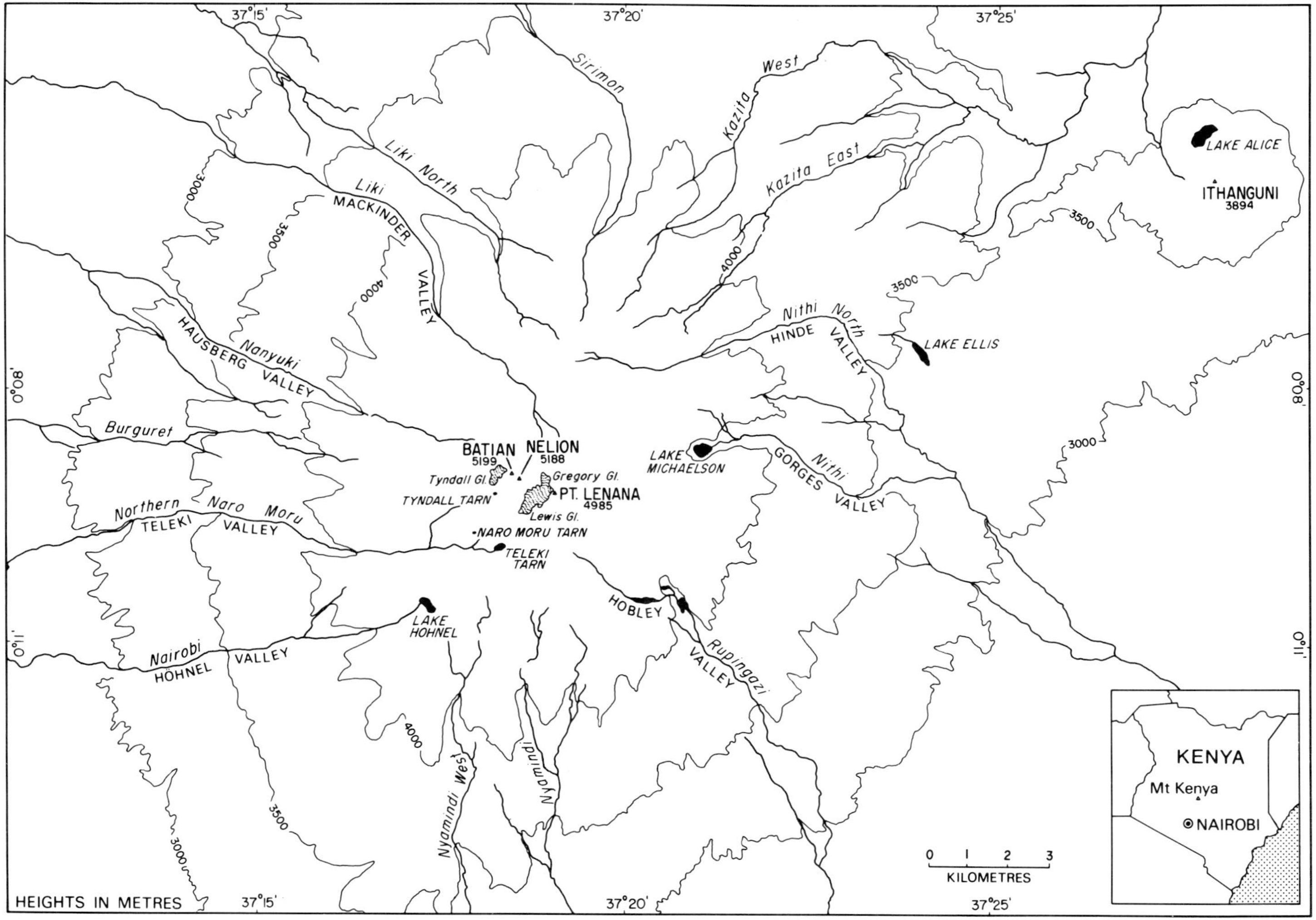

Figure 1.2: Radiating drainage patterns on Mount Kenya and Ithanguni above 3000 m.

Figure 1.3: The Central Peaks (Batian to the left; Nelion to the right), between the Tyndall and Lewis glaciers, as viewed from the south above Teleki Valley (photo by W.C. Mahaney, June 1984).

Figure 1.4: Teleki Valley from the 3900 m level. Tall woody growth of *Senecio keniodendron* and *Lobelia* in the foreground. This is the route followed by Count Teleki and J.W. Gregory in the 19th Century. The bowl-shaped depression at the head of the valley was considered an explosion crater by Count Teleki.

with respect to understanding the glacial history of the mountain, and the effect of paleoclimatic variations on the present floristic composition and plant distribution. Unfortunately, all of Gregory's plant specimens were later lost on the Tana River upon his return to England (Coe, 1967).

During this first expedition, Gregory climbed the mountain along the Naro Moru track (Teleki Valley) to the Central Peaks (Figure 1.4) where he collected and described for the first time some of the common rock suites (Gregory, 1896). In 1896, Dr. Kolbe explored the eastern moorland collecting insects, but there is almost no published information from his expedition (Coe, 1967).

Just as the twentieth century was about to appear, H.J. Mackinder climbed the mountain along the Nairobi Drainage (Figure 1.3) to Lake Höhnel. On September 13, 1899, he climbed Batian (5199 m) while opening up the first climbing route, which would be followed later by several generations of mountaineers. Although Mackinder made many new observations on the biology of Mount Kenya, most of his specimens were lost on his return trip to Europe. His notes on the floristics of the mountain, like Gregory's before him, remain as significant early descriptions of enormous importance to science (Mackinder, 1900). Above all, it was Mackinder who proved that the nepheline-syenite complex discovered by Gregory (1894b) extended to the summit of the Central Peaks.

In the early twentieth century, little scientific research was carried out, although local settlers explored in various areas, establishing some of the first tracks and huts. The most notable of these explorations include Orde Brown (1918); Arthur (1921, 1923); and Dutton, (1929). Earlier, in 1908, Ross (1911) and Hutchins (1909) completed a forest inventory, around the lower moorland, in which they mentioned former glaciation in the Hausberg Valley. They photographed the Cesar and Tyndall glaciers showing that they were formerly more extensive than in the early twentieth century.

After exploring the moraines in Gorges Valley in 1927, Eric Nilsson returned in 1932 to map the entire area. His published account in 1935 described the large terminal moraine near the Nithi Waterfall (Mahaney, Barendregt, Churcher and Spence, 1986) as delimiting the maximum extent of Quaternary glaciers. His observations of "patches" of older till below the waterfall were attributed to an earlier glaciation (Nilsson, 1931). Nilsson's observations extended from the earliest deposits to the Neoglacial as he was the first to describe the moraine sequence around Harris Tarn on the northeastern flank (into which the Kolbe Glacier descended in the first few decades of the twentieth century). It has since disappeared. Moreover, he correctly surmised that, since the moraine sequence on Mount Kenya was very similar to published reports from Mount Kilimanjaro, Mount Elgon, and the Ruwenzori Mountains, it was quite probable that they were all the product of the same paleoclimatic changes. Nilsson's map of moraine positions in Hobley Valley (Figure 1.2) correlates closely with the Gorges Valley sequence, except that he failed to recognize a broad arc of moraine close to and below the 3100 m contour (Mahaney, 1986a).

The first glaciological work was carried out by Carl Troll and K. Wien in 1934, which resulted in a photogrammetric survey (1949) and the first flow measurements. During the 1930-1936 year period, R. Hook and H. Slade stocked some of the mountain streams with fish and improved several existing maps. P.C. Spink (1945, 1949) monitored continued recession of glaciers in the Central Peaks. In 1947, F.E. Zeuner visited the mountain and described frost soil structures, amongst other geographical and geomorphological features. During the I.G.Y., various reconnaissance studies were carried out in glaciology, glacial geology, vegetation, meteorology and hydrology (Baker, 1967). A long-term mass and energy balance study on the Lewis Glacier is documented in Hastenrath (1984). The U.K. Geological Survey carried out a 5-year mapping project (1981-1986) involving bedrock and surficial geological investigations along the northern flank of the massif. Surficial geological studies carried out under the aegis of this project

(Charsley, 1989) provided important information on the origin and age of outwash bodies emplaced following episodic glaciation in the Afroalpine zone.

The Mount Kenya Quaternary History Project, from its inception in 1970, has provided a wide range of important information on the extent and timing of Quaternary glaciations on Mount Kenya, on the related nonglacial environment, and on paleoclimatic-ecologic changes. The results of this research are documented in *Quaternary and Environmental Research on East African Mountains* (Mahaney, ed., 1989) and in this volume.

CHAPTER 2

ENVIRONMENTAL SETTING

2.1 Bedrock Geology

The Central Plateau of East Africa, comprising Kenya, Tanzania and Uganda (Figure 2.1), has undergone intense tectonic activity since the Miocene, with fractures following zones of weakness established earlier in the Cretaceous. Towards the late Tertiary a system of rift valleys – Western and Eastern Rifts – with their enormous and magnificent lakes had formed as manifestations of the immense endogenetic forces at work in the crust. Major episodic faulting occurred again in the early and middle Pleistocene, reversing some rivers, such as the Kafu and Kagera, causing them to drain into Lake Victoria (Coetzee, 1967). The floors of some rift valleys were downdropped several hundreds of meters below sea level, while others remained several hundreds of meters above sea level. The escarpments surrounding the Eastern and Western Rifts rose to hundreds of meters, creating an imposing landscape of rather ethereal appearance. In some areas of the Eastern and Western Rift valleys large-scale faulting occurred even in the Upper Pleistocene (Bishop 1962a, 1962b, 1965; Isaac 1967a) which serves to substantiate Cooke's (1958) earlier conclusion that high-magnitude faulting will occur again. From the work of Cooke (1958) and Furon (1963) it seems quite likely that tectonic activity will recur, so that the region should be considered tectonically unstable.

Because the Rift Valleys are formed primarily as a result of fractures in the earth's crust, recent immense lava flows and active volcanoes, such as in the Virunga Mountains (Western Rift) and Oldoiyo Lengai (Eastern Rift) (Figure 2.1) are common. Many extinct volcanoes, including Mount Kenya (5200 m) of late Pliocene age, Kilimanjaro with Mawenzi Crater (5124 m) dating from the late-Pliocene/early Quaternary, Kibo (5895 m) from the middle Pleistocene, and Shira Peak from an even older episode (Cooke, 1958), tend to align themselves along these major fracture zones.

In the Western Rift, non-volcanic mountains of Precambrian rock are found in the Ruwenzori Massif (Figure 2.1) which rises to 4600 m, and is of pre-Miocene age (Coetzee, 1967). Occurring between Lakes Albert and Edward, the Ruwenzori is the product of upfaulting in relatively recent times, and it is surfaced with extensive glacial deposits emplaced during the last glaciation (Osmaston, 1965). Mount Elgon, almost mid-way between the Eastern and Western Rifts, is another non-volcanic mountain composed principally of meta-sedimentary rocks uplifted in the late Tertiary and surfaced with a succession of glacial deposits. To the east of Mount Elgon, but still outside the Eastern Rift, lies the Cherangani Hills – another imposing range, the remnant of an older land mass, and product of several erosion cycles (Miller, 1956). The northern half of the Cherangani Hills drains into Lake Turkana, while the southern half lies within the Lake Victoria Basin.

The geology of the volcanic pile of Mount Kenya is summarized from Baker (1967) in Figures 2.2 and 2.3. Outcrops of basement rock occur on the southern, eastern and northern flanks of the mountain. Precambrian inliers, beneath the Simbara basalts and agglomerates, occur in several drainages, and even project above the volcanic rocks in a few places, particularly at Marua. From the distribution of basement rocks it is apparent that the Precambrian crust forms the underpinnings of the volcanic massif itself.

Metamorphic blocks are found high on Mount Kenya in the trachytic agglomerates west of Inthanguni and as isolated large clasts on moraine and fluvial/colluvial deposits (Mahaney and Charsley, 1986). These large boulder and cobble sized clasts on the southwestern flanks of the mountain are mainly of biotite gneiss composition. Additional outcrops of pegmatite,

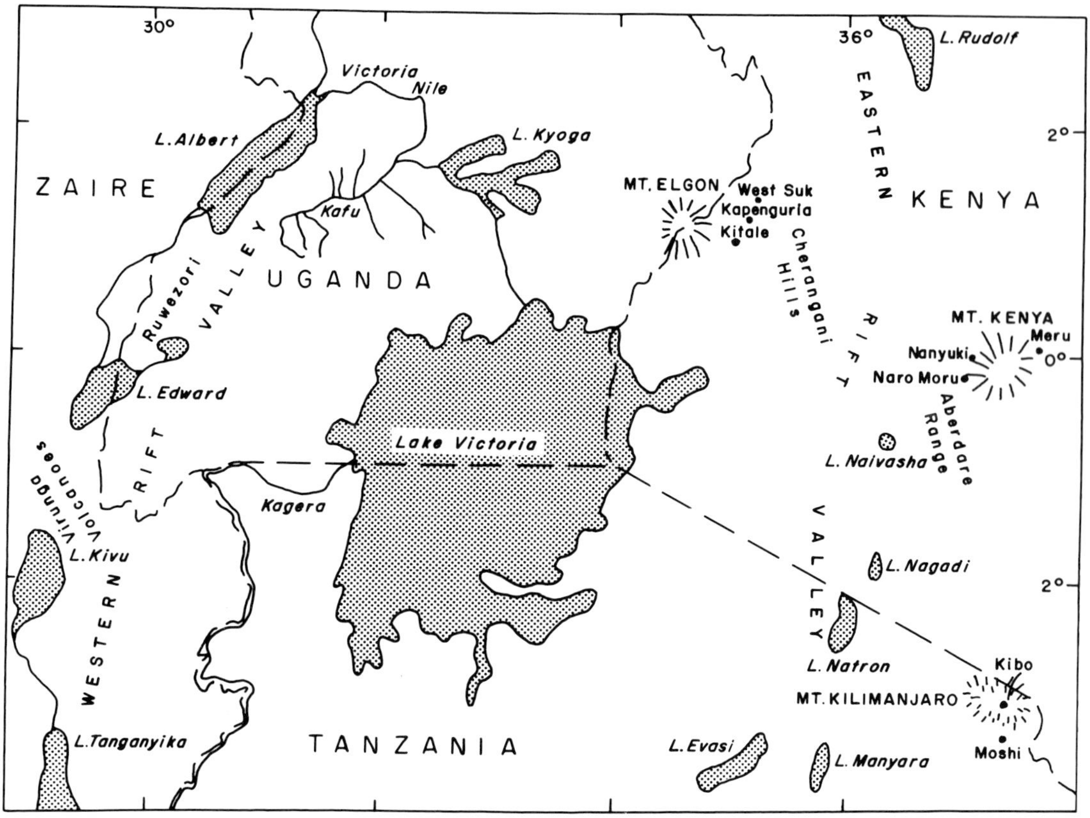

Figure 2.1 Mount Kenya Massif and the surrounding summits of Mount Kilimanjaro, Mount Elgon, the Ruwenzori and Virunga Mountains.

quartzofeldspathic granulite and muscovite-biotite-graphite schist (Baker, 1967) point to the chemical variability of the basement inliers.

The Simbara Series, of probable Miocene age, and its correlative unit to the north – the Samburu Series – represent the oldest volcanic rocks east of the Eastern Rift Valley (Shackleton, 1945; Baker, 1967). On the southwestern flanks of the mountain they overlie the basement complex and are, in turn, overlain by Nyeri Tuff. The Series thins towards the north and is replaced by the Nyeri Tuff in the area around Kiganjo. The Nyeri Tuff, with its grey color and friable texture, is easily recognized as a stratigraphic marker, and can be seen in numerous quarries where it is mined as an important building stone (Shackleton, 1945; Baker 1967).

The Laikipian Basalt, of Pliocene age, first described by Gregory (1921), and later studied by Shackleton (1945), occurs on the southwestern flank of the mountain where it forms the upper part of a dissected lava flow, that in places has thick soils formed on it. The overlying Mount Kenya Suite (Shackleton, 1945; Baker, 1967) includes all eruptive rocks from Mount Kenya and its parasitic cones, excluding the Thiba Basalts and basaltic pyroclastic rocks of the Nyambeni Series. The Mount Kenya Suite covers an estimated area of ~4300 km^2 with an approximate elliptical shape (see Landsat imagery – Figure 2.9), and a diameter of ~100 km. The original volume of the Mount Kenya Suite before erosion is estimated at 3700 km^3; the volume after erosion at the present day is ~2400 km^3. Since these calculations (Baker, 1967) are based on original *estimates* of the height of the Mount Kenya volcano, they may in fact be somewhat high.

Age	Series	Subdivision	Unit	MAX. OBSERVED THICKNESS (M)
RECENT			SURFICIAL DEPOSITS (mostly nonglacial)	~20
PLEISTOCENE	GLACIATION		LIKI TILLS	~20
	MT. KENYA VOLC. SERIES	PARASITIC VENTS	NYAMBENI VOLCANICS THIBA BASALTS	> 150
	GLACIATION		TELEKI TILL	~10
	MT. KENYA VOLC. SERIES	PARASITIC VENTS	ITHANGUNI & OTHER TRACHYTIC VOLCANICS & NECKS	c. 365
	GLACIATION		PRE-TELEKI TILLS	>40
	MOUNT KENYA VOLCANIC SERIES	PARASITIC VENTS	AGGLOMERATES	>180
		FISSURE ERUPTIONS	RIEBECKITE TRACHYTES	25
			OLIVINE BASALTS	210
			MUGEARITES, OLIVINE TRACHYTES	245
		MAIN ERUPTIVE EFISODES	NEPHELINE SYENITE & PHONOLITE of the PLUG	–
			KENYTES, PHONOLITES & PYROCLASTICS	>550
			PHONOLITES & TRACHYTES	>490
			RHOMB PORPHYRIES	335
			PHONOLITES	0-120
			UNEXPOSED VOLCANICS	c. 2900
			LOWER BASALTS	c. 45
PLIOCENE			LAIKIPIAN BASALTS olivine basalts, basanites, olivine nephelinites	c. 120
			NYERI TUFF trachytic tuffs & agglomerates	30
MIOCENE (?)			SIMBARA SERIES basalts & agglomerates	>75
PRE-CAMBRIAN			BASEMENT SYSTEMS micaceous & graphitic gneisses, amphibolites, quartzo-feldspathic gneisses, & pegmatites	–

Figure 2.2: Stratigraphic sequence of crustal rocks, late-Tertiary and Quaternary volcanics, and Quaternary glacial and nonglacial rocks and deposits (after Baker, 1967).

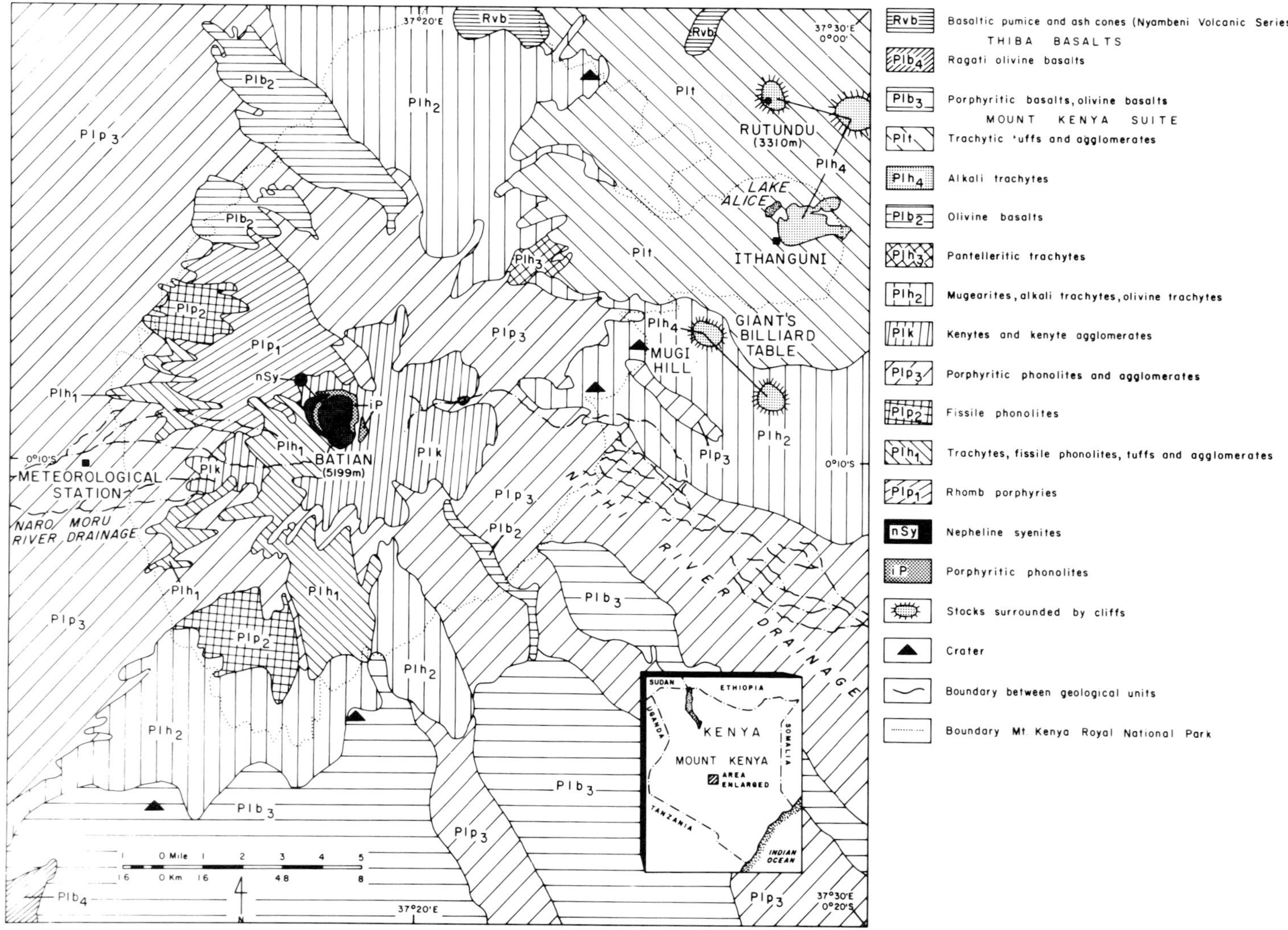

Figure 2.3: Bedrock geology of the Mount Kenya Massif showing the symmetrical distribution of extrusive rocks around a volcanic plug of nepheline syenite and porphyritic phonolite (after Baker, 1967).

Figure 2.4: Outcrop of diamict capped with tuff in Teleki Valley at 3750 m. Some of these diamicts have been described as tills; however, all diamicts studied on Mount Kenya gave random fabrics suggesting they have mass wasting/volcanic origins. None appeared to be welded as the overlying tuff.

Thicknesses of fills in lower valleys, and valley depths, do not appear great enough to substantiate the removal of 1300 km^3 of rock over the last ~2.0 my.

The Mount Kenya Suite consists of a complex of rock divided into volcanics from the main eruption and from satellite craters. The main eruptive episode emplaced the lower basalts, phonolites, rhomb porphyries, trachytes and kenytes from the main vent now closed by a plug of nepheline syenite. The second period of vulcanicity began later when lavas and pyroclastics were erupted from parasitic cones and fissures on the volcano's flanks.

The main eruptive episode led first to the emplacement of the lower basalts, which are evident on the road between Naro Moru and Kiganjo, and later to the eruption of rhomb porphyries and phonolites which have a source vent near Polish Man's Tarn high on the northwestern flank of the mountain. These rocks are overlain by porphyritic phonolites and kenytes erupted from the main vent, and by trachytes and fissile phonolite erupted from another satellite vent. The vast majority of the Mount Kenya Suite is composed of phonolites, kenytes, and agglomerates (Figure 2.3) which outcrop extensively on the western and eastern slopes, and which occur beneath younger rocks on the other flanks of the mountain. Because they are symmetrically distributed around the main vent, it is considered to have been active throughout much of the life of the Mount Kenya volcano. The youngest agglomerates, erupted during the time the Mount Kenya Suite formed, contain phonolite, kenyte and fissile phonolite which suggests that no additional volcanics likely occur below the surface, and that the vast majority of unexposed rock is of phonolitic composition.

The phonolite and kenyte lavas, and the nepheline syenite that forms the plug (K-Ar age of 2.64 my, Everden and Curtis, 1965) in the central vent are chemically similar; the main difference lies in texture and grain size reflecting different cooling times. Kenyte (glassy variety of porphyritic phonolite) outcrops are limited mainly to the central vent area; major transitions from phonolite to kenyte can be found high up on the flanks of Mount Kenya. Porphyritic phonolite and agglomerate (Figure 2.3) are laterally equivalent to the kenytes which were all erupted from the main vent over a long period in the early stage of the evolution of the volcano.

Younger volcanic rocks range from alkali trachyte with pantelleritic similarities, and olivine-alkali-trachyte to oligoclase basalt (mugearite). These rocks occur in three major areas: a) on the northern slopes where they unconformably overlie rocks from the early eruptions; b) at Hall Tarns where suites of trachytic rocks were erupted from a volcanic neck; and c) on the southern slopes where extensive and thin flows penetrated to a considerable distance below the present forest boundary.

Dikes and sills representing all the major lava suites are present on Mount Kenya within 5 km of the Central Peaks. With the exception of Polish Man's Tarn (Hausberg Valley, Figure 1.2) where numerous dikes of rhomb porphyry are located (Baker, 1967), no dike swarms are present. On Ithanguni three trachytic dikes are visible on the southern flank, and two of these are known to feed adjoining lava sheets (Baker, 1967). Confirmation of the eruptive sequence comes from dike-rock relationships where dikes of fissile phonolite, basalt, and olivine-trachyte cut the older rocks previously described.

Younger volcanic rocks formed by parasitic vents are somewhat similar to the Mount Kenya Suite in composition but different in age and mode of eruption. These satellite vents include Mugi Hill (near Lake Ellis, Figure 1.2), and six trachytic necks of which Ithanguni (Figure 1.2) is the largest. Giant's Billiard Table (south of Ithanguni, Figure 1.2) and Rutundu (north of Ithanguni, Figure 1.2) are two of these imposing volcanic necks composed of alkali trachyte rocks with prominent and widely spaced columnar joint patterns

2.2 Climate and Atmospheric Circulation

Both temperature and precipitation are dependent on elevation and both factors are interdependent to a certain extent. Temperature and precipitation control the distribution of vegetation on Mount Kenya, and considerable variations occur from one flank to another in response to temperature lapse rates and presence or absence of rain-bearing winds. Data on the climate of Mount Kenya is available from the I.G.Y. Expedition, from observations taken by various scientists over the years, and from Kenya Ministry of Works. Troll and Wien (1949), Hedberg (1964), Coetzee (1967) and Mahaney (1980; 1985) provided details on the diurnal temperature patterns.

On most days Mount Kenya is characterized by clear conditions in the early morning with the glaciers clearly outlined against the backdrop of a perfectly blue sky (Figure 1.3). The Afroalpine zone is subjected to high inputs of solar radiation which are intensified in the rarefied air. At lower elevations (~2000 m) intense radiation creates convection, and anabatic winds move upslope, producing clouds in the montane forest belt. During the late morning or early afternoon, clouds cover most of the high moorland area, which increases the albedo and reduces the amount of incoming radiation. By night the air at high altitude cools from reradiation and begins to produce katabatic winds which flow down the valleys. These katabatic winds are not strong, as indicated by the development of stratus clouds above and below the timberline (Troll and Wien, 1949). Walter (1960) argued that stratus cloud layers marked the top of the troposphere (*e.g.*, tropopause), the timberline therefore coinciding with the middle tropopause. While the tropopause is probably higher in the atmosphere, it is interesting that daily cloud formation over the montane forest causes maximum increases in precipitation, above which rainfall declines into the Afroalpine zone (Mahaney, 1985).

Diurnal meteorological patterns are superimposed on the macroclimatic seasonal rainfall patterns. Two rainy seasons are prevalent on Mount Kenya with two maxima – in April-May (long rains) and November-December (short rains) which come just after the high sun. This is evident in the rainfall data provided by Hedberg (1964). Thompson (1966) pointed out that with respect to annual precipitation at high elevation on Mount Kenya, most of it occurs during the southeast monsoon (April-May). At low elevations during the southeast monsoon a well-developed trans-equatorial ridge is in place across Kenya emanating from the subtropical anticyclone. This ridge produces subsidence with little precipitation below ~2000 m (~850 mb level), while above, at the 700 and 500 mb levels, the ridge is not in place and rainfall may occur.

Precipitation curves drawn by Coetzee (1967) from data in Table 2.1 are presented in Figure 2.7. Stations with the highest precipitation are at ~3000 m in the bamboo forest/Hagenia woodland ecotone, approximately 200 m below timberline. Using the data in Figures 2.5 and 2.6 (isohyet maps) it is evident that the highest precipitation occurs in the bamboo forest, decreasing above in the Afroalpine moorland.

The precipitation distribution shown in Figures 2.5 and 2.6 depicts a southeasterly maximum resulting from the classical monsoon, and a secondary maximum on the western side which is a product partly of daily thunderstorm activity and penetration northward of the Southern Hemisphere high pressure cell to the 2000 m level. Because rain gauges on the upper slopes are never monitored (Mahaney, 1986), rainfall distributions can only be inferred from Figures 2.5 and 2.6. Certainly there is lower precipitation in the upper Afroalpine and nival zones compared with the lower moorland and upper forest zones. Data in these figures are derived from the I.G.Y. Expedition in December 1957, and January 1958, during which most valleys on the mountain were instrumented with rain gauges. This precipitation data was compiled later by R. Jones of the Hydrology Department of the Kenya Ministry of Works. An earlier set of precipitation data (Figure 2.7) for the Naro Moru and Sirimon tracks, collected by the Ministry

TABLE 2.1: MEAN MONTHLY AND ANNUAL PRECIPITATION (MM) ON MT. KENYA DURING 1951, 1952
(East African Meteorological Dept., Nairobi)
(a) Naro-Moru Route (West slope)

Altitude (m)	2348 Montane Rain Forest Zone		3048 Bamboo Zone		3658 Lower Afro-alpine Zone		4267 Upper Afro-alpine Zone	
	1951	**1952**	**1951**	**1952**	**1951**	**1952**	**1951**	**1952**
Jan.	25.40	25.40	127.00	25.40	50.80	25.40	50.80	25.40
Feb.	25.40	12.70	127.00	12.70	76.20	25.40	50.80	25.40
Mar.	101.60	25.40	127.00	38.10	101.60	43.20	152.40	12.70
Apr.	381.00	177.80	304.80	177.80	381.00	165.10	203.20	114.30
May	76.20	177.80	152.40	266.70	127.00	165.10	50.80	101.60
June	177.80	0.00	228.60	12.70	152.40	50.80	127.00	12.70
July	76.20	12.70	88.90	177.80	101.60	63.50	76.20	25.40
Aug.	88.90	50.80	76.20	76.20	76.20	63.50	76.20	38.10
Sept.	76.20	50.80	101.60	114.30	76.20	76.20	50.80	50.80
Oct.	101.60	76.20	177.80	127.00	152.40	127.00	101.60	76.20
Nov.	330.20	203.20	304.80	251.50	241.50	152.40	279.40	76.20
Dec.	254.00	88.90	266.70	78.74	215.90	63.50	101.60	38.10
Year	1714.50	901.70	2082.80	1358.94	1752.60	1021.10	1320.80	596.90

(b) Sirimon Route (Northwest slope)

Altitude (m)	2348 Montane Rain Forest Zone		3048 Bamboo Zone		3658 Lower Afro-alpine Zone		4267 Upper Afro-alpine Zone	
	1951	**1952**	**1951**	**1952**	**1951**	**1952**	**1951**	**1952**
Jan.	25.40	0.00	50.80	31.175	50.80	6.35	–	25.40
Feb.	12.70	6.60	25.40	6.35	12.70	0.00	–	0.00
Mar.	50.80	25.90	76.20	25.40	50.80	0.00	–	50.80
Apr.	76.20	34.00	127.00	101.60	101.60	63.50	–	88.90
May	228.60	144.00	254.00	152.40	355.60	101.60	–	114.30
June	38.10	21.60	63.50	44.45	38.10	63.50	–	76.20
July	63.50	42.90	101.60	95.25	76.20	38.10	–	57.15
Aug.	76.20	31.20	63.50	38.10	63.50	50.80	–	63.50
Sept.	50.80	66.00	76.20	152.40	76.20	95.25	50.80	127.00
Oct.	127.00	66.80	177.80	95.25	76.20	88.90	63.50	63.50
Nov.	165.10	79.00	156.20	95.25	114.30	127.00	88.90	88.90
Dec.	101.60	5.30	123.20	38.10	107.95	25.40	95.25	139.70
Year	1016.00	523.30	1295.40	876.30	1123.95	660.40	–	895.35

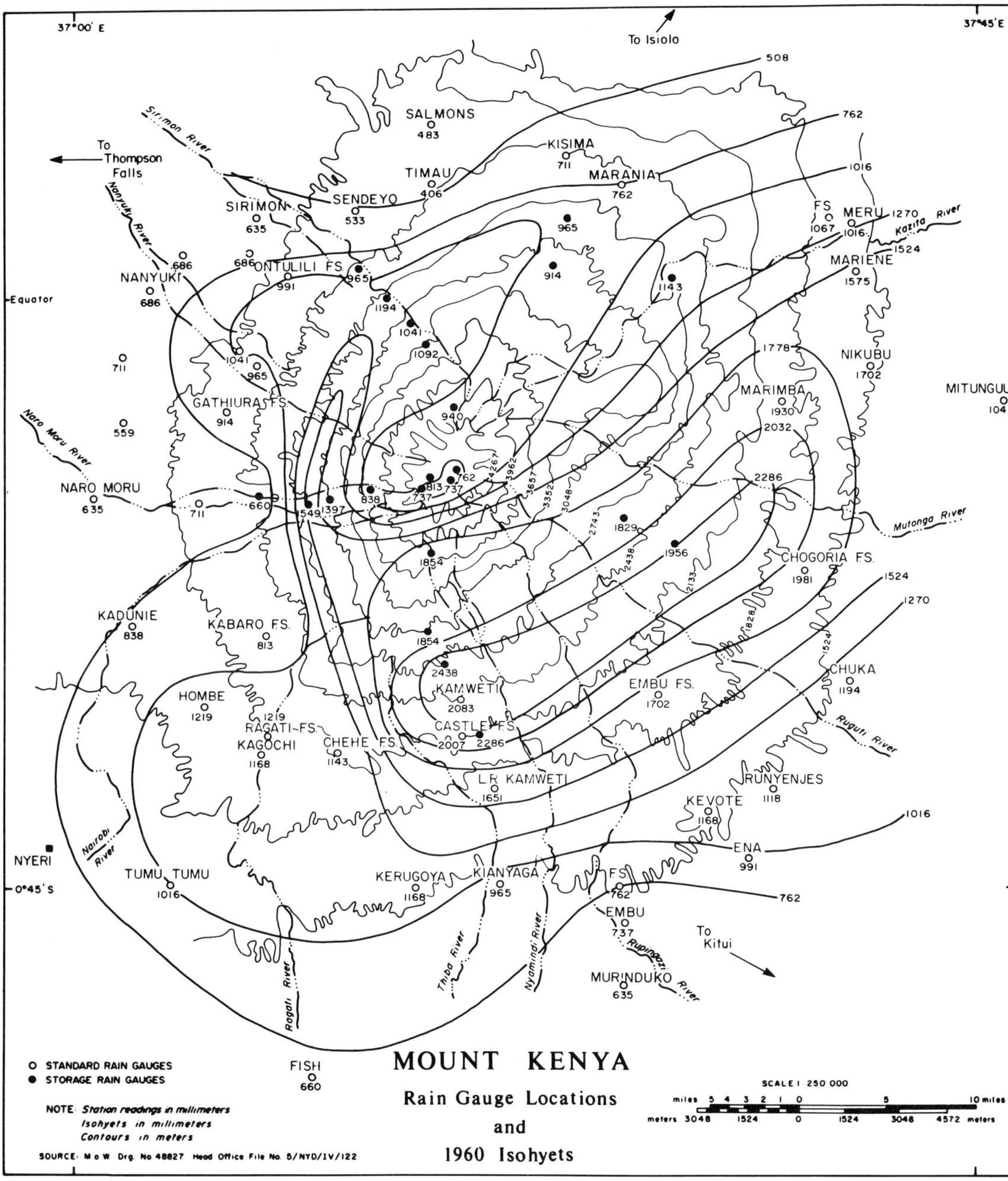

Figure 2.5: Isohyets for 1960 drawn from rain gauge data collected at stations along the Naro Moru, Kamweti, Chogoria and Sirimon tracks (after Coetzee, 1967).

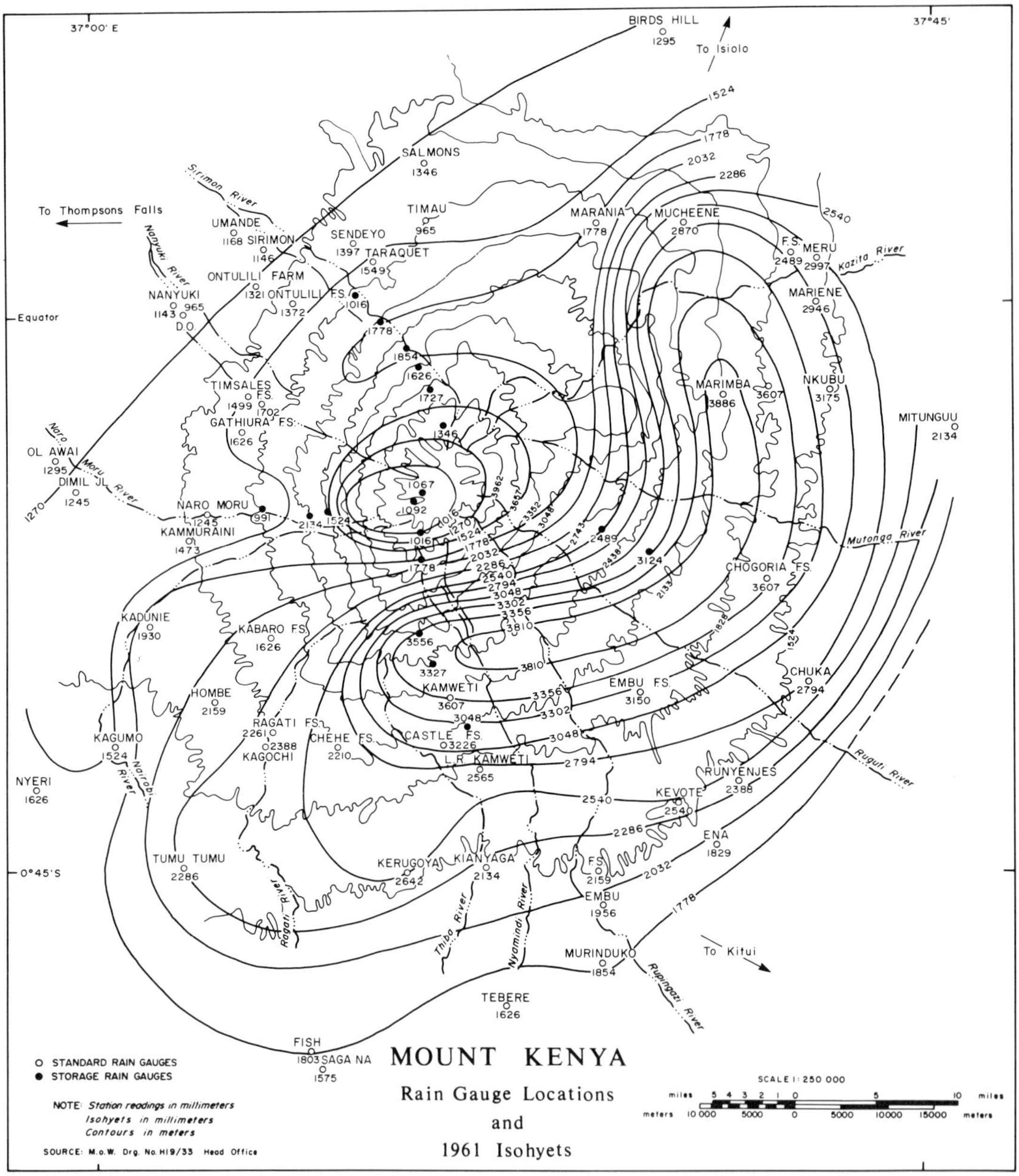

Figure 2.6: Isohyets for 1961 drawn from rain gauge data collected at stations along the Naro Moru, Kamweti, Chogoria, and Sirimon tracks (after Coetzee, 1967).

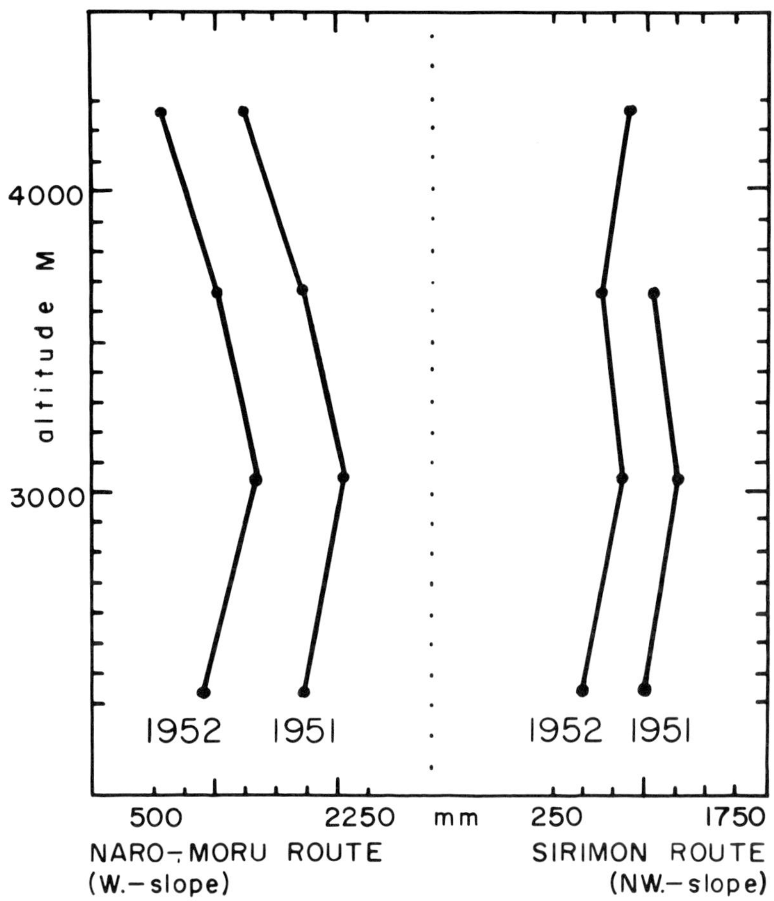

Figure 2.7: Variations in precipitation with different elevations along the Naro Moru and Sirimon routes in 1951 and 1952 (after Coe, 1967).

of Works in 1950-1952, was discontinued during the Mau-Mau Emergency. As Coe (1967) pointed out, however, there are some major items of importance in this data set:

a) Rainfall is bimodal occurring in two very well defined seasons, which parallel the rainfall pattern in lowland areas.

b) Even though the Sirimon Track is in the northern forest where it thins out towards the drier northern flank, there is a close parallel between the precipitation regime there and along the Naro Moru Track.

c) The two annual wet peaks in the precipitation regime are considerable when compared to the drier months. However, in recent years the rains have diminished considerably and even stopped entirely over periods of a few years.

The marked differences in receipt of insolation between the eastern and western flanks of Mount Kenya were pointed out by Troll and Wien in 1949. As the sun approaches its zenith during the morning, it quickly melts the snow on the eastern side; on the western side fresh snow does not melt every day owing to midday and afternoon cloud cover, and albedo changes, which create a different microclimate. This radiation pattern accounts for the distribution of the glaciers which tend to cluster on the western side where the microclimate is more favorable for their existence. Moreover, this radiation imbalance is reflected further in the asymmetry of the snowline, which is lower on the southwestern side, and higher in the eastern valleys. To the south, radiating glaciers from the Kibo Icecap, on the southwestern side of Mount Kilimanjaro, reach to 4700 m; on the northeast side they descend to 5700 m (Coetzee, 1967). At present, Mount Kenya has eleven glaciers and three of the largest – Lewis, Tyndall, and Darwin – have southwestern aspects. Osmaston (1967) mentions that snowlines on the Ruwenzori peaks lie farther down on the eastern flanks, suggesting that meteorological patterns are different in the mountains of the Western Rift.

While long-duration temperature records from different altitudinal belts and aspects are lacking on Mount Kenya, some short-term measurements from the upper Afroalpine have been made by Hedberg (1964), Coe (1967) and Mahaney (1980). These workers give the following above-ground temperatures identified in Table 2.2 (Hedberg, 1964).

Using data collected during the I.G.Y. and from the East African Meteorological Department, Coe (1967) summarized the temperature data at 3000 and 5000 m elevations on the mountain (Figure 2.8). Mean annual temperatures fluctuate only slightly (~ 2°C), while mean maximum and mean minimum temperatures exhibit greater amplitude fluctuating by as much as 6°C on a diurnal basis.

The I.G.Y. Expedition members took temperature readings from December 19, 1957, to January 17, 1958, (Coe, 1967) and provided the following data between 3000 and 4770 m (Table 2.3).

During the period over which the temperatures in Table 2.3 were recorded, the difference in mean air temperature between 3048 and 4770 m was 9.3^{o}C, giving a lapse rate of 5.4^{o}C/km. This value compares reasonably well with the International Standard Atmosphere which is 6.5^{o}C/km.

2.3 The Vegetation

The five vegetation belts in and around Mount Kenya – basal plateaux, montane forest, ericaceous belt (heather belt in Figure 2.10), Afroalpine and nival zones – are all clearly delimited on Figure 2.9; the altitudinal extent is shown in Figure 2.10. The basal plateaux vegetation belt rises from 1600 m to ~ 2100 m with precipitation varying from ~ 700 to ~ 1000 mm. On the western flank basal plateaux vegetation of *Themeda* grassveld (highland grassland) gives way to scattered *Acacia* savanna elsewhere (Coetzee, 1967). In Figure 2.9 this vegetation zone gives light tones on the western, southern and eastern flanks, showing very dark rectangular patterns in the agricultural area around Timau on the northwestern approaches to the mountain. The boundary delimiting the lower limit of montane forest is clearly shown on the southern and eastern flanks; on the western flanks the extent to which agriculturalists have made inroads into the forest is clearly demarcated where broad rectangular-shaped plowed land lies astride ribbons of residual riverine forest. The lower montane forest of *Olea africana, Ilex mitis, Juniperus procera, Podocarpus milanjianus, Podocarpus gracilior*, some of which has been partly reforested with *Pinus rigida,* contains a large number of epiphytes and an understory of ferns, which according to Cotton (1930) is indicative of higher precipitation. On the southeastern and northeastern flanks true montane rainforest persists in places.

In the upper montane forest fewer species are encountered than in the lower belt (Figures 2.11 and 2.12). *Podocarpus milanjianus* dominates along with *Ilex mitis, Olea africana, Rhamnus prinoides, Hypericum lanceolatum, Rapanea rhododendroides, Senecio johnstonii, Agaurea salicifolia* and *Hagenia abyssinica*. The vast undergrowth of fern and numerous epiphytes, so common below, are conspicuously absent here. As a result of numerous fog patches, many trees are covered with *Usnea barbarta* – the common fruticose lichen described by Coetzee (1967).

TABLE 2.2
Mean Air Temperatures and Mean
Diurnal Air Temperature Amplitude in Teleki Valley

Site	Period	Mean Air Temperature (°)	Mean Diurnal Amplitude
4200 m on[a] valley floor	Two 7-day periods in August 1948	3.1° and 3.6°[c]	8.2° and 16.4°
4200 m on[b] Rock Glacier	5 days, August 1986	-4.1°[d]	–

a. After Hedberg (1964)
b. After Mahaney (1980)
c. Mean annual, air temperature (Hedberg, 1964)
d. Mean minimum air temperature

TABLE 2.3
Temperature Data from Three Stations on Mount Kenya
From December and January 1957/58

Site m	Mean Temperature	Mean Minimum	Mean Maximum	Maximum Temperature Recorded	Lowest Temperature Recorded
3048[a]	7.4	16.2	1.7	19.4	-1.7
4191[b]	1.7	5.3	-3.6	11.0	-6.7
4770[c]	-1.9	1.1	-3.9	5.0	-8.3

a. Met Station
b. Mackinder's Camp near rock glaciers in Teleki Valley
c. Top Hut

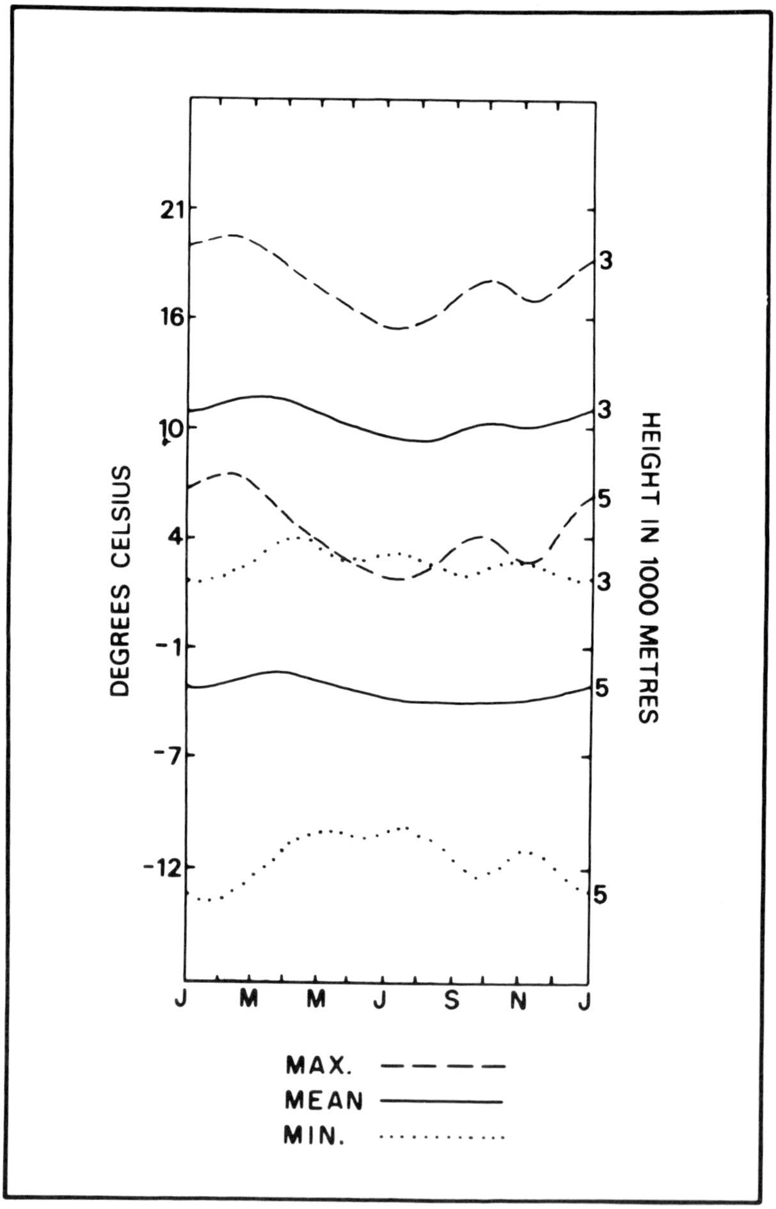

Figure 2.8: Mean monthly temperature trends at 3000 and 5000 m; mean monthly maximum and minimum temperatures for the same elevations. The mean maximum and minimum temperatures vary relative to the mean monthly temperatures (after Coe, 1967).

Figure 2.9: Satellite photo of the Mount Kenya Massif. Glaciers in the Central Peaks area appear white, radiating drainages appear dark, while higher interfluves give lighter tones. Wide U-shaped valleys are visible above a prominent break in slope that is evident at ~3000 m (lower limit of Quaternary glaciation).

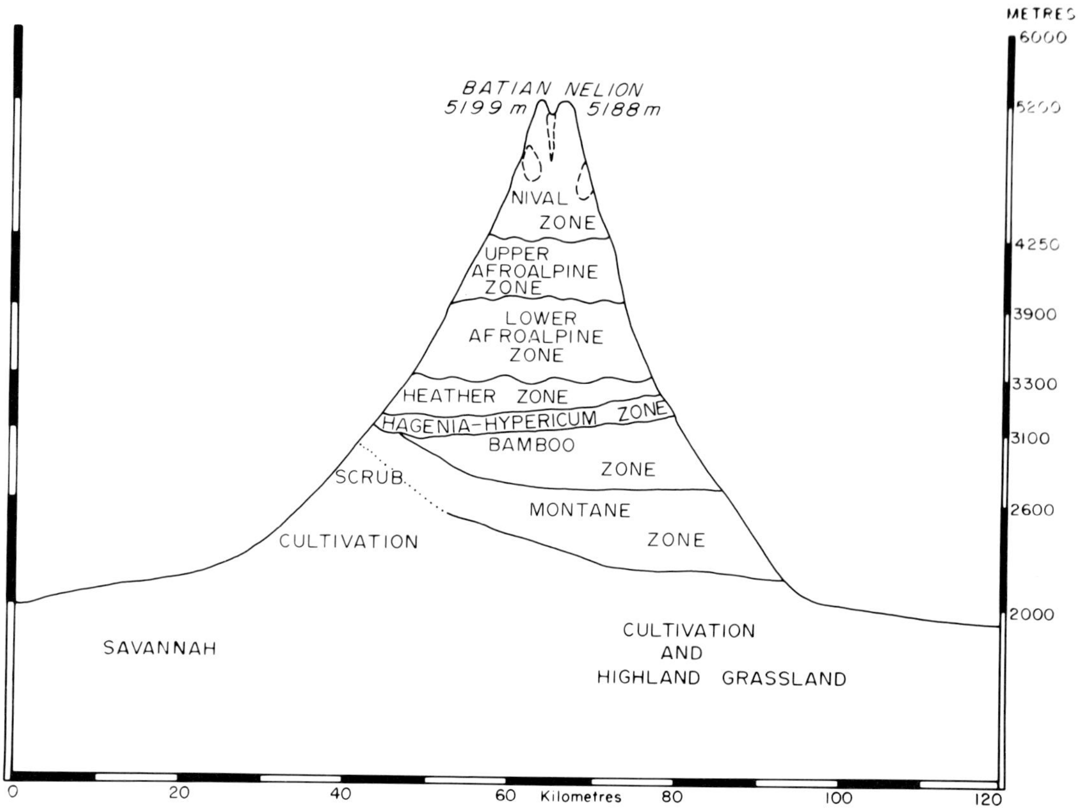

Figure 2.10: Schematic representation of the major vegetation belts on Mount Kenya (after Coe, 1967).

Above the montane forest a belt of bamboo, more extensively developed on the eastern than on the western side (Figure 2.9), stretches nearly around the mountain, thins out, and disappears to the north. Dominated by *Arundinaria alpina* it grades slowly upwards into Hagenia Woodland where *Hagenia abyssinica* and *Hypericum* trees are found festooned with *Usnea barbarta* and mosses, also beautifully reproduced in the photographs of Elliot Porter (*The African Experience*, 1974), and in Figures 2.13 and 2.14. Some of these *Hagenia* trees grow to gigantic proportions and almost all are carpeted with different varieties of lichen and moss.

The timberline ecotone at ~3200 m is easily located on Figure 2.9 by a thin, dark band clearly shown on the eastern, southern and western sides of the mountain, disappearing on the northern flank. Also several valleys on the eastern, southern, and western sides appear as deep and dark-toned trenches on Figure 2.9, reflecting the lowermost extent of glaciation which extends to immediately below the present-day timberline. Above the upper limit of timber, in the ericaceous belt, xerophytic shrubs of *Erica* and *Philippia* dominate along with inliers of

Figure 2.11: *Podocarpus* forest at ~2600 m on the Naro Moru Track, western approach to Mount Kenya.

Figure 2.12: Second-growth vegetation in upper *Podocarpus* forest at the bamboo forest ecotone. Original vegetation was bamboo (*Arundinaria alpina*). Boulders exposed along the roadcut are from the basal till in the Met Station Section.

Afroalpine grasses, often forming small to large tussocks, such as in Figure 2.15 (in Teleki Valley). The ericaceous belt is narrow on the western flank of the mountain, reaching from ~3150 to ~3300 m, widening on the eastern flank from ~3050 to ~3500 m. Often shrouded in fog, the eastern ericaceous belt produces some magnificent scenes of *Erica* and *Protea* vegetation in the Nithi drainage, just below the Gates (Figures 2.16 and 2.17).

The lower Afroalpine vegetation zone, characterized by *Dendrosenecio keniensis* and *Festuca*, is severely disturbed by tourist traffic (*e.g.* along the Naro Moru Track (Figure 1.1) creating many paths through the tussock-forming grasses (Figure 2.18). This high vegetation belt has given rise to an isolated and unique flora. Many endemic forms in this zone have been extensively investigated by Hedberg (1951, 1957, 1964, 1968); on Kilimanjaro, Salt (1954) studied a similar belt of vegetation. Among the flora, giant *Lobelia*, *Alchemilla*, and the tussock-forming grasses – *Festuca* and *Agrostis* – dominate along with numerous species of *Carex*. This broad belt of the lower Afroalpine extends along valley bottoms and interfluves (Figure 2.9) between ~3200-3500 m to ~3750 m, where it merges with the upper Afroalpine vegetation (Figure 2.19), characterized by *Dendrosenecio keniodendron*. In some drainages on the western side, it may flourish and develop into dense, tall stands of *Dendrosenecio* forest (Figure 2.19). Quite high and extensive today, these broad vegetation belts migrated down the mountain, and onto adjoining interfluves, as a result of forcing by deteriorating climate several times during the Quaternary. Glacial climates witnessed the growth of large valley glaciers which radiated outward from an ice cap that covered almost all of the Central Peaks area and even flowed over the top of Lenana, at 4985 m, the third highest summit (Figure 1.3) in the Central Peaks area.

On fresh glacial surfaces it appears to take upwards of ~1000 yr for upper Afroalpine vegetation to colonize and achieve a cover and floristic composition somewhat similar to the surrounding older exposed moraines dating from the Liki III Glaciation (~12,500 yr BP) (Coe, 1967; Spence, 1989). The foreland in front of the Tyndall Glacier (Figure 2.20) provides a sequence of Neoglacial deposits (*e.g.* Lewis and Tyndall advances) and Liki III moraines. The youngest deposits dating from the Lewis advance exhibit fresh boulder surfaces lacking *D. keniodendron* and *Usnea*, but with a developing grass and sedge community of plants forming on very thin Entisols. Beyond this an apron of till deposited during the Tyndall advance (Mahaney, 1985) shows higher plant cover with *Usnea* appearing, but without appreciable cover of *D. keniodendron*. On the oldest surface, of Liki III age, *D. keniodendron* achieves a higher cover along with other floristic components of the upper Afroalpine vegetation.

The nival zone (Figures 2.21 and 2.22) is more or less limited to cirque forelands where "boulderfield" communities of plants have developed on moraines emplaced by the Lewis advance during the late Neoglacial. When Gregory visited the mountain during the 1890s, the Lewis and Tyndall glaciers were at their maximum Little Ice Age positions. With each recession of these glaciers over the last ninety years, it has been relatively easy to document their positions with historic photographic evidence, thereby eliciting information on plant colonization. Plant succession data for both the Lewis and Tyndall moraines (Figures 2.23A and B) show striking similarities. The species present are very much similar in both cirques and some like *Senecio keniophytum* grow right up next to the present-day ice limit. Most plants of the nival zone appear within ~50 to ~100 yr; only percentage cover increases thereafter as more weathering ensues and soil formation occurs. Thereafter the boulderfield community develops into the upper Afroalpine complex within ~1000 yr (Mahaney, 1985).

On older surfaces of Liki I and II age, at the lower limit of glaciation (~3200 m), vegetation from moraine crests, and slopes, to bog-covered ground near streams has been described by Coe (1967). As shown in Figure 2.24 (map view) and Figure 2.25 (cross-section) a variety of different habitats are located in relatively small areas. On high and well-drained surfaces and on moraine slopes *Protea* and *Erica* (heather) tend to dominate. In swales between moraine crests and next to small tributary streams, tussock-forming grasses, such as *Festuca* and *Agrostis*,

Figure 2.13: Hagenia Woodland (*Hagenia abyssinica*) at ~3100 m in lower Teleki Valley. Boulders, from the Teleki Glaciation, in 25-30 cm of loess are in the foreground.

Figure 2.14: Hagenia woodland near the timberline along the Naro Moru Track.

Figure 2.15: Timberline ecotone of the Naro Moru Track with tussock-forming grasses (mainly *Agrostis trachyphylla* and *Festuca pilgeri* and taller, almost tree-like forms of *Philippia*).

Figure 2.16: Waterfall below "The Gates" in the Nithi River Basin at 3100 m. Some *Erica* and burned *Protea* appear shrouded in cloud.

Figure 2.17: Outlier of *Podocarpus* in the ericaceous belt in the Nithi River Basin at 3075 m.

Figure 2.18: Lower Afroalpine vegetation belt dominated by *Dendrosenecio keniensis* with tussock-forming grasses and *Lobelia* in the vertical bog at ~3500 m (Naro Moru Track).

Figure 2.19: Upper Afroalpine vegetation belt with *Alchemilla argyrophylla* in the foreground and *D. keniodendron* in the middle ground (upper Teleki Valley).

Figure 2.20: Boulderfield vegetation zone on Neoglacial and Liki III moraines below the Tyndall Glacier. Upper Afroalpine vegetation belt.

Figure 2.21: Nivation scree and fresh snow in upper Gorges Valley at 4600 m. Hall tarns are visible on the volcanic neck to the left of "the notch" that leads to Lake Michaelson. Thick stratus clouds from the southeasterly trade winds are well developed to ~3200 m.

Figure 2.22: Typical early morning weather over the Central Peaks. Lewis Glacier is to the right; Tyndall Glacier is to the left. View is from the south.

make up most of the vegetation. Closer to large streams, such as the Nithi River in Figure 2.24, a complex mosaic of sedges, grasses, and giant *Senecio (D. johnstonii)* reflect the change from xeric and mesic sites to the hygric conditions of boggy habitats.

The vegetation belts of the East African Mountains and their altitudinal extent are shown in Figure 2.26. The three belts of Hedberg (1964) – montane forest, ericaceous zone, and Afroalpine – are found on Mount Kenya, Kilimanjaro and the Cherangani Hills (Figure 2.1); however, on Mount Kenya the montane forest belt can be subdivided further into broad areas of Hagenia – Hypericum, bamboo forest, and montane rainforest. Because these vegetation belts are quasi-permanent features of the present day environment, one must consider that during the last glaciation each vegetation belt fluctuated over as much as 1500 m, displacing the Afroalpine species into a relatively larger area (perhaps twice the present extent) under an Afroalpine paleoclimate drier than today, and with mean annual temperatures ~ 7.0° C colder than today. It is within the context of this environmental setting that the flora of Mount Kenya evolved over 2.0 my to its present extent and composition.

2.4 Geomorphic Overview

Mount Kenya, one of the principal high mountains in East Africa, lies on the equator about 125 km north-northeast of Nairobi. Averaging about 100 km in diameter, it rises from surrounding plateaux areas to 5199 m (Figure 1.1). The twin peak area – Batian (5199 m) and Nelion (5188 m) – represent the plug of the extinct volcano, the present day summit of which is probably below the crater wall. Just how much of the summit area has been eroded is subject to considerable controversy. Gregory (1900) suggested that the original height might have been ~ 6000 m. Most workers have accepted this assessment (Baker, 1967; Coe, 1967); however, I (1986a) argued for a lower original summit elevation of ~ 5500 m, using the thin fills in lower drainages, and ~ 500 m deep valleys above the limit of glaciation as evidence for considerably less erosion.

With its deeply dissected tributary valleys, Mount Kenya stands in stark contrast to Mount Kilimanjaro, 300 km to the southeast. Drainages radiating from Mount Kenya vary considerably in depth and shape depending upon their erosional history. Glaciated valleys tend to have considerable depth and U-shaped profiles; fluvial valleys tend to have V-shaped cross-sections and rather shallow depths. Most interfluve ridges on the western and eastern flanks were covered with ice during the last glaciation and hence have smooth surfaces. But some interfluves, such as Höhnel Ridge, and the ridge at Two Tarn in the upper Burguret Drainage, have well weathered summit areas with highly altered and exposed bedrock, covered with deep, cavernous weathering pits, suggesting at least that ice did not cover all surfaces during the last glaciation.

Near the lower limit of glaciation (~ 3000 m) all glaciated valleys change from a U-shaped to a V-shaped profile, and while erratics are present in fills below this elevation, stone shapes become progressively more rounded with increasing distance downvalley. Greater glacial erosion and sediment transport on the eastern flank of the mountain have given rise to thicker outwash fills there on the order of ~ 100 m in some drainages. On the western flank fills sometimes reach ~ 100 m, but are usually considerably thinner, suggesting somewhat less transport. On the other hand, the absence of well-defined terminal moraines in lower Hausberg and Teleki valleys suggests that rapid decantation of meltwater has removed much of the glacial sediment, possibly in a jökulhlaup-type system with upper-flow regime velocities. The style of deglaciation which is known to be rapid, together with valley glaciers with meltwater confined to pre-existing glaciated valleys, and tropical-type geophysically warm glaciers might create a setting in which jökulhlaup transport might be expected.

One outstanding feature of glacial erosion on Mount Kenya is the abundance of ice-covered lake basins of variable size. Teleki Tarn, Lake Michaelson, Oblong Tarn, Hausberg Tarn, Naro Moru Tarn, and Lewis Tarn are striking examples of the effectiveness of tropical glaciers in

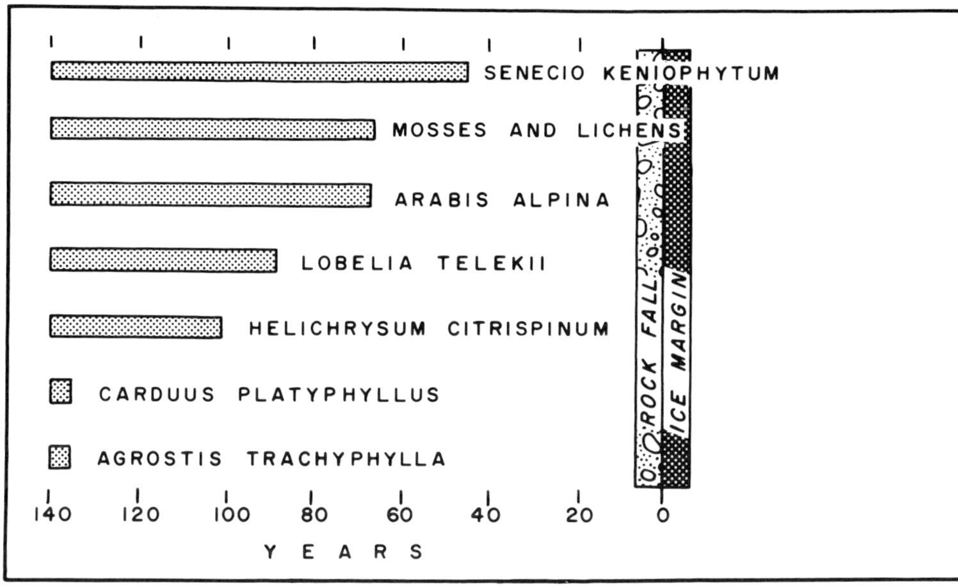

Figure 2.23A: Relative time required for the colonization of moraines by several species of plants in Tyndall Cirque (after Coe, 1967).

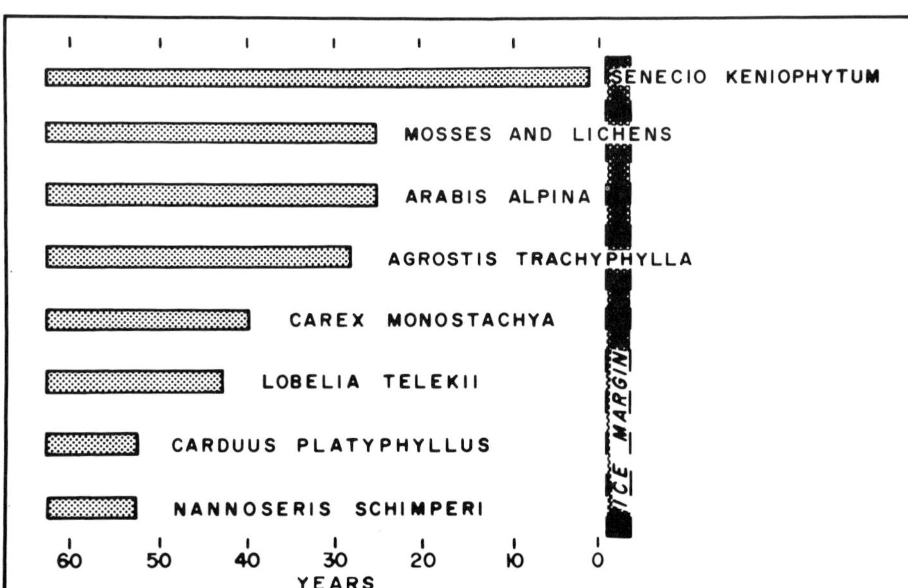

Figure 2.23B: Relative time required for the colonization of moraines by several species of plants in Lewis Cirque (after Coe, 1967).

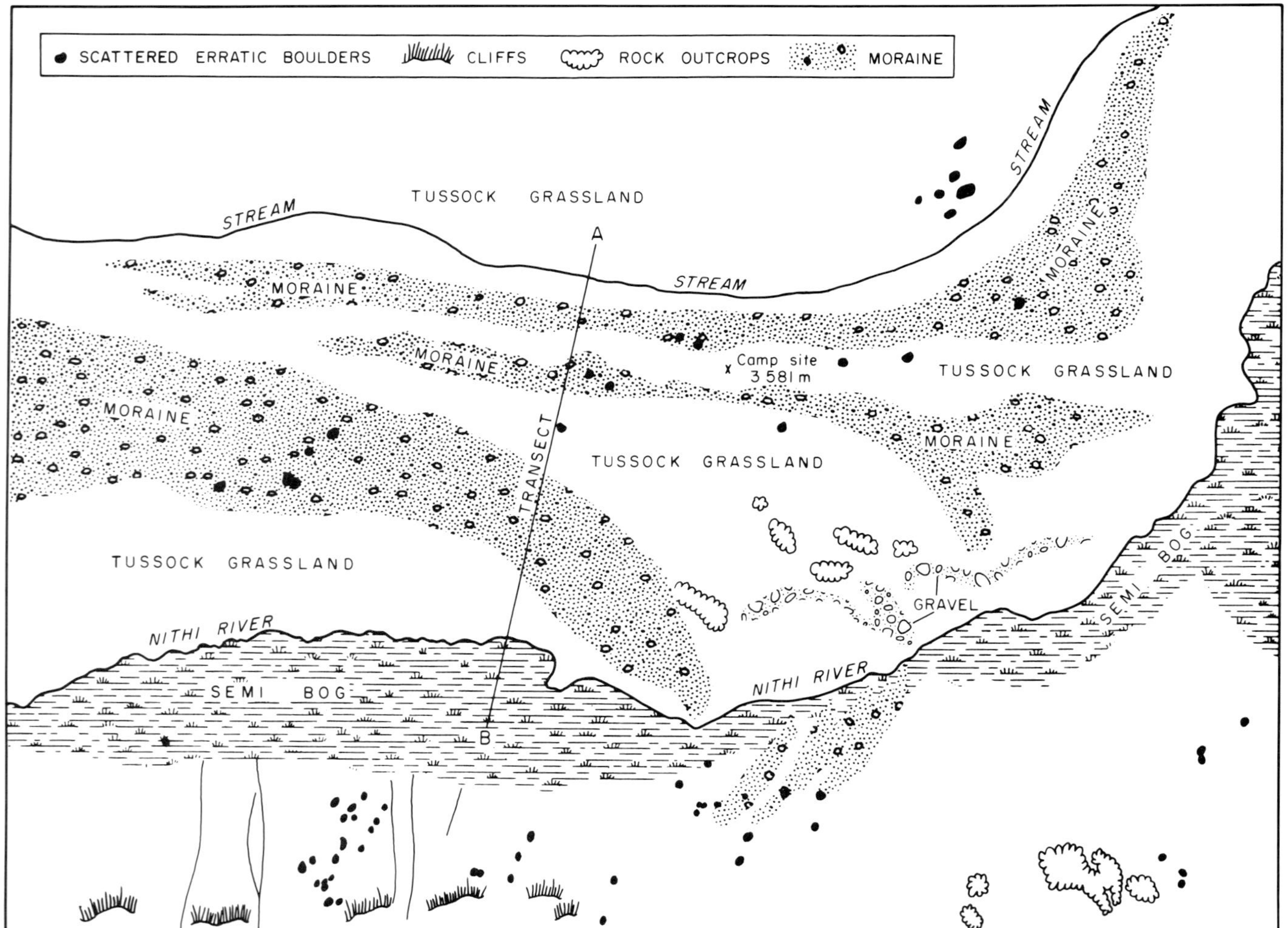

Figure 2.24: Transect from semi-bog across moraine crests to a tributary stream; (from Coe 1967).

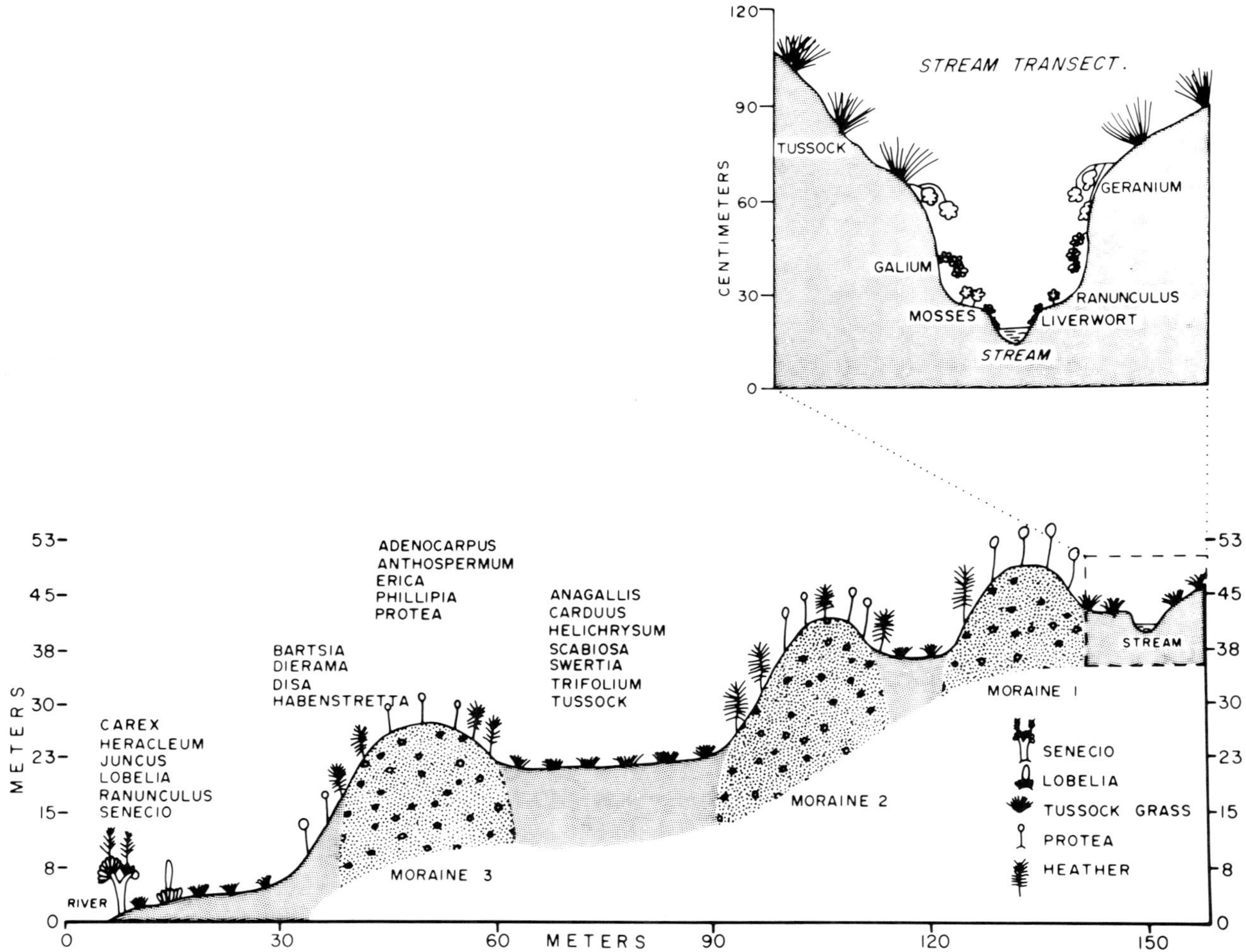

Figure 2.25: Vegetation assemblages along transect shown in Figure 2.24. Hygric forms of *Ranunculus, Carex* and *Juncus* give way to mesic and more xeric forms of *Erica* and *Protea* on moraine ridges and tussock-forming grasses in swales between moraine crests; (from Coe, 1967).

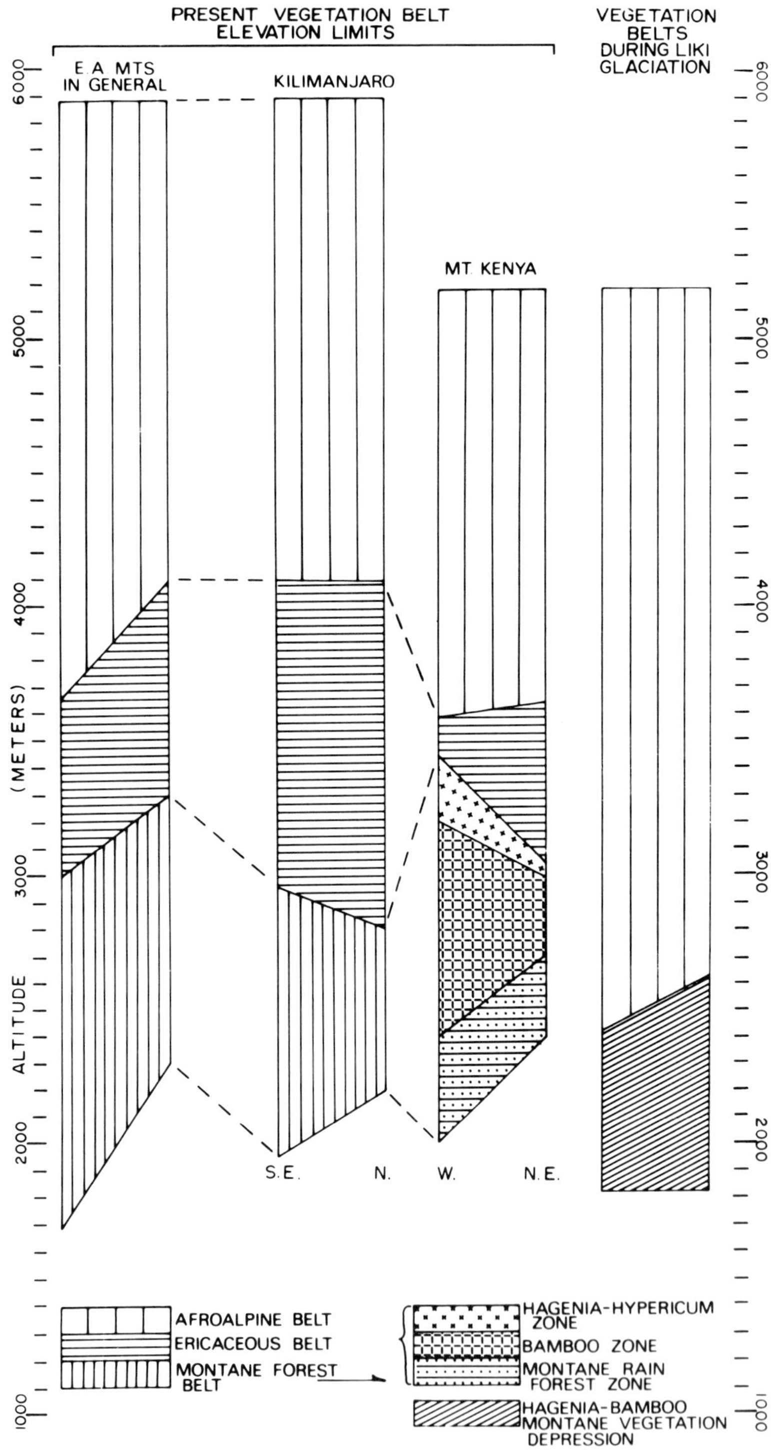

Figure 2.26: Vertical vegetation gradient on the East African Mountains. While the ericaceous belt is narrower on Mount Kenya, the Hagenia woodland and bamboo forest zones are more extensive (from Coetzee, 1967).

creating rock basins that later filled with water. Other lakes created by impounding water behind end moraines are present as well – Tyndall Tarn, Harris Tarn, Thompson Tarn are good examples. Still other water bodies, such as Hall Tarns, are found in small bedrock depressions fed by direct rainfall and dew. Because the only moisture outlet is through evaporation, these tarns often exhibit considerable fluctuation in water level during the year.

In addition to macroscale geomorphology, there are many smaller-scale landforms on Mount Kenya that will be touched upon later in this volume. These include needle-ice forms, soil stripes, and mud polygons, all of which are closely related to the microclimate, flora and fauna, soil cover, and variations in slope.

2.5 Stratigraphic Framework

The Quaternary stratigraphic framework is outlined in Figure 2.27. There is at present enough stratigraphic evidence to subdivide the older glaciation of Nilsson (1935) into four sub-units: Gorges, Lake Ellis, Naro Moru and Teleki tills (Mahaney, 1983a, 1988a). Gorges and Lake Ellis tills are known only from the eastern drainages. Naro Moru Till is known from several locations in Teleki and Gorges valleys (Mahaney, Barendregt and Vortisch, 1989). One section of Lake Ellis Till in Hobley Valley (HOB19) gave reversed magnetism suggesting an age older than 0.73 my. In the same section, loess overlying till gave normal magnetic polarity suggesting a later Pleistocene age. At site TV23 in Teleki Valley a buried paleosol formed in Naro Moru Till and younger loesses provides evidence for considerable weathering, and a wetter paleoclimate, during the Naro Moru/Teleki and Teleki/Liki interglacials. While the time lines for these interglacials are uncertain at present, the increase in clay size material, along with variations of gibbsite and hematite, are taken to suggest a lower intensity of weathering relative to the interglacial following the Teleki Glaciation.

Teleki Till is widespread on Mount Kenya and three sections are known on Ithanguni (Derbyshire and Mahaney, 1989). Nearly every drainage on Mount Kenya contains Teleki drift, but in Teleki Valley till overlies a buried paleosol formed in Naro Moru Till, loess and tephra (Vortisch, Mahaney and Fecher, 1987). Because Teleki Till is often covered with Acheulian hand axes, and contains well-weathered surface boulders and cobbles, it is considered to be of middle Pleistocene age (> 0.1 my), and hence older than the last glaciation.

Deposits from the last glaciation are most plentiful in the < 35 ky time frame and most information about them comes from bog and slope sediments, not directly from the tills. All end moraines (Liki I and II) in the lower valleys at ~ 3200-3400 m appear to date from the last advance of Liki ice. Radiocarbon ages from bog-bottom sediments on these deposits suggest that ice retreat was well underway by ~ 15,000 yr BP (Mahaney, 1988b). Cover sediments on slopes below the lower limit of glaciation bury older paleosols that radiocarbon date between ~ 21 and ~ 25 ky, suggesting that Liki glaciers advanced into the lower valleys, creating slope instability, some time before ~ 21 ky. After recession from the Liki I/II positions, ice recession was rapid, and without stillstand, until glaciers reached ~ 4000 m (Mahaney, 1985) where they halted to construct small recessional moraines that are usually double crested (as in Teleki Valley), and sometimes have triple crests (as in Gorges Valley). After a short interval of perhaps a few hundred years, ice recession began again and most moraines were free of ice by about ~ 12,500 yr BP. In some cases large lakes formed from meltwater provided by receding ice (*e.g.* Hobley and Gorges valleys). Liki III ice in Teleki Valley, although larger than in other valleys (mirroring the present-day climatic situation), existed as two distinct valley glaciers – one flowing from Tyndall Cirque, and the other from Lewis Cirque – constructing an "interlobate moraine" westward from the Midget and Point John (Figure 2.28). Following the late glacial episode, Liki III ice either receded into the cirques occupying positions similar to today, or it disappeared.

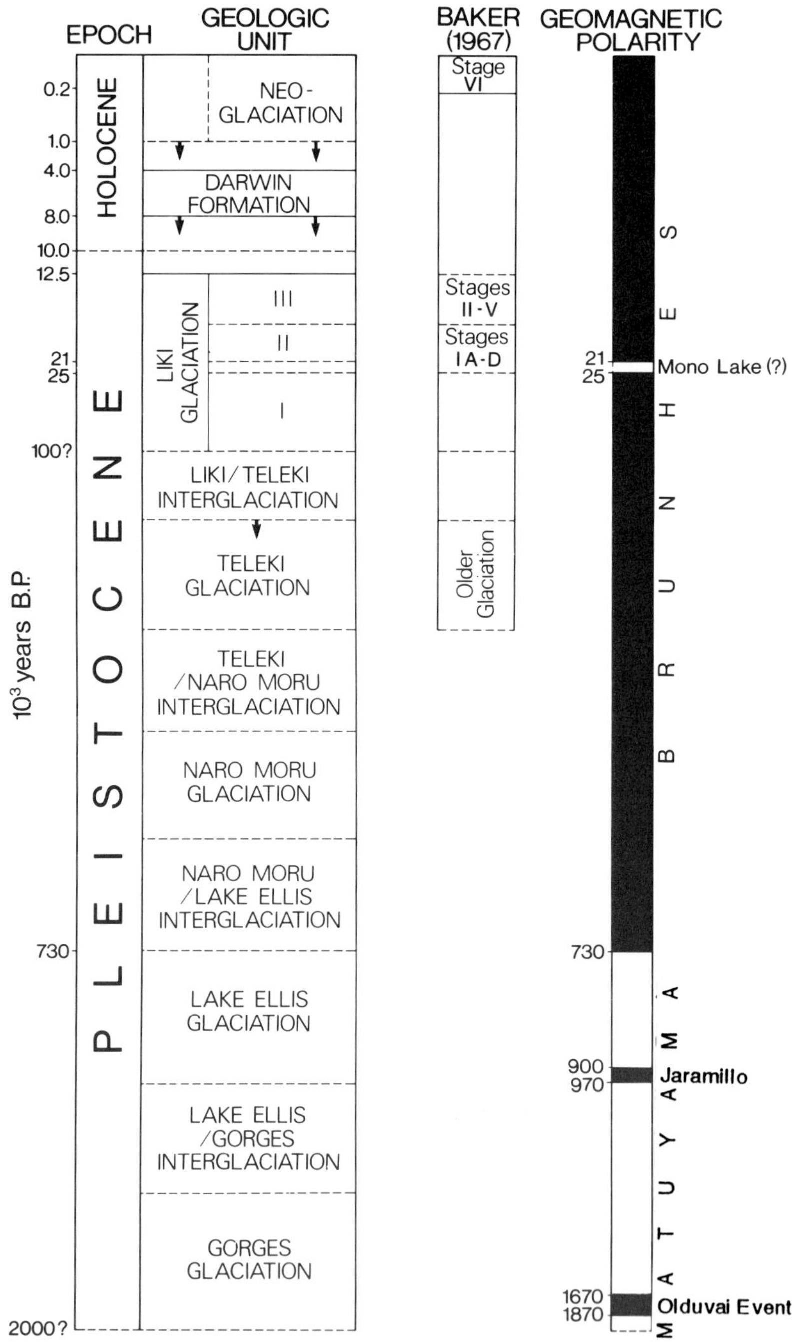

Figure 2.27: Quaternary stratigraphy and paleosols for the Mount Kenya area.

Figure 2.28: Late glacial moraines below Point John and The Midget in the Central Peaks area.

After the retreat of ice in late glacial time, surfaces stabilized in the Afroalpine zone and soils formed in them. There is no geomorphological evidence for early or middle Holocene ice advances, but there is considerable evidence for a period of alluvial and colluvial fan construction between ~8000 and ~4000 yr BP. These deposits, collectively termed the Darwin Formation (Mahaney, 1985), are widespread on Mount Kenya, contain a multitude of buried paleosols, and are bracketed by radiocarbon dates. Following this generally warmer and wetter phase of middle Holocene fan activity, paleoclimate became progressively cooler and drier, culminating first in dune construction on the northeastern side of the mountain, and later in the Neoglacial (period of glacier resurgence during the later Holocene; ~1000 yr BP). Neoglaciation on Mount Kenya is considerably different than in other middle- and high-latitude areas (Mahaney, 1978, 1984b; Mahaney and Spence, 1984; Davis, 1985) in that it is limited to the last 1000 years and consists of only two advances: Tyndall (older) and Lewis (younger). For the last ninety years all glaciers on Mount Kenya have been slowly receding, as in other parts of the world, in response to overall warmer temperature. Five glaciers have disappeared since Gregory's visit in 1893, and with projected increases of CO_2 in the atmosphere, it is possible that increased warming may melt the remaining eleven glaciers over the next few decades.

CHAPTER 3

DATING AND ANALYTICAL METHODS

3.1 Introduction

The use of multiple relative (RD) and absolute (AD) dating methods on Mount Kenya and elsewhere provides important chronological tools for Quaternary stratigraphers. RD methods used on Mount Kenya include: changes in topographic position (morphostratigraphy), superposition, weathering characteristics, lichen size and cover, soil characteristics, amino acid racemization in paleosols, and paleomagnetic signature. AD methods are confined to radiocarbon, potassium-argon, thermoluminescence, and lichenometry (from the lichen growth curve). Because recognizable pyroclastic layers are present only in the older deposits and paleosols (> 100 ky), it is not possible to use tephras to differentiate late Pleistocene sediments. Moreover, younger Holocene sequences are above timberline, making it impossible to use dendrochronological methods. Uncontaminated buried organic-rich sediments are rare, and the few radiocarbon dates available document only the Liki interstade, end of the Liki Glaciation, and the Holocene deposits.

The AD and RD methods outlined above have been used at many field localities to date deposits in the geologic succession. Of these many methods, radiocarbon and thermoluminescence have the greatest utility for AD age determination because they cover all, or most, of the late Quaternary, which contains a near-complete sequence of deposits. Weathering parameters and topographic position, although crude, give relative ages for surface deposits, which can be refined by the use of soil-stratigraphic relationships. Lichen characteristics of cover and maximum size provide an important means of dating Neoglacial substrates, while use of the lichen-growth curve on southern exposures can be considered an AD method. Used individually, each method suffers from limitations that hinder refinement of the stratigraphic record. The use of several methods at each locality increases dating precision and helps strengthen the chronology. In this chapter, the dating and laboratory analytical methods are outlined and discussed. In later chapters, dating methods are applied and evaluated.

3.2 Topographic Position and Superposition

Since there are many sequences of deposits on Mount Kenya, a simple means of differentiating units is by their surface position relative to one another in different drainages. Recognizing the relative age of deposits by their topographic position is a variation of the morphostratigraphic method, and provides a means of establishing at least a tentative sequence. Such methods have been employed by numerous workers for decades; Blackwelder (1915), Nilsson (1931, 1935), and Baker (1967) all used morphostratigraphic techniques.

Because many sequences of deposits and paleosols are "stacked", that is superimposed over one another, it is important to point out that relative age can be determined by relative position in various sections.

3.3 Weathering Characteristics

Several different weathering parameters were measured on samples of 50 or 100 stones at each field locality. Counts were made of boulder frequency; that is, the number of boulders over 0.3 m diameter in unit areas of 10 m x 10 m (100 m^2) or 5 m x 5 m (25 m^2), following the pioneering work of Blackwelder (1915) and Nelson (1954). Usually the area was paced out in the field and total counts were made of all stones without regard to lithologic differences. The results allow differentiation between deposits of glaciation rank, but are not generally useful for differentiation at the substage, or lesser level (Mahaney, *et al.*, 1981; Mahaney, *et al.*, 1984a).

The ratio of fresh to weathered stones can be determined by recording the sound produced by boulders struck with a hammer. A sharp ring and strong rebound of the hammer indicates a fresh boulder, whereas a weathered rock produces a dull sound and weaker hammer recoil. Although this method is admittedly not particularly accurate, and does not work well on phonolite, it is quickly done in the field. Alternatively, one might use the fretting ratio of Sharp (1969), where a weathered stone is defined as one where more than 50 percent of the stone surface is weathered to a depth greater than the average grain diameter. Care was taken to select stones of similar rock type, on a single valley side, to avoid lithologic and microclimatic effects. Data produced as weathering ratios allow differentiation among deposits at the glacial stage rank (Liki *vs.* Teleki glacial deposits), but are not generally useful for differentiation at the substage or stadial level (Mahaney, *et al.*, 1984b).

Weathering rinds are indicative of the degree to which iron-bearing minerals oxidize and discolor the outer periphery of clasts. The thickness of rinds measured perpendicular to stone surfaces is related to time since deposition. This method has been described by numerous workers, including Birkeland (1973), Mahaney (1974, 1978, 1981) and Kiver (1974). Most workers only measure the maximum thickness of discoloration, neglecting the irregularly developed rind commonly found on many stones. Some investigators undoubtedly measure discoloration along fracture faces, which may amount to two or even three times the rind thickness. My practice was to ignore fracture faces and to measure both the maximum and *minimum* rind on the stone periphery. I have also measured the degree of internal discoloration of stones, but find it is controlled more by lithology than time, and difficult to quantify (Mahaney *et al.*, 1984a).

Weathering pits on boulders, cobbles or rock outcrops can be measured with the aid of a dial or vernier caliper. Depth and width of pits reveal differences between sets of deposits, provided that some care is taken to avoid porphyritic rocks, and in choosing only one lithology (*e.g.* basalt). Accurate width measurements are reasonably easy to obtain, but determining depth is more difficult. Measurements were made by fully extending the caliper depth and placing it in the bottom of the pit perpendicular to the pit rim. The main beam was then brought to a position level with the top of the pit and a measurement was made. Fifty measurements were taken at each study site and the means calculated.

The data in later chapters show the degree to which deposits can be differentiated using measurements of boulder frequency, weathering ratios, rinds, and pits. The data allow differentiation at the glacial stage level, while differentiation between stades is not possible. The presence/absence of some minimum rind development is useful in distinguishing Neoglacial deposits from deposits of late-Liki age.

3.4 Lichenometry

Maximum diameters and percent cover of the dominant crustose lichens – *Rhizocarpon* section *Rhizocarpon*, *Buellia* and *Lecidea* – are used to differentiate deposits in the Neoglacial succession. The method rests on the following assumptions:

1. the largest lichen thallus represents the oldest and/or fastest growing lichen on a substrate,

2. lichen growth (including percent cover) is indicative of elapsed time since deposition of the substrate,

3. the substrate falls within the life span of an individual lichen species, and,

4. effects of climatic change are smoothed out over time.

Deposits in all cirques on Mount Kenya were sampled to determine the maximum diameter of each crustose lichen. Percent cover was estimated from percentage diagrams in Oyama and Takehara (1970). Measurements of individual thalli were made to the nearest millimeter. Only the largest elliptical-shaped thalli were measured, and only maximum diameters were used to assign relative ages. Thalli with irregular shapes and those found in depressions were bypassed to avoid problems associated with "interfingering" by individuals of the same species (Mahaney, 1973) and snowkill (Curry, 1969), respectively. Additional sampling restrictions followed by other workers (Beschel, 1957, 1958; Benedict, 1968; and Andrews and Webber, 1964) including standardization of rock type, prevailing wind, exposure to ice crystal blasting, aspect, available moisture, and surface stability were followed in this study.

Lichenometry allows the separation of Lewis and Tyndall advances which are indistinguishable on the basis of surface morphology and only marginally identifiable by weathering characteristics. Lichenometry can also be used as an AD tool, but it must be emphasized that the growth curve in Chapter 8 was established using historic and radiometric controls, and it is based on only two control points. If the data for maximum thalli diameters and percent cover between correlative units on Mount Kenya are compared some differences emerge, the apparent result of climatic variations between cirques. Thus, a separate growth curve must be made for each cirque if lichenometry is to be used as a precise AD method.

3.5 Paleosol and Soil Stratigraphy

Deposits in the geologic succession can be separated using soil morphology, particle size distributions, clay mineralogy, and selected soil-chemical parameters. Individual pedons are assigned to soil stratigraphic units[1] on the basis of stratigraphic position and morphological features which may be continuously recognized and mapped (Mahaney and Fahey, 1976; Mahaney, 1978, 1982a). Such units form, *in situ*, in deposits by the action of pedological processes working downward on fresh unaltered materials. Soil stratigraphic units are assigned informal names coinciding with the age of the deposits in which they form (e.g. use of the prefixes "post" and "pre" to avoid a proliferation of geologic names). A few examples illustrating the use of soils and soil properties in differentiating Quaternary deposits follow.

Moraine sequences have been described on all the continents, but differentiation of glacial deposits by degree of soil development is a method that was developed in the western United States (Richmond, 1965; Mahaney, 1978; Birkeland, 1984; Mahaney *et al.*, 1984a, 1984b).

Study sites were selected mainly on topographically high and well drained deposits, except when toposequences were being evaluated. Soil samples were collected from each horizon described in the field following the nomenclature of the Soil Survey Staff (1951, 1975), and Birkeland (1984).

Particle size distributions commonly allow differentiation of soils in geologic successions. If the amount of clay in the parent material of each soil is fairly uniform, then changes in clay content in B horizons of soil-stratigraphic units should reflect relative amounts of weathering over time and/or airfall influx. Not all of the clay accumulates by *in situ* weathering, as many soils on Mount Kenya have thick loess caps, suggesting large airfall influxes of silt plus clay during Quaternary glacial episodes. Many soils display upward fining sequences in the profiles, the result of loess

---

1 Soil-stratigraphic units as they are used here are surface pedons of young and old age. As with its use elsewhere (Morrison, 1978), the term paleosol refers to both surface and buried soils.

being translocated downward into the solum by percolating water. The older soils have thicker loess caps and the overall silt content is higher than in younger soils. This condition is attributed to a longer time for weathering and greater amounts of loess blown in from nearby glaciers.

The centers of gravity (mean phi) for the B horizons of some profiles were calculated as follows:

$$\bar{x}\,\varnothing = \frac{\Sigma 25th + 50th + 75th}{3} \text{ (percentiles)} \qquad (1)$$

to determine if mean phi shifts with increasing age (Mahaney, 1978).

3.6 Amino Acid Dating

Many varieties of fossils are now routinely analyzed to determine the D/L (D to L isomers) ratios of amino acids for relative and absolute age dating of sediments. The method has been particularly useful when fossils have experienced similar temperature histories and diagenetic conditions. Marine molluscs have been widely and successfully utilized for age determination, whereas freshwater and terrestrial molluscs offer promise (Andrews *et al.*, 1979, Belknap and Wehmiller, 1980; Miller and Hare, 1975; Rutter *et al.*, 1980; Wehmiller, 1982). Bone and wood have been useful, although problems have been encountered and discussed by numerous workers (Bada, 1972; Bender, 1974; Rutter and Crawford, 1984). In addition, investigations involving amino acid racemization age dating have been carried out with marine sediments (Bada *et al.*, 1970; Kvenvolden *et al.*, 1970), foraminifera (King, 1980; Wehmiller and Hare, 1971), coral (Wehmiller *et al.*, 1976; Wehmiller and Hare, 1970) and with paleosols (Limmer and Wilson, 1980; Mahaney, Boyer and Rutter, 1986).

Racemization (or epimerization) of amino acids in geological specimens is a complex process involving a number of factors. The rates of interconversion processes are dependent upon the type of amino acid, ambient temperatures, types of materials in which the amino acids are embedded (shells, bones, wood, *etc.*), and the state in which the amino acids occur; *i.e.*, whether they are terminally bound, internally bound, or free amino acids. In addition, other external diagenetic factors such as moisture content, acidity, and oxidation-reduction conditions can play important roles. The accuracy of D/L determinations depends upon the care of preparation and the reliability of the preparation methods employed. These factors contribute to the apparent lack of interest in amino acid D/L ratio determinations of paleosols. The complex nature of soil genesis and diagenetic changes that take place at the surface and after burial compound the problems already inherent in the method. It will be seen, however, that in the present work there is a tendency for D/L ratios, of aspartic acid and others, to increase slowly with age of the paleosol.

Amino acids were determined at the University of Alberta and at York University. Once in the laboratory, samples were prepared for D/L ratio determinations using the following procedures. Large samples of between 1 and 2.5 grams of organic material were weighed directly into clean vials. It was not possible to clean the samples thoroughly by the usual method of rinsing and sonicating in HCl and double distilled water because whole-soil samples were utilized.

The samples were hydrolyzed for 24 hours at 108°C in 5.5M HCl. The hydrolyzate was decanted and then dried under vacuum in a rotary biodryer. The dried residue was then taken up in double distilled water and placed on a cation exchange resin. Samples were eluted with 3M

NH_4OH and again dried under vacuum. Esterification was achieved with acidified isopropanol and heating for 15 minutes at 100°C. After drying, the samples were acetylated with pentafluoropropionic anhydride in methylene chloride. The samples were then heated for 5 minutes at 100°C and evaporated to remove any excess PFPA. After being washed with CH_2Cl_2 the samples were dried over K_2CO_3. Later samples were filtered and the filtrate injected into a 25 meter x 0.3 mm Chriasil-Val capillary column installed in a Hewlett-Packard 5840A Gas Chromatograph equipped with a FID detector. Chromatographs and analysis of typical Mount Kenya paleosols follow in Chapter 10.

3.7 Paleomagnetism

Paleomagnetic samples were collected from various sites on the mountain. After extensive excavation and removal of overburden, undisturbed sediments were found which proved suitable for paleomagnetic sampling in 2 x 2 x 2 cm plastic cubes. Azimuthal orientation of the samples was made in the field by magnetic compass, each sample being aligned parallel to the vertical face of the outcrop, and spaced from 2 to 10 cm apart. Pilot specimens were selected from samples in several valleys, and these were subjected to progressive alternating – field demagnetizations at levels of 0, 5, 10, 15, 20, 25, 30, 40, 50 and 60 millitesla (mT) using a conventional three-axis tumbler. After each successive demagnetization, remanence measurements were made using a Schonstedt DSM-1 magnetometer. Qualitative assessment of changes in magnetic directions and of the decrease in magnetization intensity led to a choice of 15, 20 and 25 mT as the standard demagnetizing fields for the remaining samples. Typical demagnetization results are shown in Chapters 4, 5, and 6, and most specimens are characterized by a single component remanent magnetization. On the basis of the low median destructive field of these specimens, it would appear that detrital magnetite is the predominant remanence carrier [magnetite is one of the principle components of most tills on Mount Kenya (Vortisch, Mahaney and Fecher, 1987)].

3.8 Radiocarbon

Samples were collected from bog and lake cores, and from buried A horizons in paleosols, for radiocarbon dating. All samples were handled with metal implements, dried on aluminum foil, and stored and shipped frozen to various laboratories. For bog and lake cores, wood and detrital plant fragments usually provided sufficient material for dating by gas- or liquid-counting methods, but on occasion humus-rich sediment without wood and detrital plant fragments yielded radiocarbon dates. One particle accelerator date was determined by Dr. R.E. Taylor at the University of California, Riverside. All other age determinations were made by: Dr. Jaan Terasmae and Mr. Howard Melville at the Geochronology Laboratory at Brock University, St. Catharines, Ontario; Drs. J. Stipp and M. Tamers at Beta Analytic, Coral Gables, Florida; and Gakushuin University, Japan.

Buried paleosol samples were collected from the middle of each horizon sampled, and meticulous descriptions of root penetration from ground soils, as well as groundwater influx from below, were made at the time of sampling. Both ground soils and buried paleosols were analyzed for leaching effects by determining depth distributions of pH, extractable Ca/Mg ratios, and base saturation to determine if contamination might have occurred. In addition, some paleosols were analyzed for their microbial compositions to determine if depth distributions of bacteria and fungi might indicate geobiochemical contamination (Mahaney and Boyer, 1986a).

3.9 Soil Microbial Analysis

The soil samples collected for microbial analyses were taken as subsamples of those collected for chemical and particle size determinations. At least 1 kg was collected in sterile bags from each profile. Following exposure of the soil, and just prior to collection, each horizon was cut back several cm to provide an uncontaminated face for sampling. Subsamples were collected immediately and placed in sterile scintillation vials in which the screw caps had been drilled and loosely plugged with cotton. The vials were stored in a carrying case which permitted air exchange. Samples were analyzed microbiologically from one week to four weeks after they were collected in the laboratory of Dr. M.G. Boyer, York University.

A number of routine methods were employed to obtain estimates of the numbers and species of fungi and bacteria. For fungi a soil extract medium (SEA) was prepared from the surface mineral horizon of a typical loam from Southern Ontario. The extract was prepared by autoclaving, for 15 minutes at 15 p.s.i., 1 kg of crushed air-dried soil with 1000 ml tap water followed by cold storage until clear. The filtered extract was then brought up to 1000 ml and the following added: 0.2 g K_2HPO_4, 1.0 g glucose, 3.3 ml of 1 percent rose bengal, and 15 g Sigma agar. The pH was stabilized at 6.0 with 10 percent lactic acid following sterilization. For bacteria, standard Difco nutrient agar was used and supplemented by 1.0 g tryptone 1^{-1}.

Soil samples to be processed were air dried asceptically, homogenized with a mortar and pestle, and processed in distilled water. The pour plate method was employed for routine counting procedures. Triplicate plates were prepared for each dilution, usually 1×10^{-2} to 1×10^{-4}, and samples were incubated at 16°C for 7 days prior to counting. Confidence limits were estimated, except in cases where zero values were obtained in one of the plates (see discussion in Mahaney and Boyer, 1986a).

3.10 Thermoluminescence

Minute amounts of energy (light) are trapped in minerals from environmental radioactive elements. This light is zeroed when a sediment is exposed to sunlight prior to burial by a geologic agent (Wintle and Huntley, 1982). After burial, TL begins to increase as light is trapped in the lattices of various minerals. By comparing accumulated TL to the incident radiation at the surface, an age can be calculated ranging from a few hundred to ~100,000 yr or older.

TL in eolian and fluvial sediments is zeroed by sunlight bleaching during transport. After burial, the accumulated TL in a size fraction (usually 2-4 μm or 2-11 μm) (Berger, 1985), separated from the bulk sample, is observed by heating under a photomultiplier tube. Along with analysis of the light released from the mineral traps, the naturally occurring radioactive elements U, Th, and K are determined, as well as the particular mineral sensitivity to beta and alpha radiation.

Samples for thermoluminescence (TL) dating were collected under tarps, in the dark, after clearing section faces back a distance of ~30 cm. Samples were placed in quart-size paint cans lined with plastic. Can lids were sealed tight, and samples were kept cool until they could be placed in a freezer. All sample cans were completely filled with sediment, together with its *in situ* water content. All samples were taken from the center of the buried A horizons in paleosols, dated independently by radiocarbon, and later complete section descriptions were made in the field. From horizons above and below the buried A's, samples were collected to determine *in situ* water content.

Two major problems with TL dating are: a) that mineral traps for TL might be saturated, and b) that minerals used might not have been zeroed by sunlight prior to burial. In this study it was found that loesses and fluvial sediments appeared to give reliable dates, while alluvial fan or slope sediments sometimes gave older ages than radiocarbon probably as a result of not being zeroed at time of burial (Dreimanis, Hütt, Raukus, and Whippey, 1985). Quartz and Ca-plagioclases were the principal minerals used to obtain TL ages for buried paleosols on Mount Kenya.

3.11 Potassium-40/Argon-40

Potassium-40/argon-40 (K-Ar) isotopic ages were determined for some recognizable tephras found associated with older deposits on Mount Kenya. Only one date indicates the time of volcanic eruption, presumably for Ithanguni, although it could be from Mount Kilimanjaro, or from other volcanoes in the Eastern Rift Valley. For these samples K-Ar ages were determined from whole-tephra samples without mineral separation because the ashes did not contain minerals that could be separated, either magnetically or by specific gravity. The samples were analyzed using a Reynolds-type MS-10 mass spectrometer with an ultra-high vacuum direct fusion system. Argon was extracted with an RF induction heating procedure to insure complete removal with minimal atmospheric contamination, which is critical in dating very young samples with little argon content. Even though all tephras are known to have Quaternary ages on the basis of their stratigraphic positions, only one sample (described by Charsley, 1989) gave a ~.3 my date. All other tephras gave anomalously old dates (e.g. early Pliocene to Miocene).

3.12 Neutron Activation Analysis

Samples comprising the < 2 mm fractions of air dried material were subsampled for geochemical analysis. They were selected to weigh between 700 and 1000 mg (2 to 3x larger than normal) to insure representative sampling. The samples were stored and analyzed in Olympic plastic flip top vials.

To determine the concentrations of elements that produce short-lived radioisotopes, such as U, Dy, Ba, Ti, Mg, Na, V, Al, Mn, Ca and K, samples were irradiated serially for 1 minute at a neutron flux of $1.0 \times 10^{11} n\ cm^{-2}\ s^{-1}$ in the SLOWPOKE nuclear reactor at the University of Toronto. After waiting 19 minutes to allow short-lived ^{28}Al to decay to acceptable levels, samples were assayed with 5 minute counts using on-site gamma-ray spectrometers (see Hancock, 1984). After this, appropriate gamma ray peaks were measured and chemical concentrations calculated.

Following this procedure, samples were later batch irradiated for 16 h at a neutron flux of $2.5 \times 10^{11} n\ cm^{-2}\ s^{-1}$; after seven days the concentrations of Sm, U, Yb, La, As, Sb, Br, Sc, Fe and Na were calculated; Na was used to crosscheck with the first analysis. After two weeks, the samples were recounted and concentrations of Nd, Ce, Lu, Ba, Th, Cr, Hf, Sr, Cs, Ni, Tb, Sc, Rb, Fe, Co, Ta and Eu were determined. Sc and Fe were used to crosscheck the second phase of the analysis (see Hancock, Mahaney and Stalker, 1988, for discussion).

3.13 Laboratory Methods

Particle size analyses (Folk, 1968) were performed on the < 2 mm fractions; coarse grain sizes (> 63 μm) were determined by sieving, whereas fine grain sizes (< 63 μm) were calculated by sedimentation following procedures established by Bouyoucos (1962) and Day (1965). Chemical analyses of the < 2 μm fractions include pH on a 1:1 paste and 1:5 solution, total organic nitrogen by the Kjeldahl method (Bremner, 1965), cation exchange capacity (CEC) after leaching with 10% acidified NaCl and steam distilled on a Kirk-type unit. Extractable cations were determined with 1N NH_4OAc. Organic carbon was determined by the Walkley Black (1934) method. Iron was extracted with sodium dithionite-citrate (Fe_d) according to procedures outlined by Mehra and Jackson (1960), and Coffin (1960). Oxalate extractable Fe (Fe_o) and pyrophosphate extractable Fe (Fe_p) were determined following procedures established by McKeague and Day (1966) using atomic absorption spectrophotometry.

The mineralogy of the <2 μm grade size was determined by agitating samples with a cell dismembrator and pretreating with H_2O_2. Clay was centrifuged onto ceramic tiles, air dried, and X-rayed following procedures established by Whittig (1965), and Jackson (1956).

Weathering intensities were determined using weathering ratios of light minerals following procedures set forth by Ruhe (1956). In the light mineral suite quartz is more resistant than feldspar. Counts of resistant and weatherable minerals for each size fraction (250-63 μm) in each soil horizon were made under a binocular microscope with an electronic point counter. For some profiles discussed at length in later chapters, sand (and clay) samples were studied with X-ray diffraction (XRD), polarizing microscope (representative grain mounts), and scanning electron microscope (SEM). The XRD results are given as peak heights and peak areas (in some cases). Textural features of the XRD-powder mounts were kept constant in most analyses so that mass attenuation coefficients varied only slightly between samples, making the absolute peak heights reliable semi-quantitative measures of the clay minerals. Corrosion and etching studies on primary minerals were done on the SEM (Vortisch, Mahaney and Fecher, 1987).

CHAPTER 4

GORGES AND LAKE ELLIS GLACIATIONS

4.1 Introduction

The Gorges Glaciation is the oldest early Quaternary glaciation on Mount Kenya for which there is considerable geomorphic and stratigraphic evidence. These deposits generally consist of complex end moraines comprising waterlain till overlain with alternating sequences of lodgement till and loesses. The available data indicate two advances of Gorges ice within the Olduvai subchron, each major advance followed by periods of loess deposition and weathering. The most recent advance left moraines on the surface that have been weathering *in situ* for 1.8 my.

A complex suite of landforms of moraine, outwash fans, glacial spillways, crevasse fillings and loess sheets are associated with the Gorges Glaciation and contrast with younger and less obtrusive deposits of Lake Ellis age (middle to late Matuyama Chron). Deposits of Lake Ellis age are known from Mutonga, Hobley and Gorges valleys and range from extensive end moraines in Gorges and Hobley valleys to isolated pockets of drift in the Mutonga area. Paleosols in these deposits are well developed and some are complex multi-story units that contain important information on glacial and interglacial paleoenvironments.

The most significant information on the ages of Gorges and Lake Ellis deposits comes from paleomagnetic studies that have been carried out (Barendregt and Mahaney, 1988; Mahaney, Barendregt and Vortisch, 1989). Paleomagnetic ages ranging from the Olduvai subchron to the later Matuyama have been assigned to various moraines. Loesses have often been shown to have either late Matuyama or Brunhes ages; in some cases hiatuses of several hundred thousand years between till and loess, suggest that loesses may be stripped periodically, and replaced by subsequent aeolian activity. As a result there is often a considerable difference in weathering between tills and loesses.

Paleosols play an enormously important role on Mount Kenya in providing information on the magnitude of weathering that occurred within glacials and between glacials and interglacial periods. Just as in Chapters 5 and 6, where it is possible to reconstruct magnitudes of weathering within the Liki interstadial, here it is possible to interpret weathering effects between advances of Gorges ice in the Olduvai subchron. The distribution of clay minerals and other weathering products in these early Quaternary paleosols provides a means of interpreting stratigraphic differences and weathering intensities, as well as paleoclimatic intensities (see Chapter 11 for a discussion of paleosols and paleoclimate).

4.2 Older Diamictons and Tills

Forty oriented diamicton and diamictite samples were studied using scanning electron microscopy (SEM) to determine if their microfabrics might produce features related to their mode of origin. Some samples appear to have properties consistent with emplacement as lahars, whereas others have morphoscopic and microfabric features consistent with emplacement in a glacial environment. It proved possible to establish diagnostic microfabric features for tills that are consistent with deposition in low energy mountain glacial environments. Glacial diamictons yield crushed grain surfaces, fractures in grains and matrix materials, fissure fillings, striated surfaces, and some grain orientation. Nonglacial diamictons emplaced by mass wasting/volcanic processes exhibit lower clast frequency, overall higher amounts of matrix material, and grain roundness. Grain orientation (microfabric) appears to be a diagnostic property useful in differentiating glacial from nonglacial diamictons.

Many older diamictons lack definitive glacial landform charcteristics making it important to study microfabric characteristics of representative oriented samples from deposits considered to have glacial origins. Moreover some diamictons have volcanic/mass wasting origins (*e.g.* lahars). Because these deposits have similar physical appearances, it is important to find microfabric properties that will allow differentiation according to mode of emplacement. The samples discussed here are from prominent diamictons and diamictites on the eastern flank of Mount Kenya.

Sand samples (250-63 μm) from diamictons (Figure 4.1) considered to have glacial (Figure 4.2A) and/or volcanic/mass wasting (Figure 4.2B) origins are shown for comparative purposes. Overall there is a larger amount of volcanic glass in the till (GOR48) sample compared with the volcanic/mass wasting diamicton. The angularity of clasts is about the same in both samples.

Microfabric samples from the GOR48 diamicton were studied to determine if grain orientation and texture could be used to determine mode of origin. Relatively thick coatings of MnO_2 + Fe_2O_3 were encountered on most subsamples (Figure 4.3) and in some cases grain orientation was difficult to determine. In Figure 4, fissure fillings in the GOR48 diamicton (Figures 4.4 and 4.5) may have been caused by terminal meltout. Both photomicrographs show representative textures of till bodies with a great range of sizes from clay to sand. Crushed surfaces[1] on some grains (Figure 4.6) provide evidence for a glacial origin.

Samples from the GOR4 diamictite show many clasts coated with $CaCO_3$-rich material as determined by X-ray microanalysis (Figure 4.7A). Enlargement of zones with coarser material and higher porosity probably indicate dewatering channels (Figure 4.7B). Subsamples of the same diamictite show somewhat more matrix material (coated with $CaCO_3$) than in the glacial diamicton and a random orientation of clasts (Figure 4.8). In addition, clasts are often separated from one another by a high percentage of tuffaceous material (volcanic glasses). A general view of a representative random microfabric is found in the GOR4 nonglacial diamicton on Mount Kenya (Figure 4.9); enlargement shows a weathered coat of $CaCO_3$ on a feldspar grain with a bimodal grain size distribution.

Several recent studies of lodgement till from a number of middle latitude localities (Li *et al.* 1984; Haldorsen, 1983) point to glacial crushing features on grains as the major morphoscopic evidence of glacial deposition. Because the microfabrics studied here show only some evidence of glacial crushing it is possible that the glaciers were thin with considerable amounts of interstitial meltwater. If this was the case, deposition processes in these alpine environments would have been dominated by subglacial and terminal meltout, with perhaps somewhat less shear stress and abrasion considered common in continental ice (Krinsley and Donahue, 1968; Vortisch, Mahaney and Fecher, 1987; Mahaney, Barendregt and Vortisch, 1989; Mahaney, Vortisch and Julig, 1988). In some samples, a preferred dip direction or inclination suggests that at least some diamictons have the character of lodgement till. This, coupled with a high frequency of matrix shearing, enhances the notion that the glacigenic diamictons were deposited mainly as subglacial or terminal till (Derbyshire, 1978; Li *et al.*, 1984).

Diagenetic changes seem to have led to the development of Fe_2O_3 + MnO_2 coatings on and between grains, which creates considerable difficulty in observing grain orientation, thus making it difficult to determine the mode of origin on some of the samples studied. Precipitation of these coatings on primary grain skeletons is particularly dense around ancient root channels and faunal cavities. In addition to Fe + Mn, Si, Al, K, and Ca have been detected in Mount

1. Crushing features may also result from frost shattering or from abrasion during release from source rocks. However, the only observed crushing surfaces were on clasts derived from glacial diamictons.

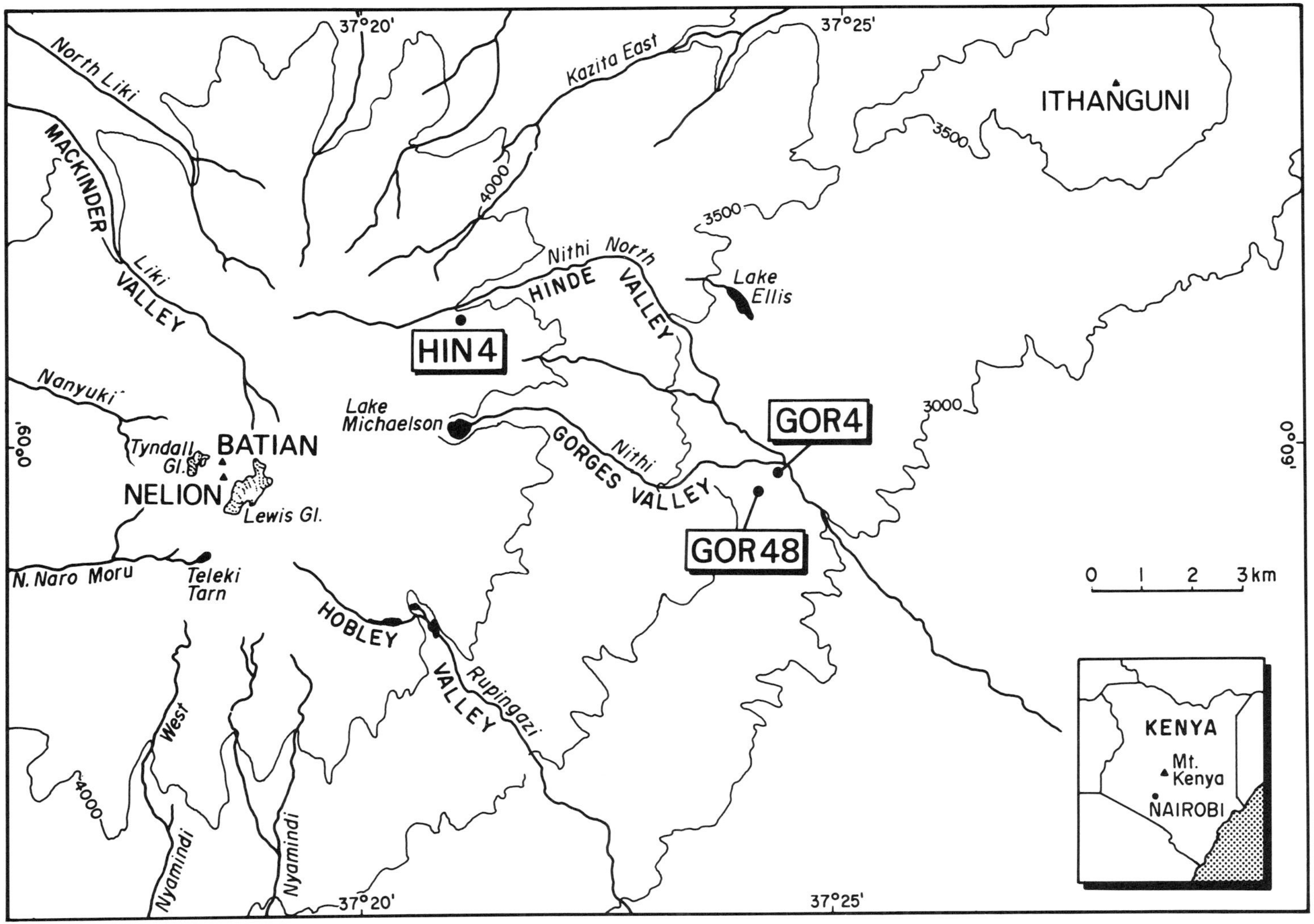

4.1 Location of glacial (GOR48) and nonglacial diamictite (HIN4 and GOR4) in Gorges and Hinde valleys.

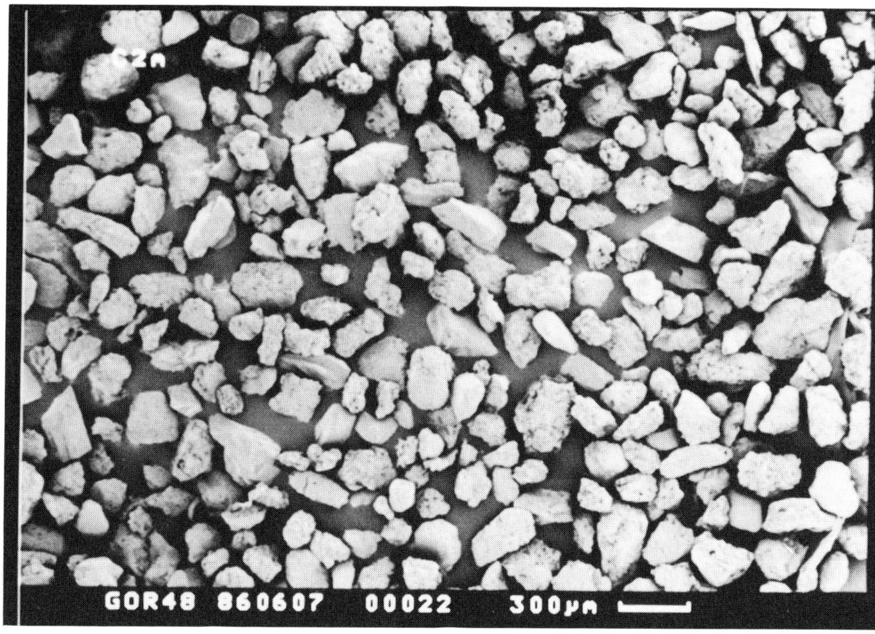

A.

B.

Figure 4.2: General views of fine and very fine sand fractions (63-250 μm) considered to be of glacigenic (A) and volcanic (B) origins. No principal morphoscopical differences are visible with the lower magnifications used here.

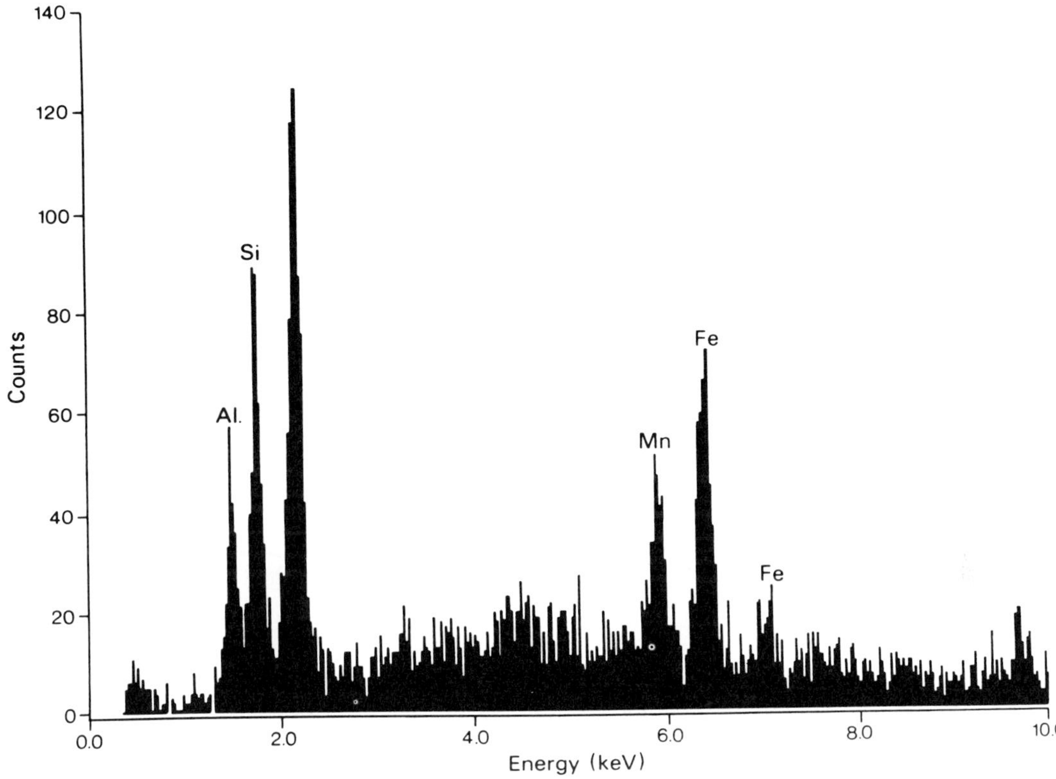

Figure 4.3: Energy dispersive X-ray printout from oriented microfabric suface of sample GOR48. From Derbyshire and Mahaney (1989).

Kenya microfabric samples (Derbyshire and Mahaney, 1989). Moreover, the matrix cement seems to increase in density with greater age.

Because of widespread precipitation on grain surfaces in the microfabric samples, the degree of clast rounding is sometimes difficult to determine. However, in some cases it proved possible to remove the cement using a microspatula, and in others the thinness of the cement coating allowed at least an impression of grain shape. Moreover, some grains appear fractured and in general grain angularity appears consistent with a glacial origin and with previous microfabric studies on Mount Kenya (Derbyshire and Mahaney, 1989). Crushed edges and corners considered indicative of subglacial deposition (Derbyshire, 1978) proved difficult to find on the photomicrographs in this study owing to the prevalent carbonate and oxide coatings.

The skeletons of all glacigenic diamictons studied showed some preferred clast orientation, which in most cases is consistent with deposition in a moraine system (Derbyshire and Mahaney, 1989). Those diamictons considered to have a volcanic/mass wasting (lahar) origin appeared to have random clast orientations and showed no evidence of glacial crushing features encountered in the glacial diamictons. While some grains appeared fractured in the nonglacial diamicton samples, we did not observe grains with crushed corners or edges. We did, however, observe a considerable increase (relative to glacial diamictons on Mount Kenya) of tuffaceous matrix

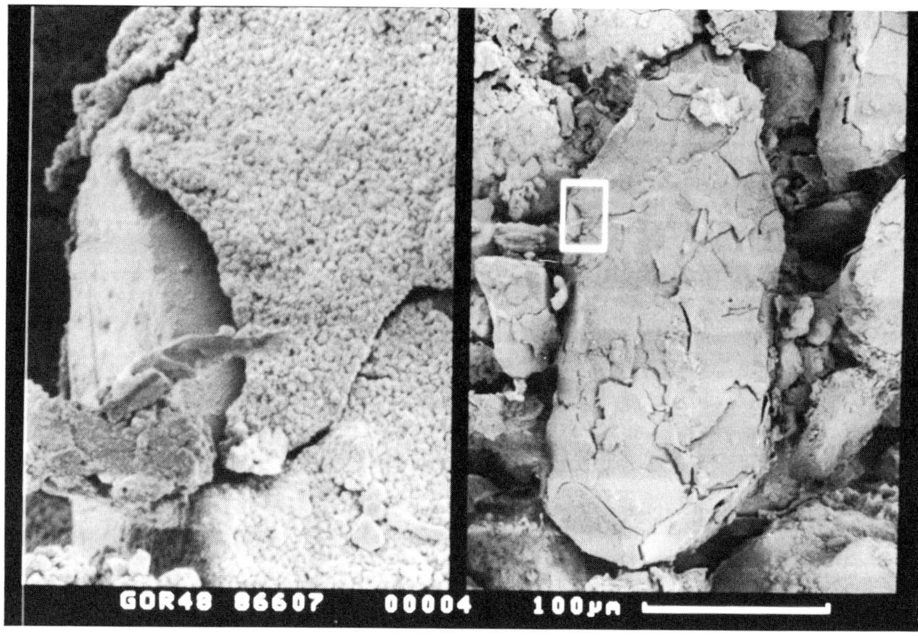

A.

B.

Figure 4.4: Typical texture of Mt. Kenya till (site GOR48). Arrow in B shows location of fissure filling in Figure 4.5. Most grains are heavily coated with MnO_2-rich material.

Figure 4.5: Fissure filling (center) showing subangular and subrounded clasts in glacial diamicton (GOR48).

Figure 4.6: Crushing features (arrows) on alkali feldspar in Mt. Kenya till probably caused by glacial action.

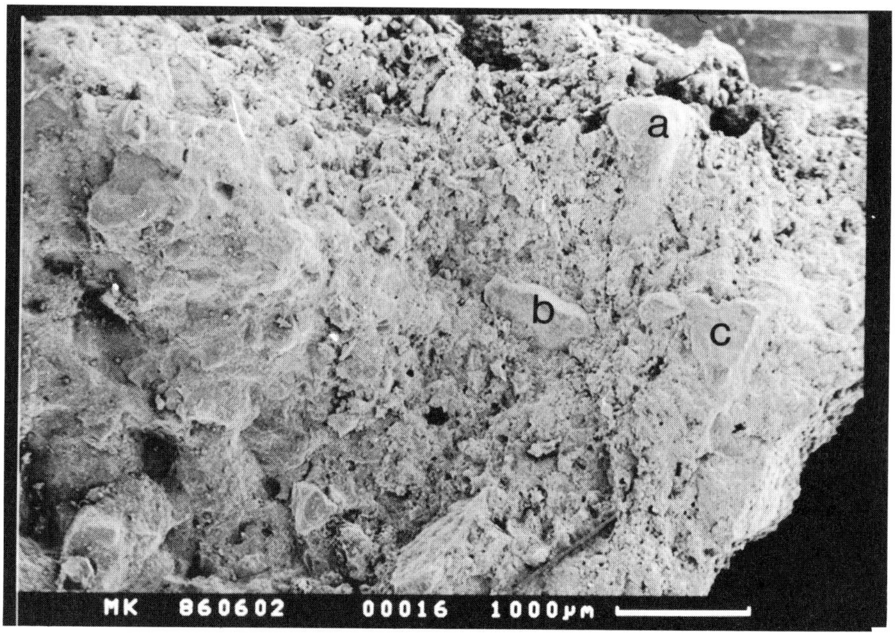

A.

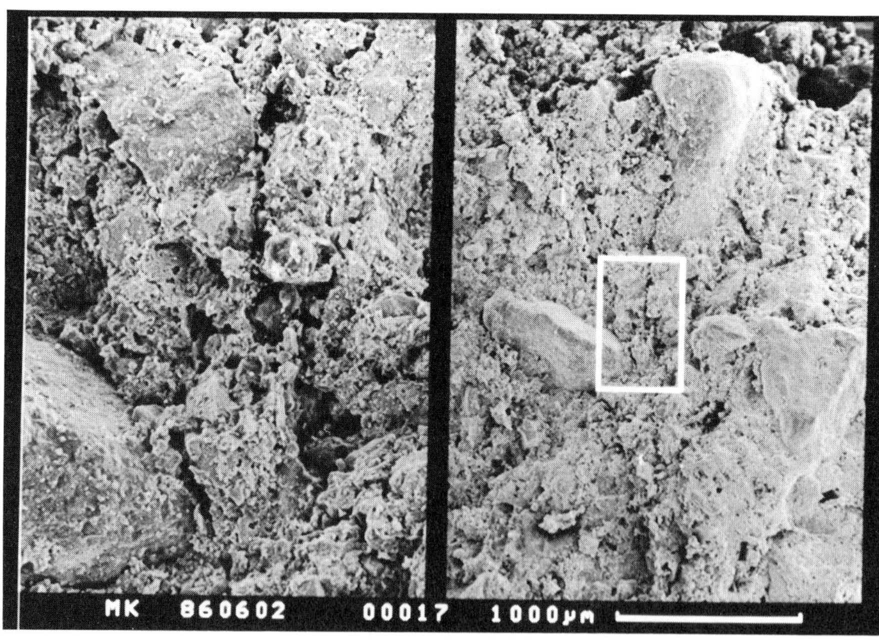

B.

Figure 4.7A, General view of Mt. Kenya lahar (diamictite in site GOR4) showing several randomly oriented clasts (a, b, c) coated with $CaCO_3$-rich material; B, Enlargement of zones with coarser material and higher porosity which may be dewatering channels.

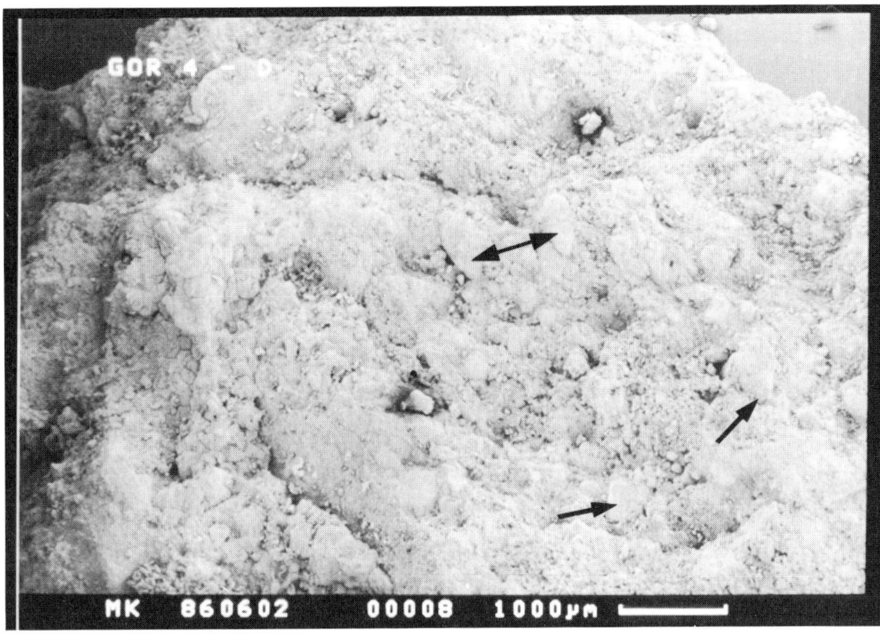

Figure 4.8: Randomly oriented coated clasts (arrows) in Mount Kenya lahar (diamictite of site GOR4).

Figure 4.9: Typical texture of Mt. Kenya lahar (diamictite in Site GOR4) showing bimodal grain size distribution and random clast orientation.

material and cement between clasts in the nonglacial diamictons. As with the glacial diamictons, clast surfaces were coated with clays and precipitates making it difficult to observe roundness/angularity characteristics. The cement in these lahar samples is usually $CaCO_3$, although occasionally Fe_2O_3 and MnO_2 dominate, suggesting some variability in the chemistry of subsurface ground and soil waters. Where cleaner grains could be observed we encountered fracture cracks, but without sharp edges.

Fissure fillings, while common in glacial diamictons, were not observed in nonglacial diamictons on Mount Kenya. Also, crushing planes which are common on glacially emplaced clasts (Vortisch, Mahaney and Fecher, 1987), were observed occasionally on subsamples of the nonglacial diamictons indicating that lahars have fewer clasts which do not abrade one another on a regular basis.

The Mount Kenya diamictons exhibited diagnostic criteria of glacial deposition resulting from low-energy conditions in tropical ice, which produced tills with coarse matrices. Post-depositional diagenesis leading to increased clay and precipitate coatings on the primary skeletons hindered observations of grain-to-grain relationships, and masked the primary features of much of the matrix and some of the clasts in all samples studied. Although volcanic clasts dominate in all the microfabric samples studied on Mount Kenya, glacial diamictons appear on average to have an overall higher clast frequency as well as higher percentages of glass.

While there are still problems in differentiating tills from lahars on Mount Kenya (and in other alpine areas), it is apparent that tills tend to produce some glacially-crushed grains, fractures in matrices and grains alike, and fissure fillings. Lahars, on the other hand, generally show random fabric orientation, yield a lower clast frequency, and have higher amounts of matrix material. The unimodal to bimodal grain size distributions of all nonglacial samples studied, show that texture is an important property useful in differentiating glacial from nonglacial diamictons. As described in section 4.3, some glacially-crushed grains in nonglacial diamictons are taken to provide evidence of a pre-Gorges (pre-Olduvai) glaciation on Mount Kenya.

4.3 Evidence for pre-Olduvai Glaciation

A scanning electron microscopic and morphoscopic study of lahars on Mount Kenya (all older than Gorges Glaciation, Figure 2.27) showed many glacially crushed sand grains. On average about 1 in 15 grains were observed to carry subparallel crushing features (including linear and conchoidal crushing features of the low frequency type). No concentic gouges were observed, but several grains showed v-shaped percussion microtextures indicating they might have been subjected to meltwater release in addition to crushing effects. Other pseudo-crushing effects were considered to represent intergrowths of different minerals such as pyroxene and amphibole, not true crushing effects.

The grains most frequently encountered in the lahars include anorthoclase (mainly K-analbite and sanidine), amphibole, pyroxene, orthoclase, apatite and volcanic glass in different weathering stages (see Vortisch, Mahaney and Fecher, 1987). Some grains were fresh and others coated, either with clay minerals, or other weathering products such as Fe-oxides and gibbsite. The composition of weathering products gives the impression that a significantly *long* period ($>$100,000 yr) of weathering ensued at relatively low elevation (3000-3300 m.a.s.l.) assuming precipitation and temperature patterns similar to today, or a short period with increased precipitation (and temperature?). Overall, it appears these weathering products formed on grains following crushing events. No quartz was observed in the samples studied and most grains followed the sequence: fractured $\rightarrow$ crushed $\rightarrow$ weathered. As usual some grains were fractured but not crushed (Mahaney, Vortisch and Julig, 1988).

The reworking of several grains in Mount Kenya lahars is evidenced by their loss of angularity and general tendency of subangularity → subroundness (Figure 4.9) and by several distinct stages of crushing and weathering on a single clast. Some grains recovered from lahar deposits on Mount Kenya showed slight tendencies toward roundness and many showed a sequence of crushing, followed by weathering, and later emplacement in a volcanic deposit. Because some grains were coated with Fe and Si, the possibility of diagenetic effects has to be considered. Fluctuating ground water or seepage of waters from the surface might produce some of the observed coatings (Figures 4.7A, and B).

For most of the aluminium-rich coatings weathering seems a reasonable explanation. Many feldspar grains, including apatite, sanidine and orthoclase are coated with gibbsite. Even nepheline syenite grains were encountered with coats of aluminium (Al:Si ratio of 2:1) indicating substantial and aggressive leaching conditions some time prior to the Olduvai subchron and following the initial emplacement of the Mount Kenya volcanic suite of rocks (see Fig. 2.27).

Because the deposits in which these old glacially crushed grains were emplaced have been destroyed by volcanic debris-flow activity, it is only possible to gain an impression of the probable sequence of events and relative age of the clasts. Moraines associated with pre-Olduvai glaciation on Mount Kenya (if they survived the lahar activity) were probably consumed by the younger Gorges Glaciation, the older clasts being redistributed in the younger drift. A methodical SEM study of the clasts in till of Gorges age might reveal reworked grains that were crushed, weathered and then crushed again.

Overall it appears that sometime between the crystallization of the volcanic plug (2.6 my) and the emplacement of Gorges Drift (~ 2.0 my), a major mountain glaciation followed by a major weathering episode (interglacial) occurred on Mount Kenya. Allowing for the emplacement of the Mount Kenya volcanic suite (~ 300,000 yr) and redistribution of volcanic debris by volcanic debris-flows (30-100 m thicknesses of lahar suggest a considerable but inestimable time), narrows the available time to between 300,000 and 400,000 yr, or most probably from ~ 2.0 to 2.4 my.

4.4 Geomorphologic Evidence for Early Quaternary Glaciations

Glacial landforms of pre-Brunhes age (> .73-.79 my) in the Nithi River Basin, on the eastern flank of Mount Kenya, provide evidence for the dynamics of early Quaternary ice. The glacial sequence, composed largely of multiple sequences of ground and end moraines, also contains associated crevasse fillings, outwash and aeolian deposits, and thin deposits of waterlain tills (Figure 4.10). These morphosequences represent two distinct glaciations beginning with the more extensive Gorges Glaciation and ending with the less extensive Lake Ellis Glaciation. The Gorges Glaciation began with deposition of waterlain till on weathered tuff followed by deposition of end and ground moraines, crevasse fillings, and aeolian deposits. The Lake Ellis Glaciation began with deposition of end and ground moraines and outwash deposits on weathered tuff, followed by emplacement of aeolian sediments.

Paleoclimatic and glacial geomorphologic inferences can be drawn from the sediments, degree of weathering, and landform relationships. Waterlain till near the margin of the Gorges Glaciation is derived from subglacial deposition of sediment in an aquatic environment. The overlying basal tills dominate in the end moraine. Hummocky end moraines represent ice stagnation during a later stage of the Gorges Glaciation that led to emplacement of meltout debris along with some lodgement till. Crevasse fillings near these hummocky moraines indicate static ice margins that eventually wasted away near the end of the Gorges Glaciation. A somewhat similar morphosequence, dating from the Lake Ellis Glaciation, began with the construction of prominent end and ground moraines, and ended with deposition of outwash

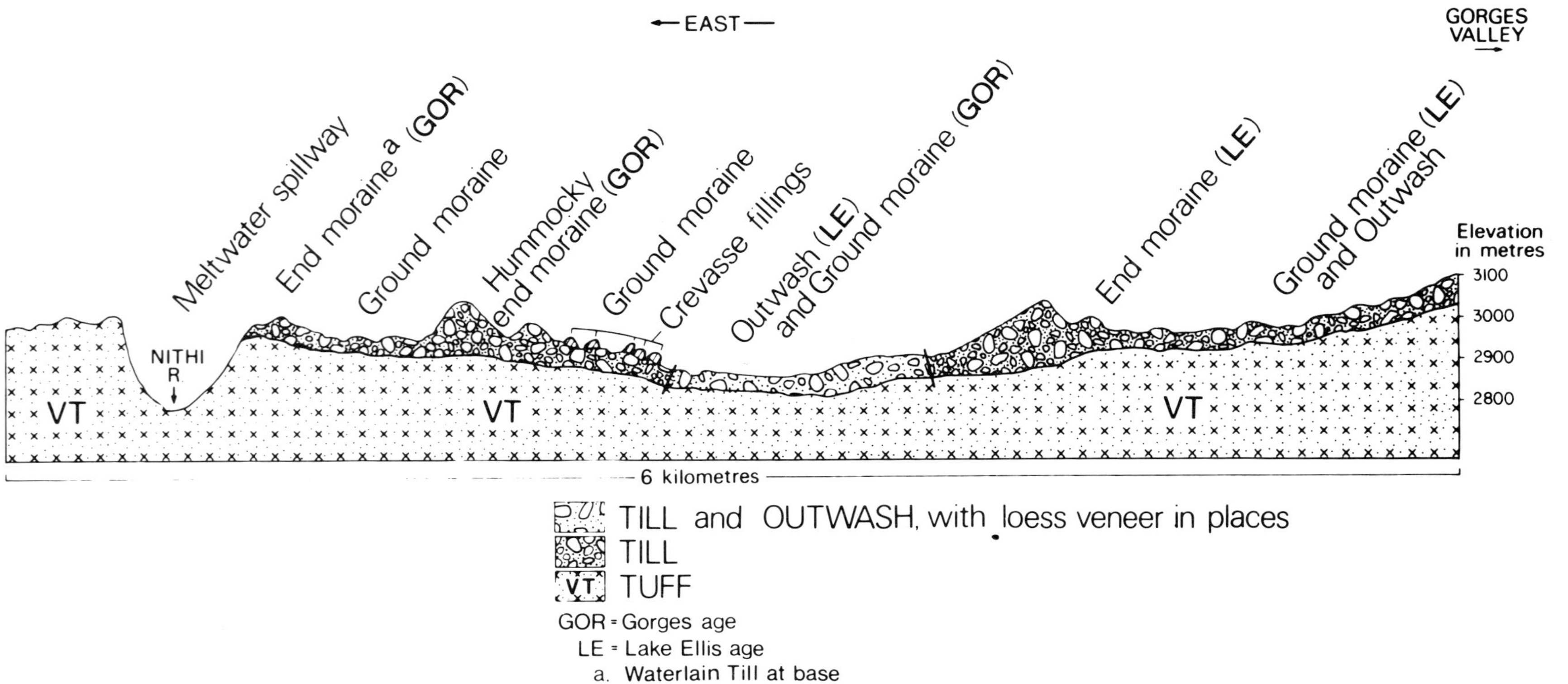

Figure 4.10: Landform profile showing morphosequences of deposits emplaced during Gorges and Lake Ellis glaciations. Glaciers moved from west to east; largest expanded mountain ice sheet occurred during the earlier Gorges Glaciation.

bodies. No crevasse fillings or hummocky moraines were observed associated with this younger set of glacial landforms.

The older pre-Brunhes glacial landforms (Plate 1 and Figure 4.11) consisting of meltwater channels, complex end moraine underlain with waterlain till, end moraines, ground moraines, hummocky moraines, crevasse fillings, and outwash plains were not recognized by previous workers on Mount Kenya (Nilsson, 1935; Baker, 1967; and Hastenrath, 1984). T. J. Charsley (personal communication, 1984) first recognized extensive moraines as low as 2850 m on the eastern flank of the mountain, which were subsequently investigated by Mahaney in 1984, 1986, 1987. This led to the identification of morphostratigraphic units for two of the oldest glaciations – Gorges and Lake Ellis – on Mount Kenya. These units comprise a sequence of well-preserved deposits as shown in Figures 4.12 through 4.15.

End moraine complexes consisting of one or two ridge systems, usually abutting outwash fan deposits or meltwater channels (Figure 4.12) are difficult to map from air photographs and satellite imagery. These moraine complexes are 100 to 200 m in width and 3 to 4 km in length. They stand only a few meters higher than surrounding ground moraine. The end moraines are covered with thin veneers of aeolian sediment; commonly two lodgement tills separated by loess, and overlying waterlain till are present in individual deposits of Gorges Till. These tills record possibly two advances of Gorges ice during the Olduvai subchron between 1.67 and 1.87 my (Mahaney, Barendregt and Vortisch, 1989). Because the lower beds in the outer Gorges moraine were deposited subaquatically along most of the length of the moraine, it is possible to make only limited inferences regarding ice dynamics and paleoclimate.

Waterlain till has been observed at four locations below the lodgement tills of Gorges age. These tills were deposited by ice moving from west to east (see Figure 4.10). All sediments in this sequence exhibit horizontal bedding, lack appreciable deformation (minor deformation is visible), and have sharp and abrupt upper contact with overlying lodgement till. These sediments were deposited as poorly sorted beds about 2-4 cm in thickness that remain remarkably consistent with respect to particle size distributions from top to bottom. No subaquatic mass wasting or diamicton structures were observed, so it is possible to rule out episodic gravity flow deposition for these sediments (such as described by Evenson, *et al.*, 1977).

Subaquatic deposition of these sediments may have occurred under thin mountain ice actively receding behind bedrock bars above the Nithi River outlet. Because the underlying weathered tuffs were preserved it is logical to assume that ice action was weak, and that sediment was being deposited subaquatically beneath floating ice, perhaps in shallow water dammed by ice in the meltwater channels leading into the Nithi River or in the main Nithi Valley itself. Because this deposit contains pebbles that could be considered dropstones, the orientation of 50 clasts was measured. The resulting fabric gave a bimodal distribution with stones oriented mainly toward 070° and 105° (nine samples gave spurious readings). The presence of a glacial fabric and minor deformation of beds in these deposits favor a glacial rather than a glaciolacustrine origin. However, additional data is needed to rule out the latter hypothesis. This type of sediment, only observed in the outer Gorges moraines, may be possibly linked to catastropic ice surges followed by rapid melting, or to sudden paleoclimatic warming followed by rapid deposition of till.

Behind prominent end moraine complexes and beneath outwash and aeolian cover sediments, are numerous areas of ground moraine (Figure 4.13). In general, they are nearly flattopped and irregularly distributed in the landscape. Some deposits are overlain with crevasse fillings that have been described as man-made mounds (see Mahaney, *et al.* 1986, for discussion). A number of exposures displayed irregular sequences of lodgement till with thin, sandy diamictons (Figure 4.14). Some may consist of crevasse fills or subglacial cavity collapse features, but most were probably formed contemporaneously with, or prior to, deposition of the hummocky

Figure 4.11: End moraine complex (middle ground) deposited during the Lake Ellis Glaciation (>0.73 x 10^6 yr) in Gorges Valley. The ridge of Gorges moraine as seen on the horizon marks the outer advance of Quaternary ice on the mountain.

Figure 4.12: Meltwater sluiceway in front of Gorges-age end moraine in Gorges Valley; left, distal moraine slope; right outcrops of tuff of the Mt. Kenya volcanic suite (early Quaternary).

Figure 4.13: Ground moraine deposited during the Gorges Glaciation (early Quaternary) in Gorges Valley.

Figure 4.14: Hummocky moraine at site GOR60 emplaced during the Gorges Glaciation in Gorges Valley.

Figure 4.15: Crevasse filling on Gorges ground moraine.

moraine complexes. Deposition occurred on level to slightly sloping topography, presumably under relatively thick ice with high subglacial water pressure.

Broad hummocky moraine occurs less than 1.0 km behind the outer Gorges end moraine complex (Figure 4.11). Several asymmetrically shaped hills (Plate 1) separate depressions filled with assorted till, fine laminated stratified drift, and ponded sediments. The hummocky moraines (Figure 4.13) lie on the margin of a bedrock bar which was eroded by meltwaters issuing from both Gorges and Lake Ellis ice. This morphosequence may have resulted from thick and active ice during the third Gorges advance, which led to eventual thinning followed by stagnation and sedimentation of a range of clast sizes (*e.g.* from boulders to clay). The later Gorges advances emplaced deposits that are clearly different from the waterlain till in the outer end moraine complex, and are possibly indicative of debris-covered ice (Lawson, 1981). Because the hummocky moraine belt is composed of thick drift (~ 3 m), deposition probably occurred under active and thick ice conditions, followed by rapid stagnation and ice thinning that led to emplacement of numerous crevasse fillings.

Lateral moraine sequences are easily identifiable on air photographs and LANDSAT imagery. They define the outer margin of Gorges ice when the expanded mountain ice sheet occupied almost all of Gorges Valley and one unnamed tributary drainage to the north (Plate 4). These lateral moraines have steeper slopes and sharper crests compared with the end moraine. They are 5-7 km in length, 50 to 150 m wide and up to 25 m high. Commonly they have single or double crests and together they account for a greater amount of sediment accumulation when compared with the end and ground moraine complexes.

Behind and around the hummocky moraine belt an irregular distribution of conical mounds (Figure 4.14) consisting of till and stratified sediments were interpreted to be crevasse fillings. Originally considered to be man-made mounds (Zeuner, 1948), these deposits appear related to thinning and eventual breakup of the expanded mountain ice sheet following the third advance of ice during the Gorges Glaciation.

Between the hummocky moraine belt of the Gorges Glaciation and the end moraines of Lake Ellis age (Figure 4.11) a system of outwash deposits, built by streams flowing from active ice to the west, forms prominent terraces and fans in the landscape. Even though some of this outwash could have been deposited following the Gorges Glaciation, the major proportion appears to belong to the younger Lake Ellis Glaciation. Unlike the older Gorges Glaciation, which left much of the underlying weathered bedrock intact, younger outwash of Lake Ellis age is found often in thin bodies (2-4 m thick) over scoured and fresh bedrock surfaces.

Wind blown sediments cap much of the outwash and till deposits (Zeuner, 1949), and in some cases (mainly around the margin of glaciation) even separate buried till sheets (*e.g.* Section GOR68). As in other drainage basins on Mount Kenya, aeolian sediments have high quartz and low ferromagnetic mineral contents when compared with tills (Mahaney, 1982; Vortisch, Mahaney and Fecher, 1987). Many aeolian deposits are thin (~ 0.5 m) and almost all are weathered giving rise to A or Ab horizons in paleosol sequences. Many contain histic or mollic epipedons of considerable thickness that were formed under Afroalpine biotic elements during deteriorating paleoclimate when major vegetation belts occurred further down the mountain a distance of at least 700 m (Mahaney, Harmsen and Spence, 1990).

The emplacement of these previously undescribed glacial landforms has led to a sequential arrangement of end moraines (with or without waterlain till), ground moraine, hummocky moraines (with crevasse fillings in the older group) and outwash that form distinctive morphosequences related to the earliest cycles of glaciation on Mount Kenya. These morphosequences provide important information on glacier dynamics and paleoclimate in the tropical Afroalpine. The landform morphosequence is considered to represent the following glacial events: (a) thin ice flow depositing waterlain till in the outer Gorges end moraine complex; (b) somewhat more aggressive ice conditions leading to deposition of two tills separated by aeolian sediments; (c) deposition of hummocky moraines during an interval when ice appears to have been capable of moving considerable debris followed by stagnation and development of numerous crevasse fillings; and (d) development of an end moraine sequence in Lake Ellis time accompanied by emplacement of significant ground moraine and outwash bodies. The Lake Ellis Glaciation appears to have resulted from the simple forcing of a deteoriating paleoclimate, whereas the earlier Gorges Glaciation left a multitude of different landforms with important paleoenvironmental ramifications.

4.5 Surface Weathering Features

Weathering features on surface stones are only of limited use in distinguishing between Gorges and Lake Ellis deposits. This is partly the result of the location of deposits under different vegetation belts – Gorges moraines are situated under Hagenia woodland at present, whereas Lake Ellis drifts are under ericaceous elements (Figure 4.16).

Also several weeks of observations in 1983/84, 1986 and 1987 showed that advection currents tend to force air masses upward to ~3000 m bringing rain to approximately the same elevation. Above ~3000 m climate is noticeably drier, the drier/humid boundary coinciding with Gorges and Lake Ellis deposits. Hence clasts on Gorges deposits are not only in a wetter microclimate, they are also older, and have been exposed to subaerial weathering for a longer period of time. The result is that clasts on Gorges deposits have weathered to produce thick (10 to 30 mm) clay and oxide-rich rinds. Stones on Lake Ellis drifts are somewhat fresher, although they can be split easily with a hammer and some can be cut with a knife.

4.6 Stratigraphy of Gorges Deposits

Early Quaternary glacial deposits were first mapped and analyzed in the outer Gorges moraine crest. These deposits (Figure 4.17) showed interesting paleomagnetic and weathering records that appear related to their age as well as relative topographic positions, and vegetation covers. Both sites are situated on the outer moraine; GOR55 is under Hagenia forest, whereas GOR58 is under tussock grass in a cleared area. Field observations of texture and relative degree of weathering suggest that GOR58 has a younger loess cover which may have been deposited some considerable time after glaciation. Site GOR55 appears to have an older loess that may have been emplaced shortly after the Lake Ellis Glaciation (Figure 4.18) during the early Brunhes Chron.

Because these paleosols are representative of weathering in other early Quaternary sections in Gorges Valley, it is important to establish firm age controls using paleomagnetism and weathering relationships (including neoformation of clay minerals) using the SEM.

The two paleosols (Figure 4.17) are Alfisols (Tropudalfs-Soil Survey Staff, 1975). Both profiles have similar thicknesses of aeolian sediment over till; however, profile GOR55 has a thicker loess cover possibly resulting from deposition over longer time (see Birkeland 1984; Mahaney, 1978; Nelson, 1954 for discussion of loess thickness *vs.* age). In general, changes in field texture – silty loam in GOR55 to loam in GOR58 – without any important change in percent clay, tend to show higher proportions of silt in the former. Colors are variable in the two profiles and reach 5YR 4 and 5 (Oyama and Takehara, 1970) in the B horizon complexes (somewhat redder in GOR55). The cemented subsoil in profile GOR58 is probably related to different microclimates, or to differential release of Si upon weathering, and its reprecipitation as a cementing agent. The relatively low amount of gibbsite in profile GOR58 supports the notion of different degrees of Si removal and leaching as discussed by Vortisch, Mahaney & Fecher (1987).

Figure 4.16: Weathering rinds from outer Gorges end moraine Nithi River Basin (R = rind weathered to clay and Fe-oxides).

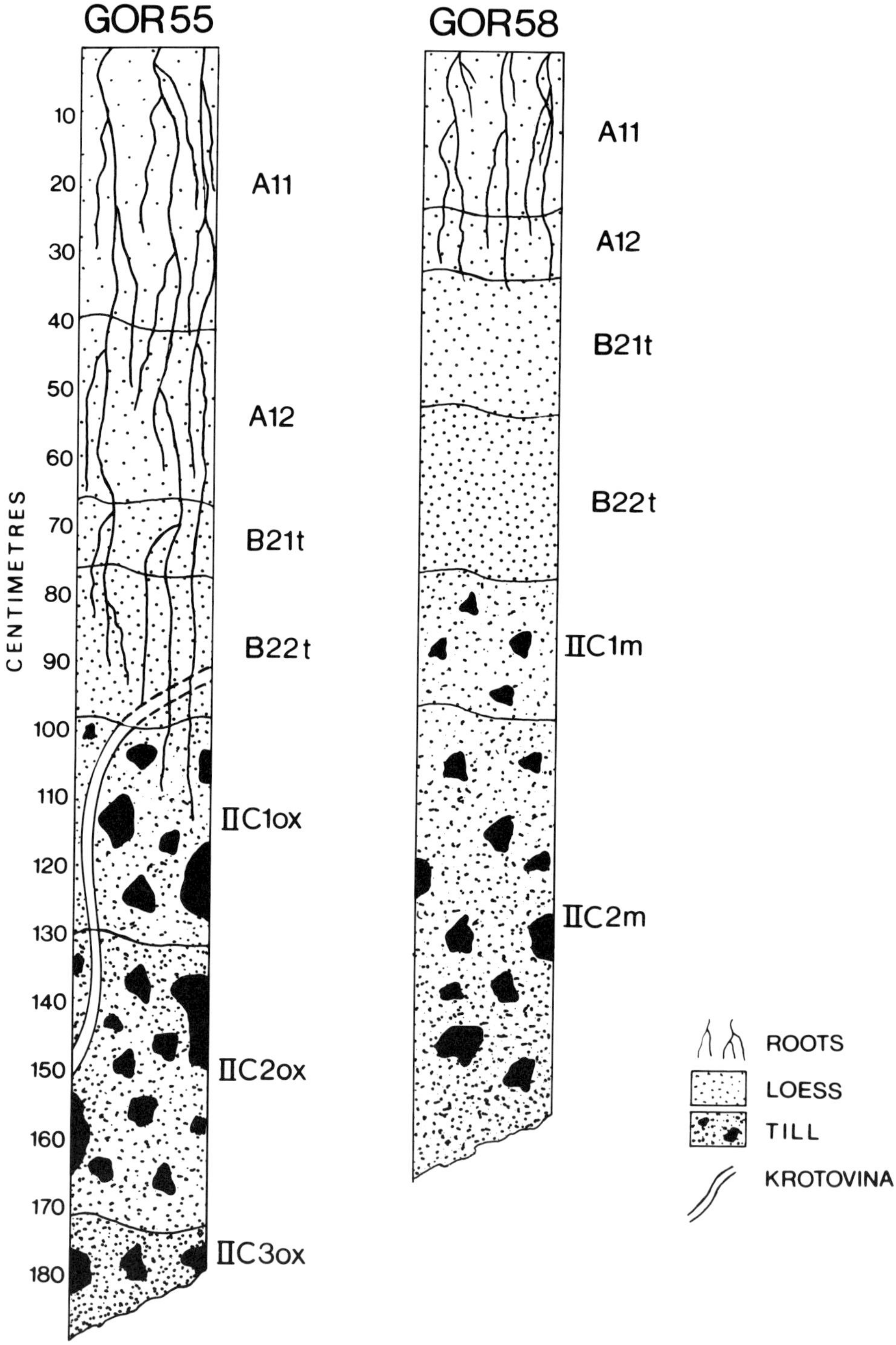

Figure 4.17: Stratigraphy of sections GOR55 and 58 in outer end moraine of Gorges age.

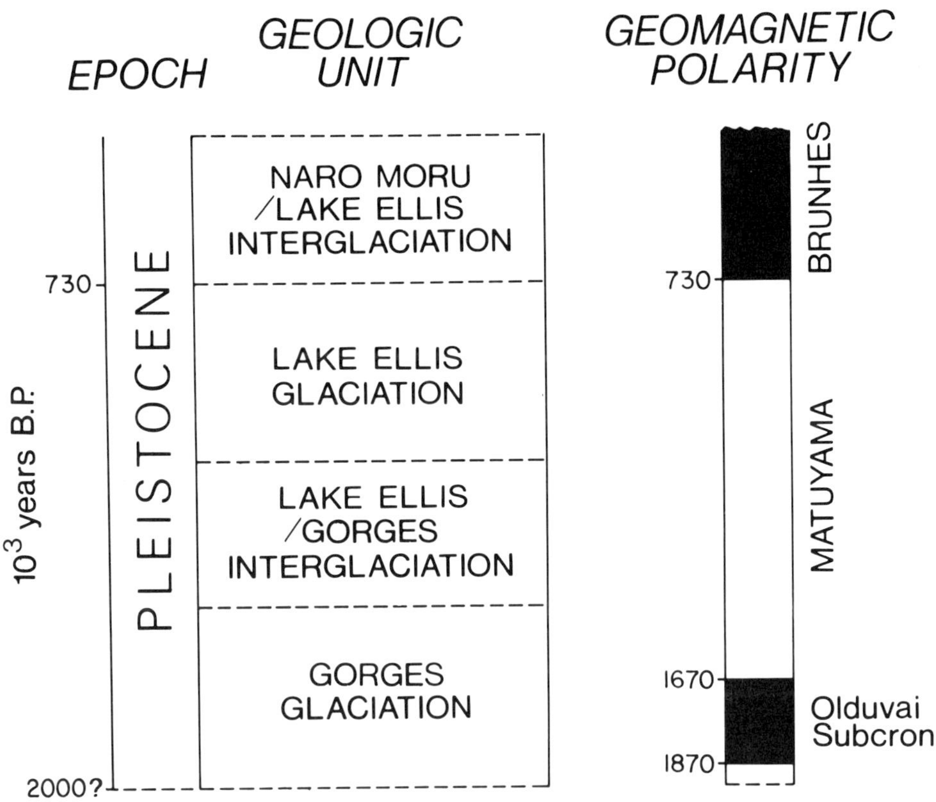

Figure 4.18: Stratigraphy for early Quaternary tills on Mount Kenya, showing relative time of emplacement of Gorges and Lake Ellis tills. Because the bottom beds in moraines have reversed polarity, while the middle beds are normal, the earliest deposition is considered to have occurred just prior to the Olduvai subchron. The youngest beds in the Gorges moraine system have reversed polarity, suggesting deposition after the Olduvai subchron. The Lake Ellis-Gorges Interglaciation may have lasted for several hundred thousand years and generally gives a reversed magnetic polarity. In some cases where the original loess cover on these deposits was stripped and later replaced, the magnetic polarity may be normal (*i.e.*, Brunhes Chron). Tills of Lake Ellis age carry reversed magnetism, and are considered to have been emplaced in the upper Matuyama Chron. Loesses on these deposits are less highly weathered and often carry normal magnetic polarity, indicating more recent emplacement during the Brunhes Chron.

Particle size trends (Figure 4.19) in the two paleosols were studied to determine differences between the tills and overlying aeolian sediments. Both profiles have somewhat different distributions of clay and silt with depth; profile GOR55 has pronounced clay films in the B horizons. In general, profiles GOR58 and 55 have similar amounts of clay in their till bodies, with evidence for translocation of clay from the A to the B horizon complexes within the overlying loessic materials. Higher amounts of clay in the A11 horizon of profile GOR58 may result from different vegetation covers at the two sites. The silt distributions are particularly revealing because of the higher percentages in the A12/B21t/B22t horizon complex of profile GOR55. Overall the higher silt percentages in profile GOR55 appear to represent emplacement sometime after the Gorges Glaciation followed by subaerial weathering over a long time period.

The mineralogy of the < 2 μm fraction in profiles GOR55 and 58 was analyzed to determine if distributions of clay and primary minerals might provide useful lithostratigraphic or weathering parameters. Within the primary mineral suite, quartz appears particularly useful in separating till from loess. In both profiles quartz is noticeably higher in the paleosol sola (A/B horizon complex) and lower in the tills (C horizons). The importance of quartz as an aeolian indicator in other drainages on Mount Kenya was first observed by Mahaney (1982b); in several subsequent studies (Mahaney, 1988c; Vortisch, Mahaney and Fecher, 1987) quartz distributions have proved particularly useful in distinguishing aeolian from glacial sediments.

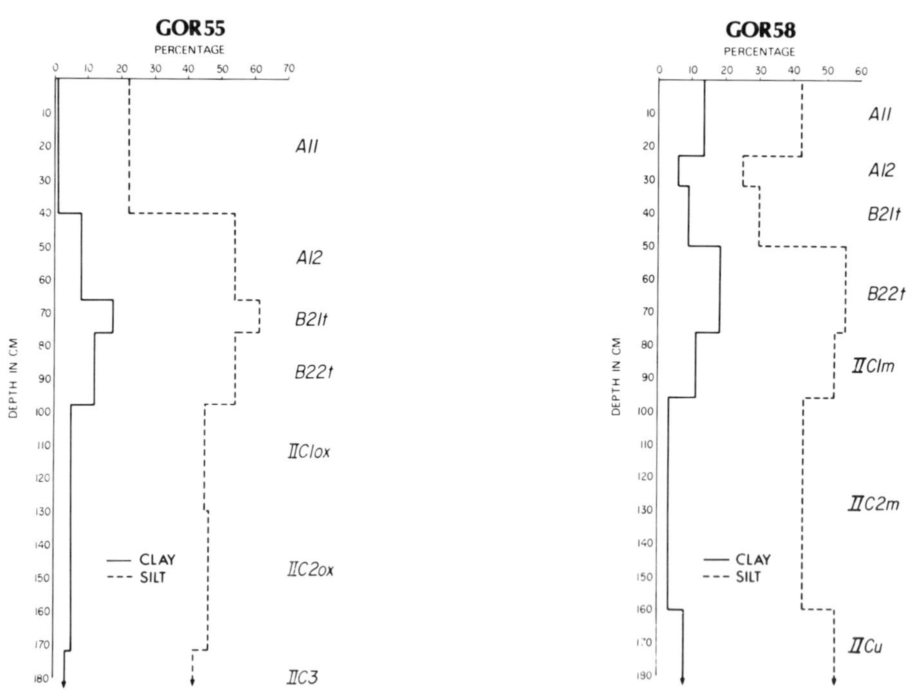

Figure 4.19: Silt and clay distributions with depth in profiles GOR55 and 58. Aeolian/till boundaries (I, II) were drawn on presence/absence of silt caps in the upper part of profile 55. Profile 58 may have translocated aeolian sediment in the B22t horizon.

The feldspar (mainly anorthoclase) distributions (Table 4.1) indicate higher quantities in tills and lower quantities in aeolian cover sediments. The distributions of magnetite, while rather uniform, suggest higher degrees of weathering in the paleosol sola.

With respect to clay minerals and other weathering products metahalloysite is found only in the lower GOR58 profile. Its absence in the GOR55 profile may be related to the presence of gibbsite, and to its paleoclimatic/drainage history (see Chapter 9). The greater amount of gibbsite in the GOR55 profile indicates more aggressive leaching conditions, which may have removed all or most Si released on weathering to the pore solutions. Because gibbsite is distributed so differently in the two profiles, it must have a paleoclimatic/weathering rather than a detrital origin. For a full discussion of gibbsite-metahalloysite distributions in other pre-Liki (pre-Würm, Wisconsinan) profiles on Mount Kenya see Vortisch, Mahaney & Fecher, (1987). The presence of illite in the IIC2ox horizon of profile GOR55 is probably the result of aeolian transport followed by glacial reworking (see Mahaney, Vortisch and Julig, 1988, for discussion).

Table 4.1: Mineralogical[a] Composition of Profiles GOR55 and 58

		H	I	G	Q	F	M
GOR55	A11	-	-	-	xx	x	x
	A12	-	-	-	xx	x	tr
	B21	-	-	x	xx	x	tr
	B22	-	-	x	xx	x	x
	IIC1ox	-	-	x	x	x	x
	IIC2ox	-	tr	x	tr	xx	x
	IIC3	-	-	tr	x	xx	x
GOR58	A11	-	-	-	xxx	x	tr
	A12	-	-	-	xx	x	x
	B21t	-	-	-	xx	x	x
	B22	-	-	x	x	tr	tr
	IIC1m	-	-	x	x	xx	x
	IIC2m	tr	-	-	x	xx	x
	IITill	tr	-	-	x	xx	tr

a. H - halloysite/metahalloysite, I - illite, G - gibbsite, Q - quartz, F - feldspar, M - magnetite. The relative amounts of minerals include nil (-), trace (tr), small (x), and moderate (xx).

A SEM study of the fine and very fine sand fractions (63-250 μm) showed that relative weathering features and neoformation of clay minerals were useful age indicators in the two profiles. In profile GOR55 every horizon was scanned from the surface loessic material to the tills at depth. The strongest weathering effects were observed in the paleosol solum in the form of solution etching, grains coated with gibbsite, decomposed feldspar grains, dissolved microliths from volcanic glass, and weathered glass with A1 coatings.

Glacially crushed feldspar grains with pronounced conchoidal fractures, and linear crushing surfaces were found throughout the profile (even in the loess), indicating that grains which survive glacial transport are often reworked by airfall processes (see Vortisch, Mahaney and Fecher, 1987; and Mahaney, Vortisch and Julig, 1988) (Figure 4.20A). Grains in the A12 horizon, with minor etching features, contain coatings with some Fe (determined by X-ray microanalysis) and alternate with other feldspar grains that show solution etching parallel to cleavage planes (Figure 4.20B). Some feldspar grains in the B21t horizon have high Mn coatings;

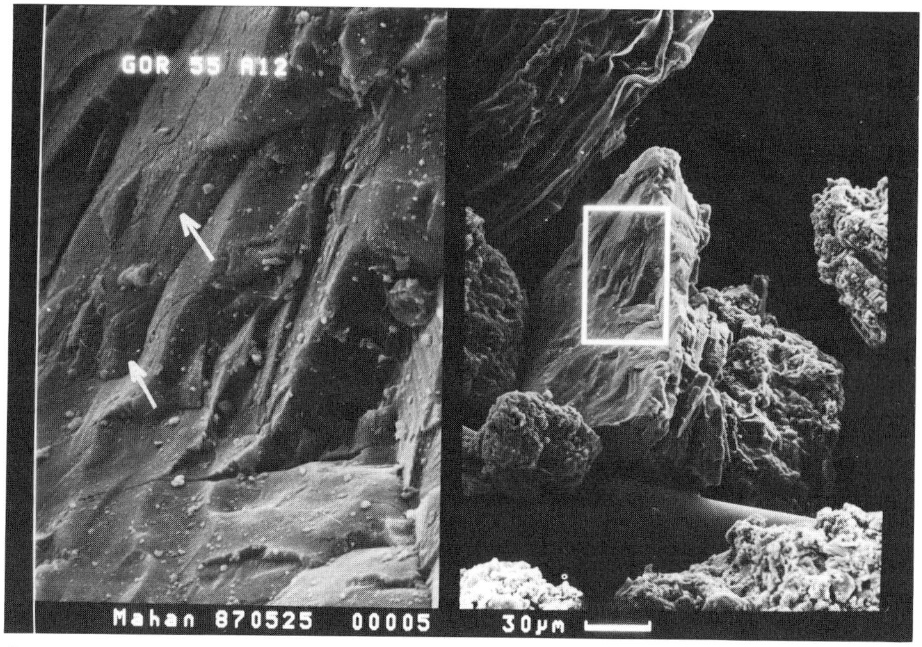

A.

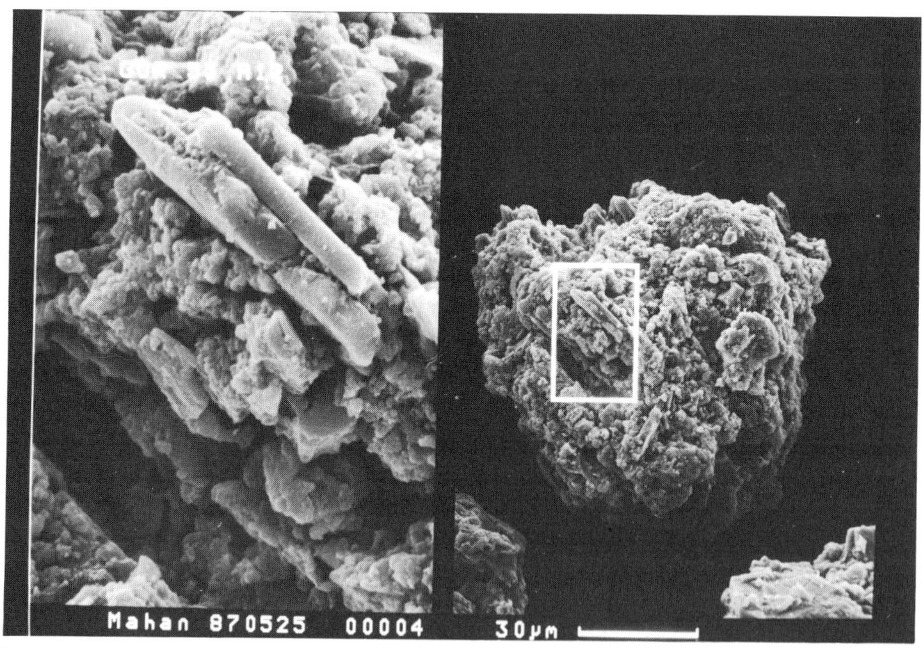

B.

Figure 4.20: A, Glacially-crushed feldspar grain (arrows indicate crushing planes) with other coated composite grains in the A12 horizon of profile GOR55; B, Na-feldspar grain in the A12 horizon showing solution etching features parallel with cleavage planes.

others have very high A1 with almost no Si. In the B22t horizon many feldspar grains have high A1; other etched feldspar grains have Si:A1 ratios of 1:1 (Figure 4.11A).

In the subsoil horizons (IIC group of profile GOR55) fresh feldspar grains (Figure 4.21A) alternate with weathered grains which often show complex weathering histories (Figure 4.21B). Glacial crushing features are more pronounced in the IIC horizons; many relatively clean feldspar grains alternate with coated grains of weathered ash. Some weathered glass particles display dissolved microliths of variable composition. Even in the IIC3 horizon at the base of the paleosol many grains are still coated with A1. In all, chemical corrosion appears slightly higher in the paleosol solum, with gibbsite coatings on more than 50 percent of grain surfaces through the entire profile. As in previous studies of older Brunhes and Matuyama age profiles (Vortisch, Mahaney & Fecher, 1987; Mahaney, Barendregt and Vortisch, 1989) most weathering occurs on volcanic glass particles.

The GOR58 profile contains somewhat higher amounts of quartz in the sand fractions of the paleosol sola (A & B horizons). In the A11 horizon compound grains of hematite (Figure 4.21A) alternate with opal phytoliths – X-ray microanalysis confirms higher Si. A *fresh* quartz grain (Figure 4.22B), with glacial crushing features, in the A11 horizon points to the recent influx of airfall sediments following glacial and aeolian transport. An alkali feldspar grain in the A11 horizon shows pronounced solution etching and a small hole probably caused by magmatic corrosion (Figure 4.23A). Glass in the A12 horizon is often rounded with dissolved microliths. These weathered ash grains alternate with fresh grains of glacially-crushed quartz (Figure 4.22B). In the B21t horizon the nodular variety of gibbsite dominates on Na-feldspar (Figure 4.23B); on other feldspar grains weathering occurs as dissolution etching parallel with cleavage planes. X-ray microanalysis showed higher Al-coatings on grain surfaces in the B22t horizon (see Table 1 for XRD data on gibbsite in the B22t horizon).

The till in GOR58 (IIC2ox and IIC3 horizons) is nearly identical to the till in profile GOR55. Glacially-crushed grains alternating from fresh to weathered occur with grains of weathered ash (Figure 4.24A and B). Some plagioclase and K-feldspar grains exhibit variable etching features depending on their susceptibility to weathering. Overall weathering is less intensive in the till bodies compared with the overlying loesses.

Profiles GOR55 & 58 show overall trends in particle size, clay, sand mineralogy, and magnetic remanence that indicate different ages for the two loess covers and similar ages for the tills. The loess in profile GOR55 is presumed to have a lower Brunhes age; weathering features on sand grains and higher content of gibbsite suggest greater chemical corrosion and more aggressive leaching when compared with profile GOR58. Normal magnetism throughout the loess cover in profile GOR58 indicates a younger Brunhes age, which is supported by a higher proportion of fresh grains and lower neoformation of clay (supported also by field observations of clay films). One interesting aspect of these different ages is that it appears loess sheets may be periodically stripped and replaced with fresh sediment, especially in areas lacking forest cover.

4.7 Paleomagnetic Dating of Gorges Deposits

Paleomagnetic dating has proven to be a valuable tool for the differentiation of sediments which were deposited during the early Pleistocene, and those which were deposited during the middle and late Pleistocene. The boundary between these two groups of sediments falls between 735 and 790 Ka (Johnson, 1982; Imbrie *et al.*, 1984) and separates the Normal Brunhes Chron from the Reversed Matuyama Chron. Although polarity subchrons lasting from 4 to 200 Ka occur within the Brunhes and Matuyama, most sequences of glacial deposits on Mt. Kenya reveal stratigraphic evidence of multiple glacial and interglacial events spanning long periods of time. On the basis of this stratigraphic evidence (Mahaney, Barendregt and Vortisch, 1989) as well

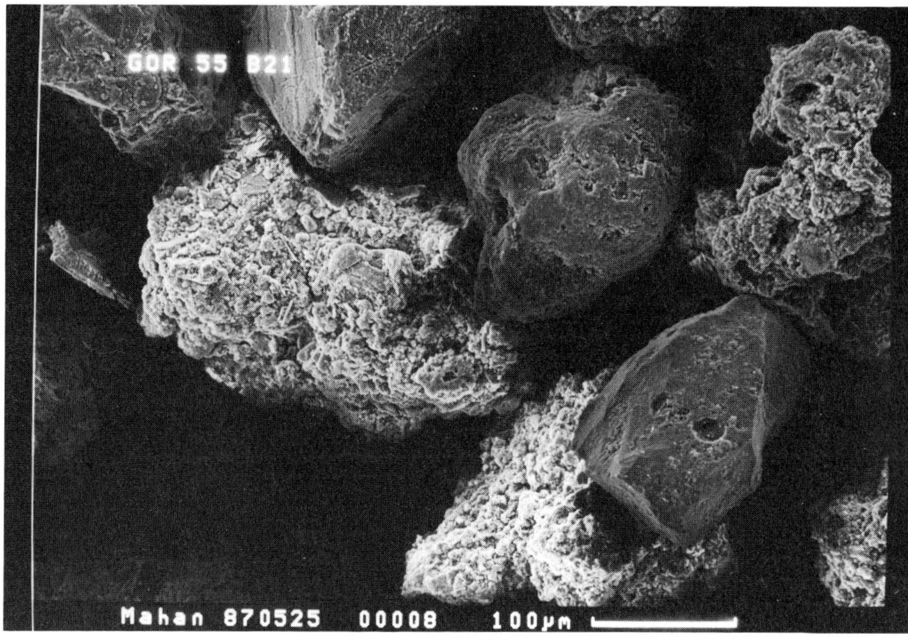

A.

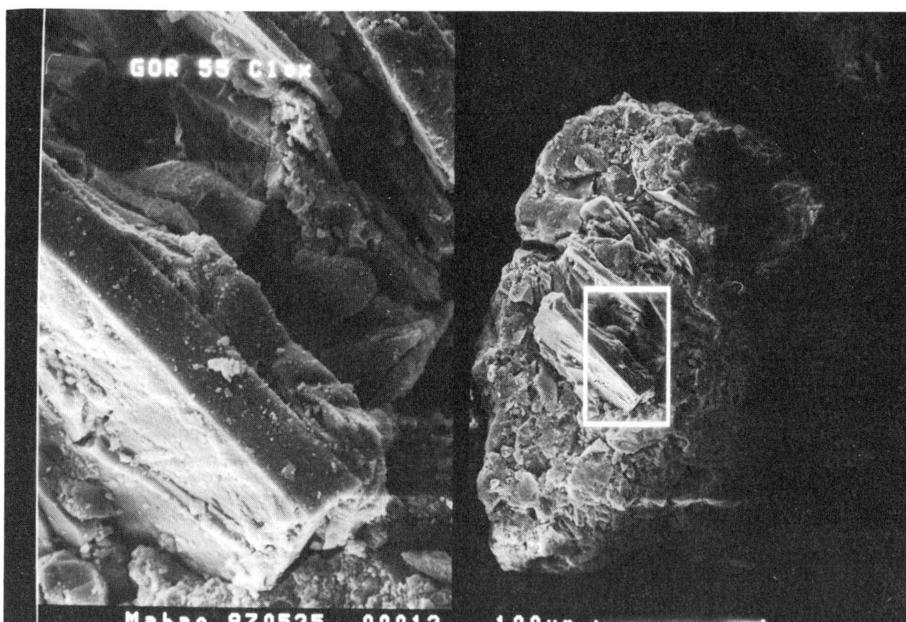

B.

Figure 4.21: A, Etched feldspar grains and volcanic glass grains with thick gibbsite coatings in B21t horizon of profile GOR55. Al:Si ratio 2.5:1; B, partially coated and etched feldspar grain coated with clays having a 1:1 Si:Al ratio. Si:Fe ratio in the IIC1ox horizon is 5:1 (Fe is very high).

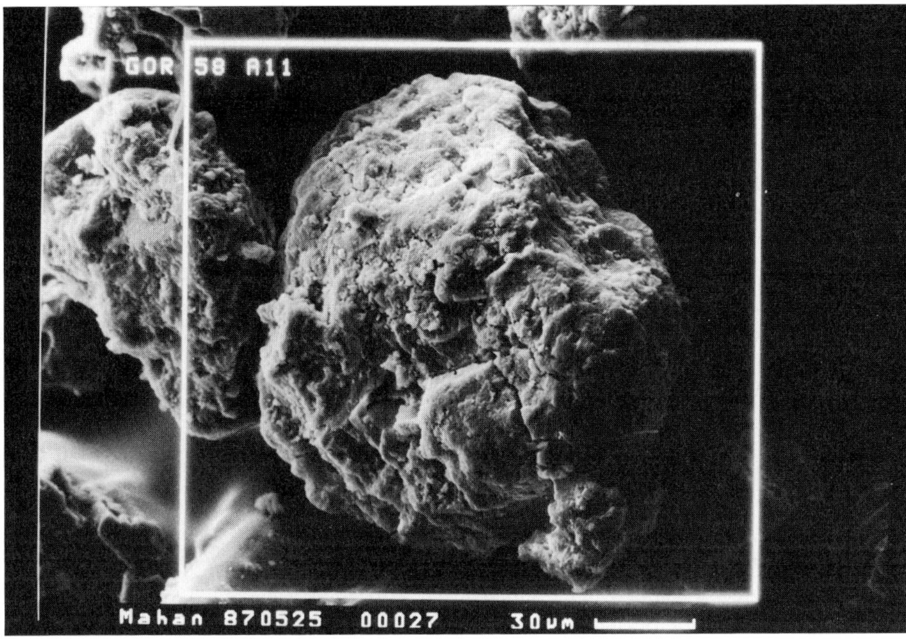

A.

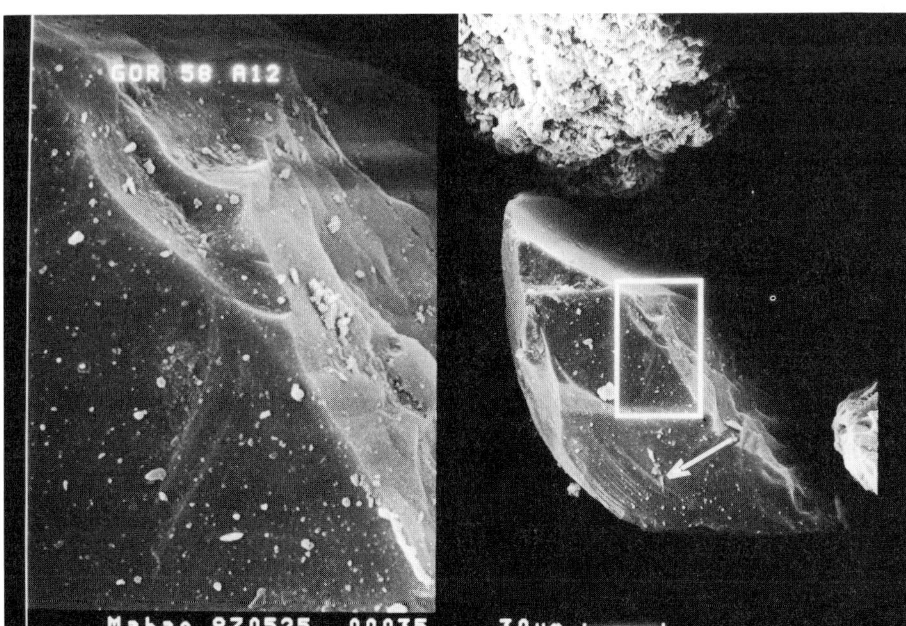

B.

Figure 4.22: A, Compound grain of Fe (probably hematite) in GOR58-A11 horizon (high Fe in paleosola of profiles 55 and 58); B, Glacially-crushed quartz grain showing subparallel crushing planes (arrow); enlargement shows step features. Only minor dissolution etching has occurred in places.

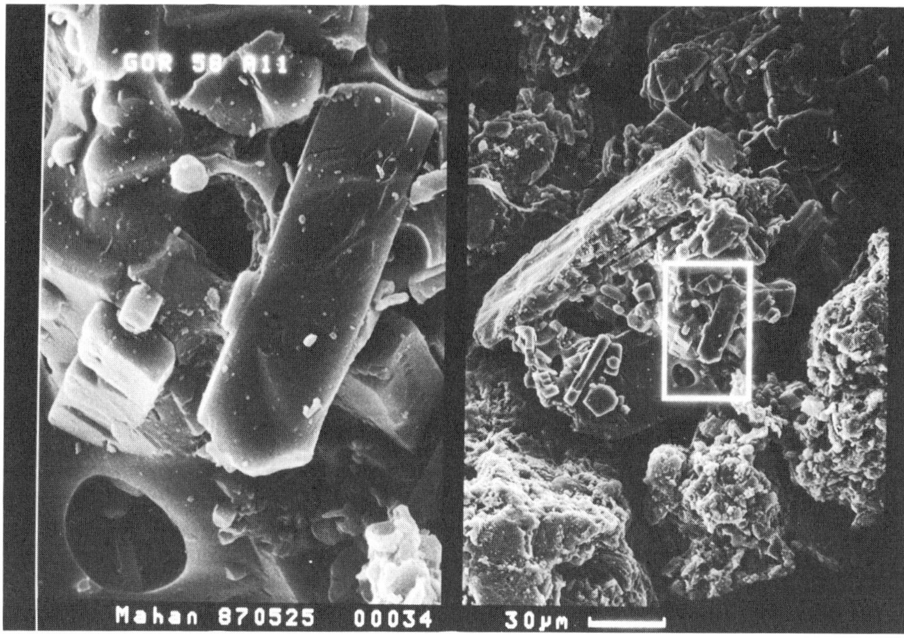

A.

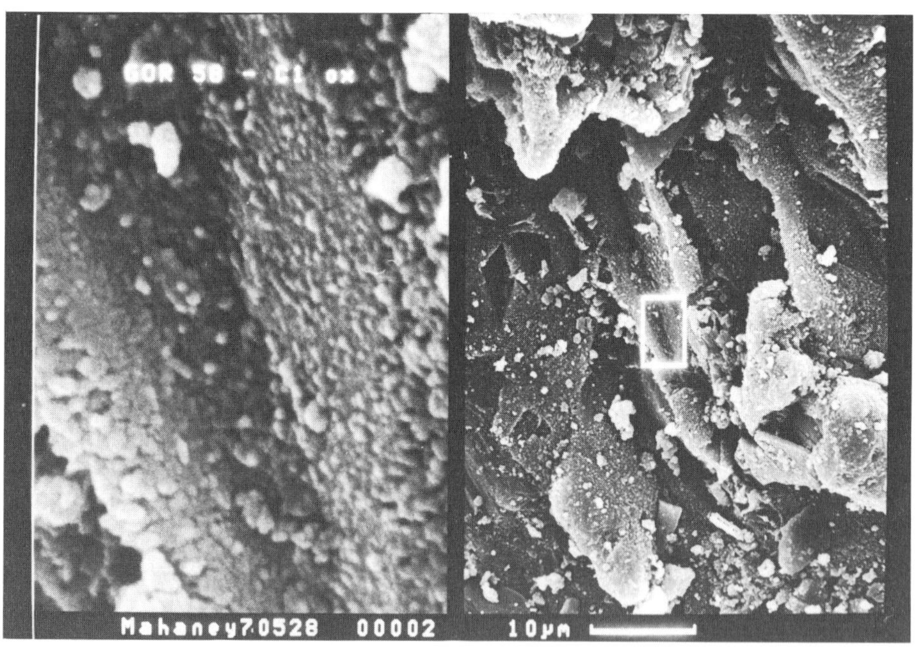

B.

Figure 4.23: A, Alkali feldspar with very fresh surfaces; small holes (arrows) are probably caused by magmatic corrosion-hole is too asymmetrical to be chemically weathered (upper hole is slightly weathered); B, Neoformed clay minerals (and nodular form of gibbsite) with Si:Al ratio 1:1 on volcanic glass grain-long furrows may have been stretched gas bubbles. Thin walls are typical of glass structures. Some Fe present as a coating on this particle.

A.

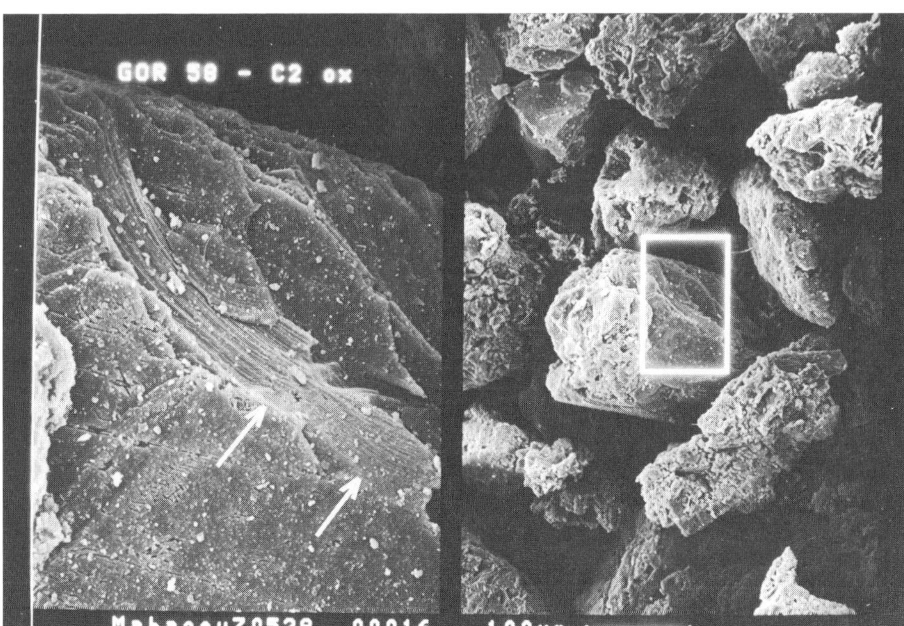

B.

Figure 4.24: A, Weathered feldspar grain with extreme etching in the IIC1ox horizon of profile GOR58. Relatively high Fe in grain coating; B, Glacially crushed feldspar grain (arrows = crushing planes) showing preweathered face (to left).

as paleosol weathering histories (this chapter), materials sampled at GOR55 and GOR58 (Figure 4.25) are considered to represent time frames which far exceed the estimated temporal extent of known excursions (subchrons). During the summers of 1984 and 1986, sites GOR55 and GOR58 located within Gorges Valley (Plate 1) were sampled for paleomagnetic analysis.

Remanence measurements and stepwise alternating field (af) demagnetization of selected pilot specimens from each lithologic unit were carried out at the levels indicated in figures 4.26 and 4.27. Qualitative assessment of changes in both direction and magnitude of the measured magnetic vectors led to the selection of a 20 mT cleaning field for the remaining samples. Typical demagnetization plots of pilot specimens are shown in figures 5 and 6.

On the basis of the low median destructive field of these samples, it appears that detrital magnetite is the predominant remanence carrier. Magnetite is one of the principal components of the tills on Mount Kenya (Vortisch, Mahaney and Fecher, 1987).

All paleomagnetic samples from GOR55 and GOR58 fall into two groups: those showing normal polarity (loesses) and those showing reversed polarity (tills). Transitional polarities were not seen.

As shown in Figure 4.25 and 4.26 all specimens are characterized by a single component remanent magnetization. A viscous component is removed by the first step of af demagnetization. No overprint histories were recorded. Table 4.2 contains a summary of the polarity data for each lithologic unit at both sites. The tills show nearly identical pole data while the loesses are significantly different (Figure 4.25). The loess cover at GOR55 reveals a pole very close to a Geocentric Axial Dipole (G.A.D.) and shows no evidence of secular variation. One can assume that the loess was deposited over a relatively short period of time, presumably during the Brunhes Normal Chron. The loess cover at GOR58 on the other hand shows evidence of at least two stages of deposition: an earlier one during which the sediments recorded a relatively high positive inclination for this sampling latitude, and a later stage when the sediments recorded a north/northwesterly declination, and showed a tenfold (order of magnitude) increase in magnetization intensity. Based on their rather different paleomagnetic properties, these two stages of deposition are likely separated by a considerable period of time. Both deposits are normally magnetized and are assumed to have been deposited during the Brunhes Normal Chron.

Table 4.2 Summary Table of Stratigraphy and Mean Remanent Magnetization Directions

Site	Thickness	Lithology	Age	Polarity	N	M	D	I	K	α95	Pole Lat.	Pole Long.
GOR55	1.00	Loess	Brunhes?	Normal	13	0.13	357	+2	186	3	86°N	329°
GOR55	0.60	Till	Matuyama?	Reversed	8	0.35	179	-5	112	5	87°S	192°
GOR58	0.25	Loess (top)	Brunhes?	Normal	7	0.06	338	+5	26	12	68°N	315°
GOR58	0.45	Loess (bottom)	Brunhes?	Normal	11	0.004	1	+17	119	4	81°N	46°
GOR58	1.00	Till	Matuyama?	Reversed	26	0.034	180	+11	10	15	85°S	42°

N = number of specimens
M = average intensity of magnetization (AM-1)
D = average declination (degrees)
I = average inclination (degrees)
K = precision parameter (Fisher, 1953)
α95 = radius of the cone of 95% confidence about the resultant vector

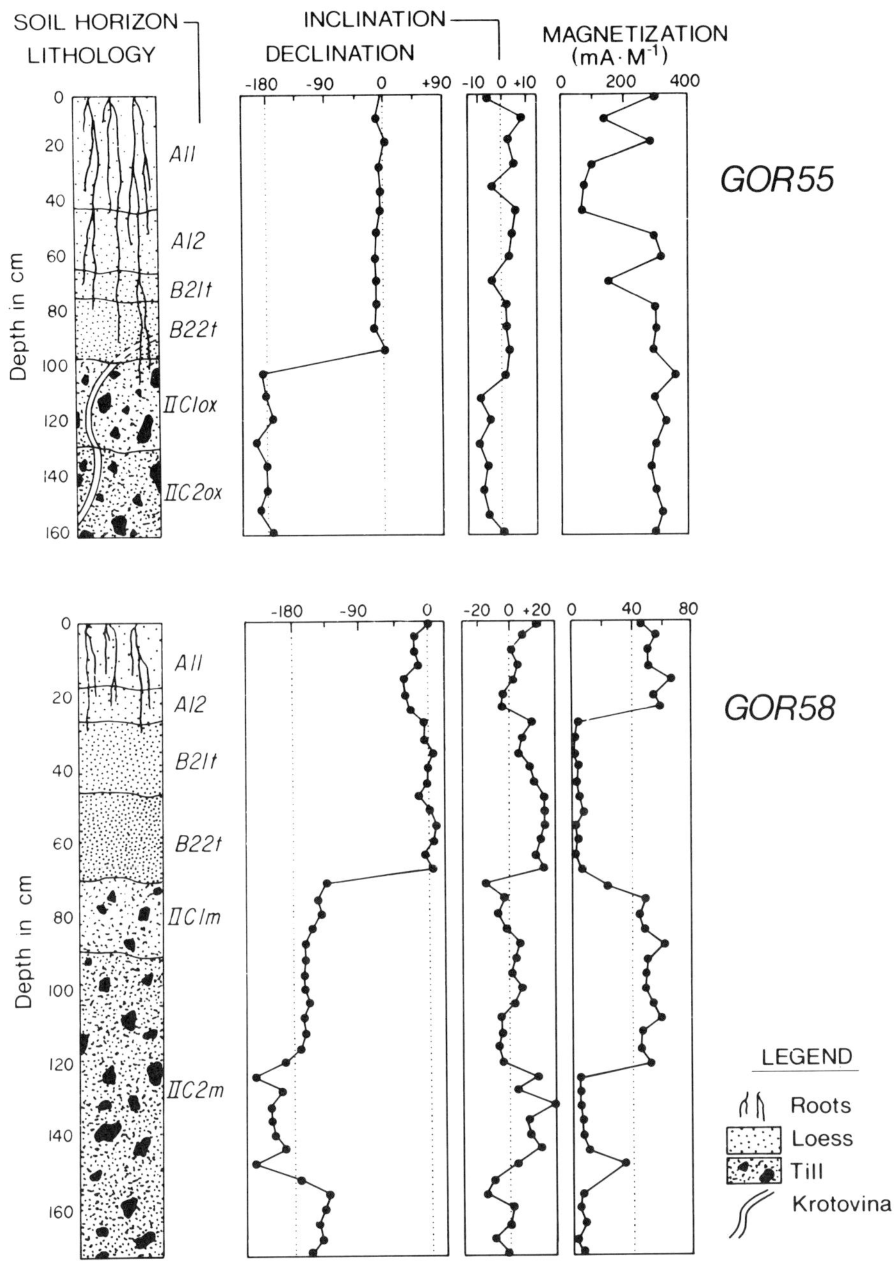

Figure 4.25: Stratigraphic and paleomagnetic profiles for site GOR55 (A) and site GOR58 (B), lower Gorges Valley, Mt. Kenya. GOR55 is located under Hagenia forest; GOR58 under tussock grass. Both sites are on the outer Gorges moraine crest. Unit I is an aeolian unit; unit II is till. The normally magnetized loess covers are assumed to have been deposited during the Brunhes Normal Chron, the reversely magnetized tills during the Matuyama Reversed Chron. Magnetization values for all samples are at 20 millitesla (mT) demagnetization. Note that a declination shift at both sites occurs at the loess/till contact. Units of declination and inclination are in degrees; units of magnetization are in milliamperes per meter.

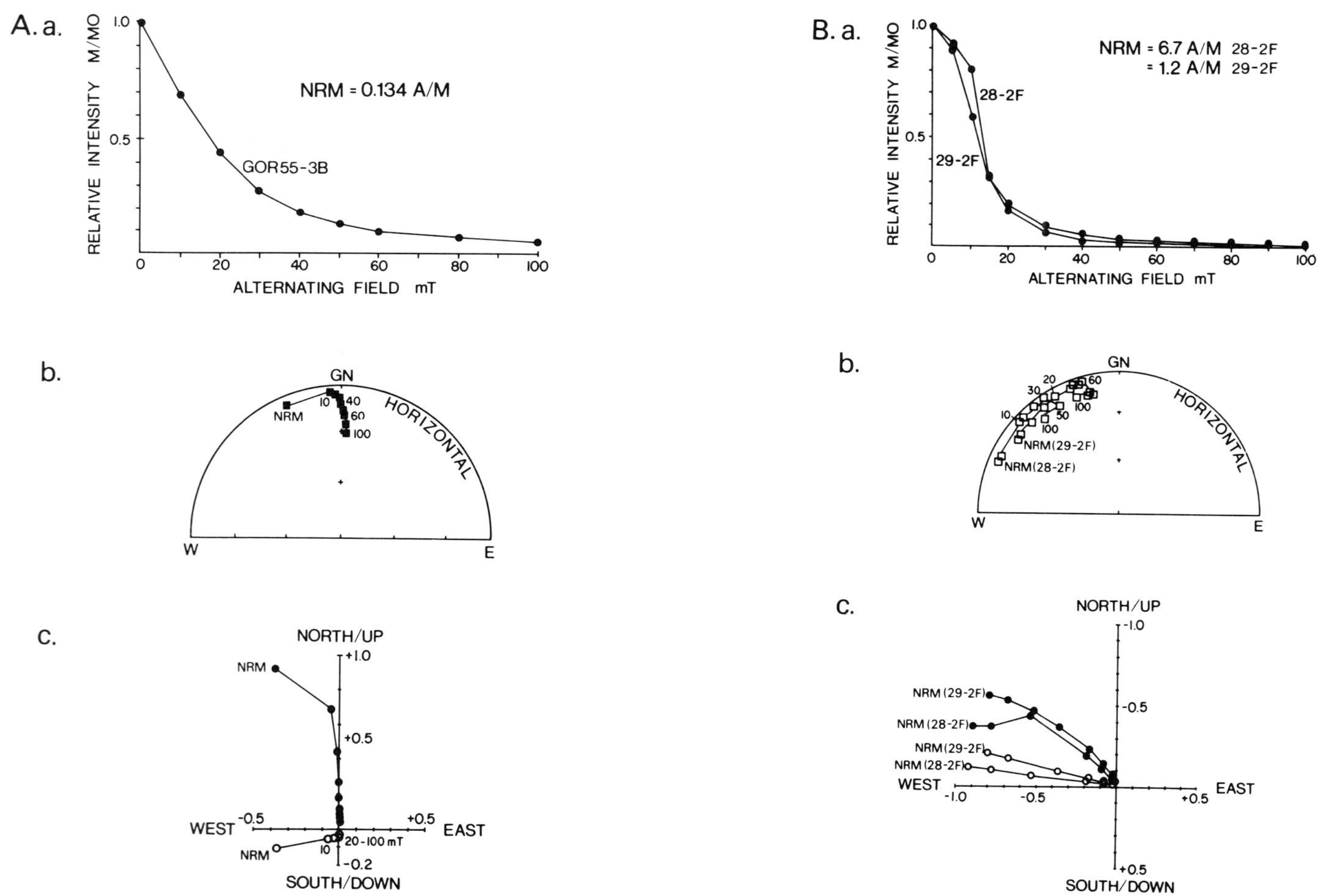

Figure 4.26: Magnetization behavior of pilot specimens from normally magnetized loess deposits at GOR55 (A) and GOR58 (B) following stepwise alternating field (af) demagnetization. Natural remanent magnetization is indicated by letters NRM. Normalized magnetization (M/MO) versus af demagnetization (a), and directional changes during af demagnetization projected on a lower hemisphere equal area stereographic projection (b) where solid squares (■) represent positive inclination and open squares (□) negative inclination, and orthogonal vector projections (c), with respect to present horizontal, where solid circles (●) refer to N/S axis (horizontal plane) and open circles (○) to Up/Down axis (vertical plane).

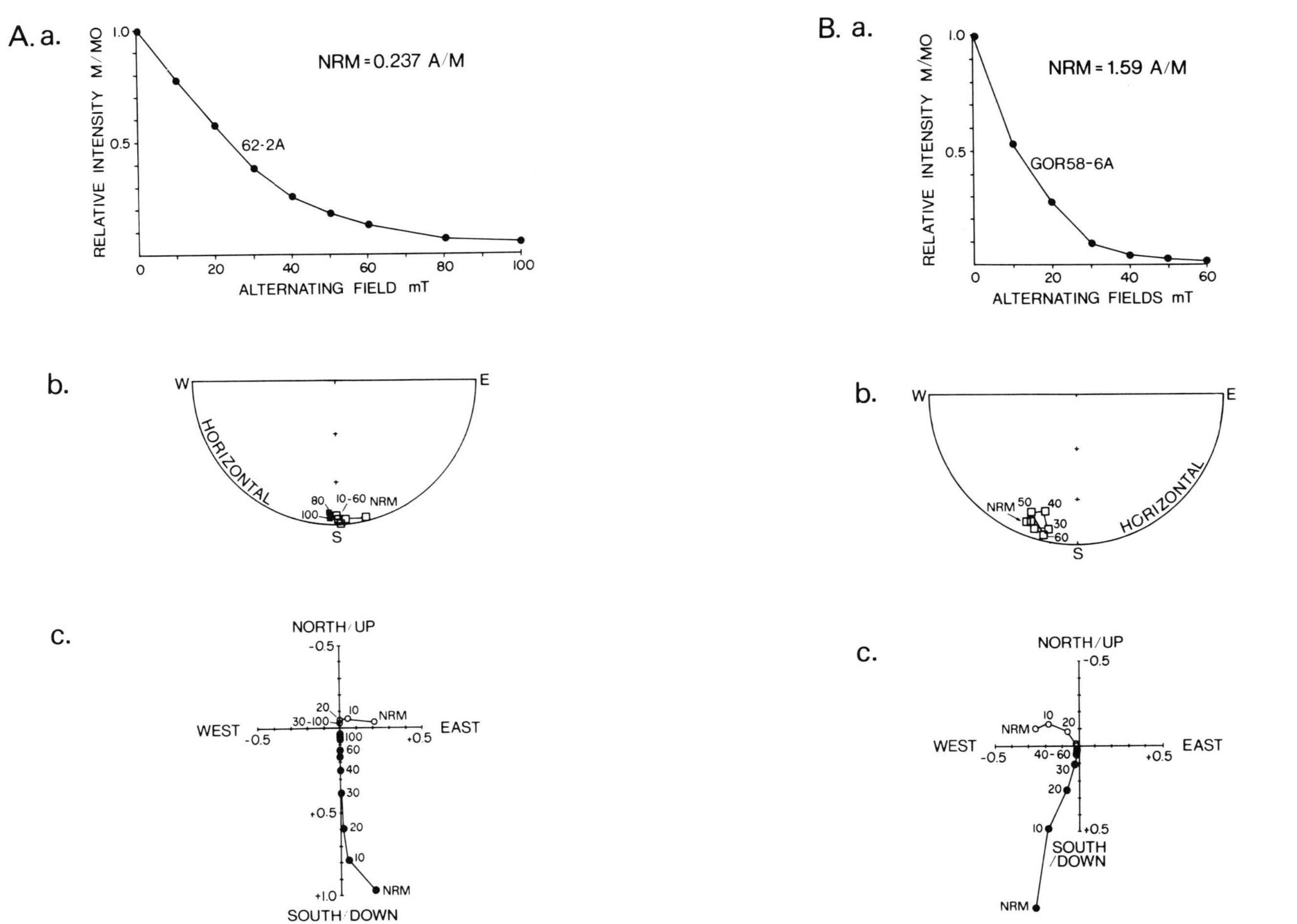

Figure 4.27: Magnetization behavior of pilot specimens from reversely magnetized till deposits at GOR55 (A) and GOR58 (B) following stepwise af demagnetization. Labelling conventions as in Figure 4.26.

In contrast to GOR55 the loess record at GOR58 reveals hints of secular variation swings. As these swings appear as incomplete cycles of secular variation, the interruptions, if real, may represent hiatuses of deposition and/or erosion and thus may also argue for the passage of considerable time.

Paleomagnetic measurements of the tills at GOR55 and 58 (Table 4.2) indicate that these deposits obtained their primary magnetic remanence at the same time, and during a period when the field was reversed. As these sites are near the margin of glaciation, and it is known from other work (Mahaney, Barendregt and Vortisch, 1989) that the earlier Quaternary glaciations descended farthest down the mountain, it can be assumed that these deposits were laid down during the Matuyama Reversed Chron and are thus older than .73-.79 my. The lack of secondary magnetizations, in the form of magnetic overprints is noteworthy (Figure 4.27). This may imply that these sediments were for the most part well drained, and that ground water movements were insufficient to substantially remobilize iron in a form that would have allowed secondary magnetizations.

Paleomagnetic measurements of the loess covers at GOR55 and 58 (Table 4.2) show significant differences in primary magnetic remanence directions. The intensity of magnetization also differs by a factor of 10. This suggests that the loesses at the two sites are of different ages and probably also of different provenance. The significant changes in direction and intensity within the loess profile at GOR58 suggest that these sediments were deposited over a considerable span of time. Furthermore, the presence of partial records of secular variation indicate that periods of erosion and/or non-deposition may account for some of this time.

4.8 Stratigraphy of Lake Ellis Deposits

Profile HOB19 (Figure 4.28) reveals a Tropudalf (Soil Survey Staff, 1975) formed in Lake Ellis till and loess, at 3150 m, in Hobley Valley. The paleosol formed in this sequence is older than the Brunhes-Matuyama boundary ($>$7.3-7.9 my). This profile was chosen for paleomagnetic study because it correlates with, and is representative of, older Lake Ellis drifts on the eastern side of Mount Kenya (Baker, 1967; Mahaney, 1982a, 1983a). Although these drift surfaces lack distinctive moraine form, the microfabric (as examined by scanning electron microscopy) shows alignment of very coarse to medium (2000-500 μm) sand grains in the downvalley direction (SE-NW) (Derbyshire and Mahaney, 1989). Such strong orientation suggests emplacement by ice. The paleosol formed in this till is capped with loess that presumably is somewhat younger than the C horizon complex. While unusually thick (1.1 m for the total pedon), the C horizons have an indurated matrix cemented with Fe_2O_3 and MnO_2 (Derbyshire and Mahaney, 1989).

A scanning electron microscopy study of sand clasts in the loess and till units of profile HOB19 revealed a number of glacially crushed grains (Figure 4.29A) in the loess. On the whole grains were fresher and less weathered in the loess, although many showed evidence of preweathering presumably during the Gorges-Lake Ellis Interglaciation. Both subangular grains (Figure 4.29A) and grains rounded by aeolian transport were found in the loess. The till in HOB19 yielded mainly subangular grains, many of which carried glacial crushing planes (Figure 4.29B) consisting of linear and conchoidal features. Some grains have V-shaped percussion cracks indicating they were moved some distance by running water.

Lake Ellis deposits in Gorges Valley consist of end moraines and outwash fans that were emplaced during the Upper Matuyama Chron. Most deposits lie on weathered tuff that belongs to the Mount Kenya volcanic suite of early Quaternary age. Presumably the bedrock was weathered during the Gorges-Lake Ellis Interglacial which may have lasted $\sim$200,000 yr. Because gibbsite dominates as a weathering product in these underlying saprolite beds, it is likely the paleoclimate that existed during its formation was one that produced high soil moisture

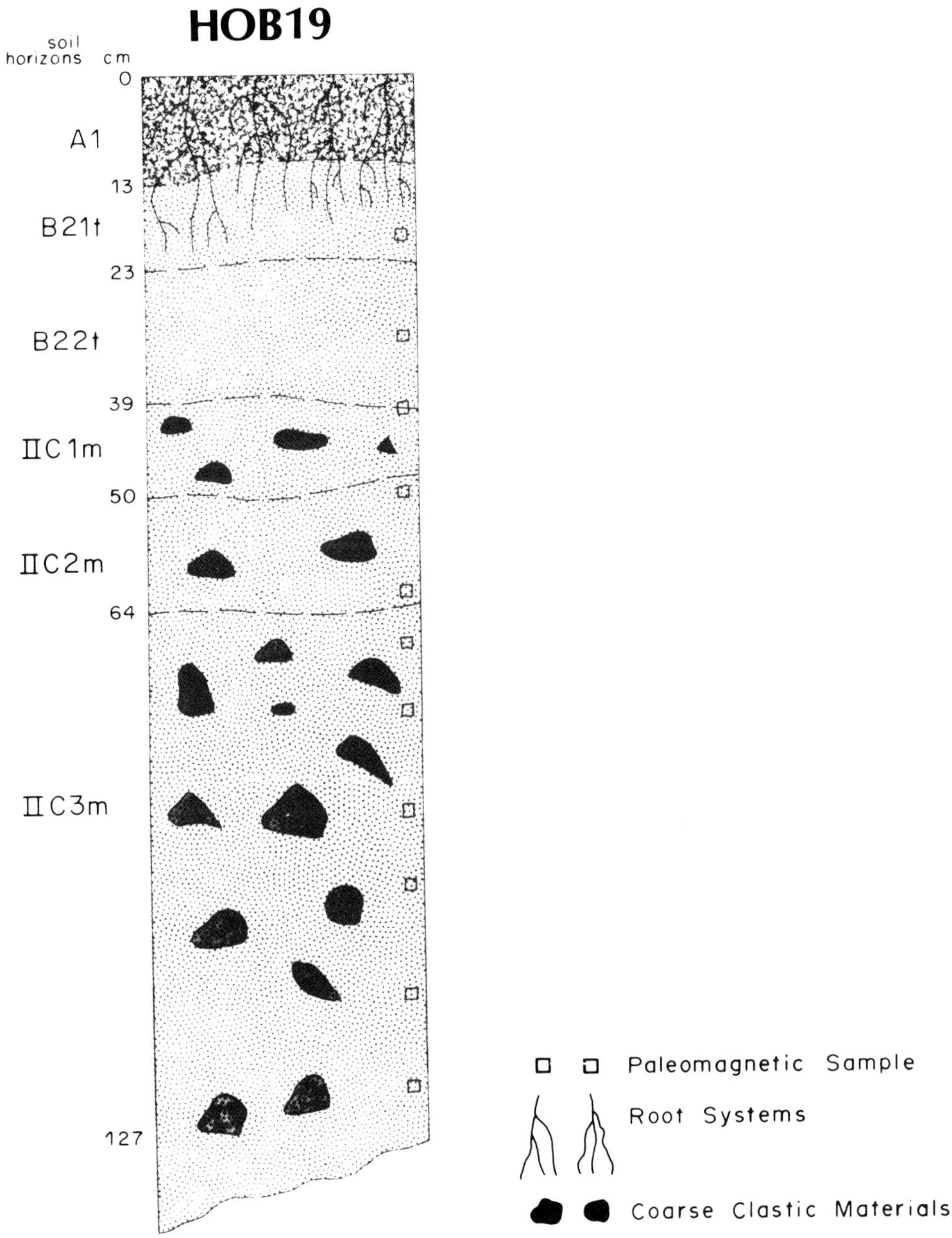

Figure 4.28: Profile HOB19 in Lake Ellis drift, Hobley Valley

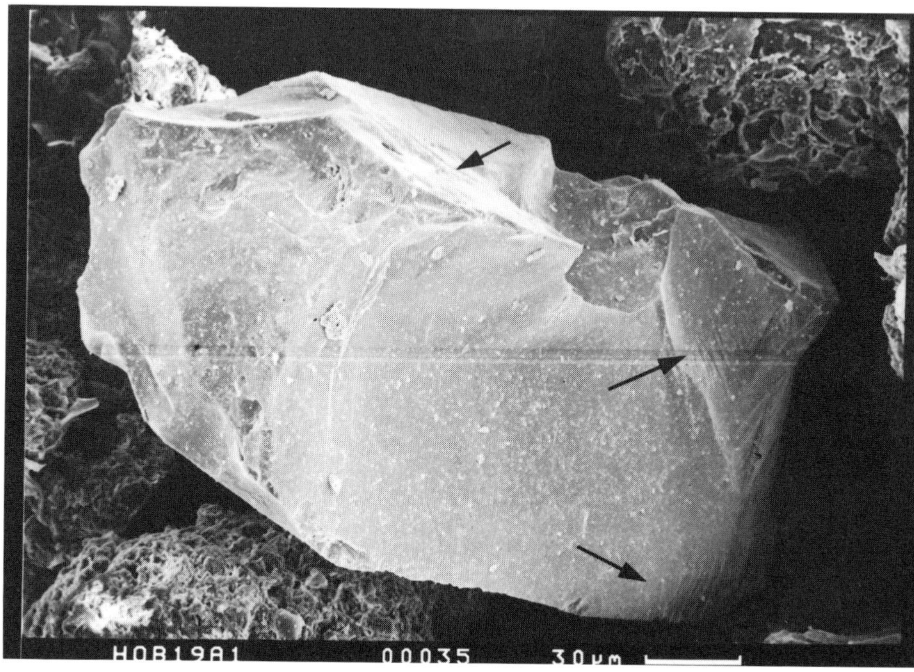

A.

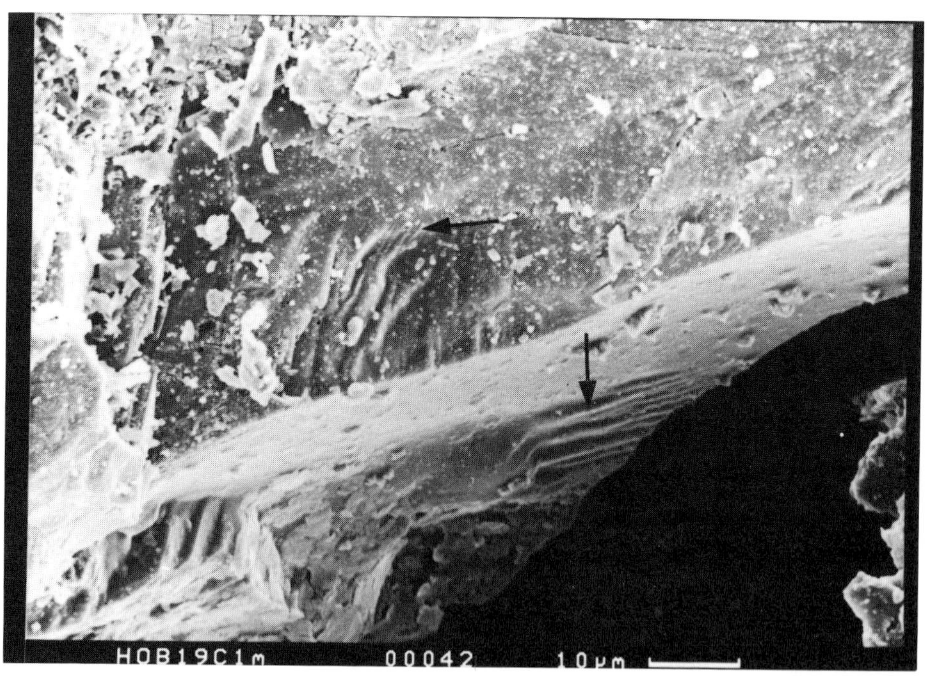

B.

Figure 4.29: A, Glacially-crushed quartz grain in the A1 horizon showing linear and conchoidal crushing planes (arrows) and an older preweathered surface probably formed during the Gorges-Lake Ellis Interglaciation; B, Glacially crushed quartz grain in the IIC1m horizon showing crushing surfaces (arrows).

and hence aggressive leaching. Therefore the paleoclimate was wetter and presumably warmer than today.

Profile GOR54 (Figure 4.30), a sequence of till overlain with outwash and loess, that is remarkably similar to profile GOR53 (see Chapter 9). Because the buried B horizons in profiles GOR53 and 54 are so well weathered it is possible they were exposed at the land surface for a long time prior to the deposition of the overlying sediments. Normal polarities in the overlying outwash and loess and relatively less weathering, make it probable they were deposited during the Naro Moru or Teleki glaciations. Thus they are likely ~250,000 to ~700,000 yr in age.

Profile GOR71, formed in loess and till of Lake Ellis age overlies a highly weathered till that is probably of Gorges age. The upper till in this section is representative of till deposits of Lake Ellis age. Weathering is not as intense as in older deposits of Gorges age, but gibbsite still dominates (in the buried till) as a weathering product indicating that the Lake Ellis-Naro Moru Interglacial paleoclimate was as wet (and presumably as warm) as the Gorges-Lake Ellis Interglacial. The lower till, with nearly obliterated clasts and high clay content, is probably a remnant from the first early Quaternary glaciation.

Near Lake Ellis (north of Hinde Valley) an end moraine remnant (to the east of the lake) records an advance of ice close to Mugi Hill (Plate 2). The till and overlying loess (Figure 4.31) show different degrees of weathering, as at site HOB19, with gibbsite dominating as the chief weathering product along with hematite and goethite.

4.9 Paleomagnetic Dating of Lake Ellis Deposits

Twelve pilot specimens from the Hobley Valley site were selected and subjected to stepwise alternating field (af) demagnetization at levels of 5, 10, 15, 20, 25, 30, 40, 50, 60 millitesla (mT) in a conventional three-axis tumbler. After each demagnetization step, the remanence was measured with a Schonstedt DSM-1 magnetometer. Qualitative assessment of changes in magnetization values following af demagnetization led to a choice of 15, 20 and 25 mT as the standard demagnetizing fields for the remaining samples. Typical demagnetization plots are shown in Figure 4.31 and most specimens are characterized by a single component remanent magnetization. On the basis of the low median destructive field of these specimens, it would appear that detrital magnetite is the predominant remanence carrier [magnetite is one of the principal components of most tills on Mount Kenya (Vortisch, Mahaney and Fecher, 1987)].

The declination, inclination, and intensity profiles for all samples are shown in Figure 4.33 and the summary of remanent magnetization directions is shown in Table 4.3.

Table 4.3: Summary of remanent magnetization directions and mean "A" pole positions (for definition of "A" pole, see Irving, 1964)

							"A" Pole	
Symbol	N	M	D	I	K	$\alpha 95$	Lat.	Long.
HOB19 (Normal)	8	5.11	358	-23	132	9	78°N	225°
HOB19 (Reversed)	4	0.85	188	-7	221	6	81°S	284°
HOB19 (Normal) Overprint)	4	2.02	309	4	11	9	39°N	310°

N = number of samples per site.
M = average intensity of magnetization (AM^{-1})
D = average declination (degrees).
I = average inclination (degrees).
K = precision parameter.
$\alpha 95$ = radius of the 95% confidence circle about the mean direction (in degrees).

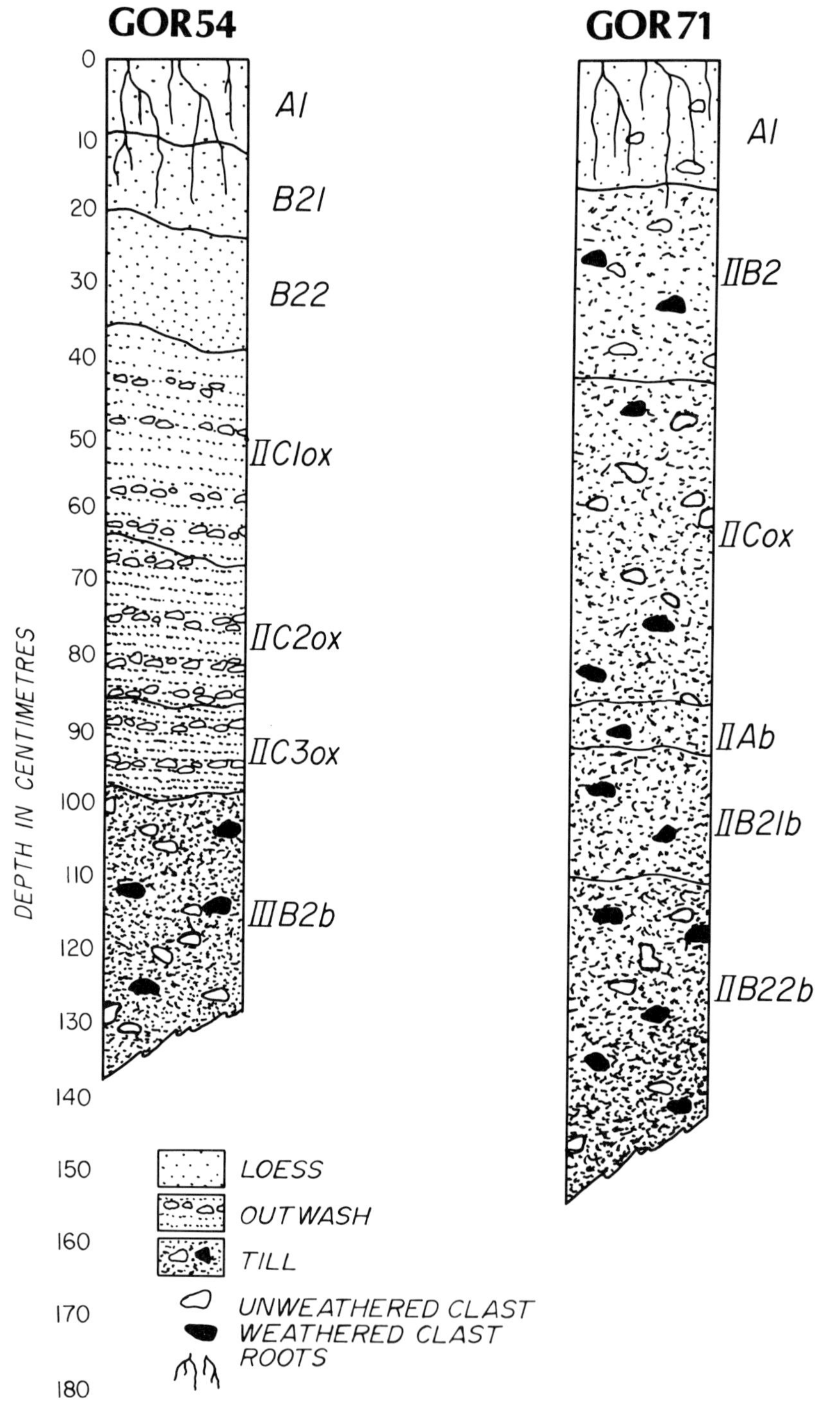

Figure 4.30: Stratigraphy of Lake Ellis Deposits (sites GOR54 and 71) in Gorges Valley.

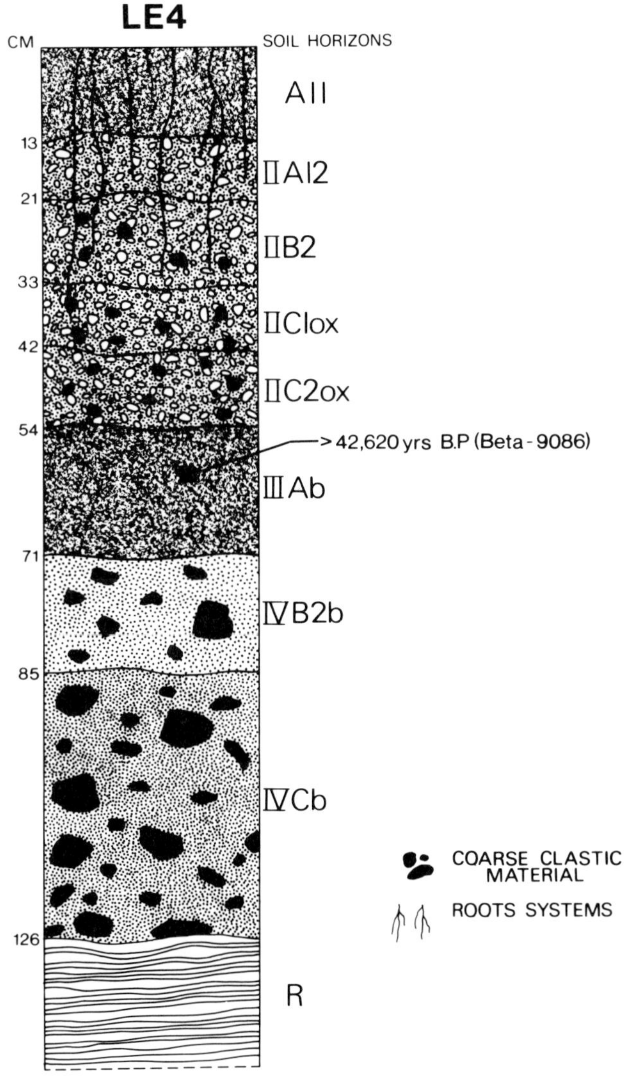

Figure 4.31: Stratigraphy of till, alluvium and loess (site LE4).

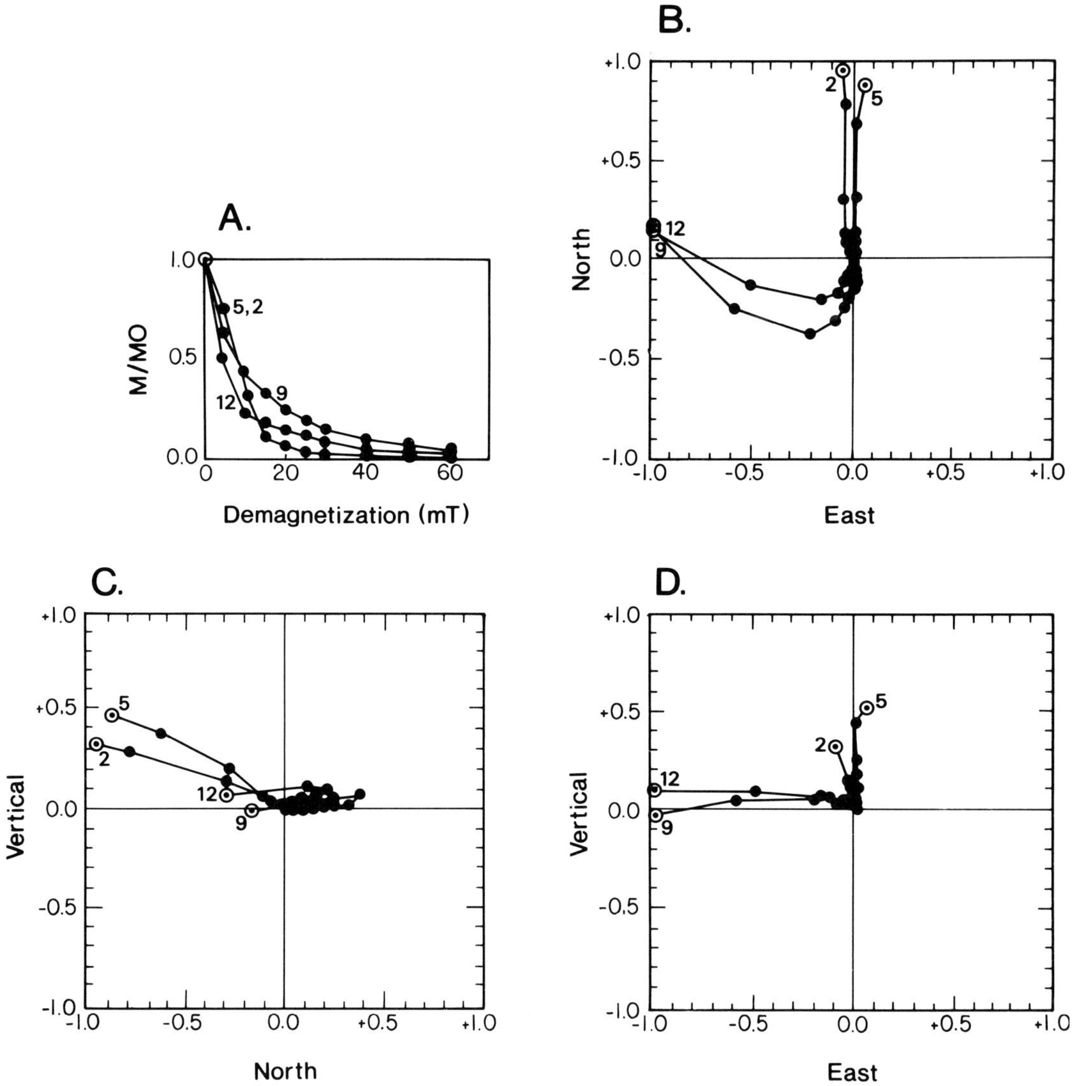

Figure 4.32: Representative demagnetization curves of samples 2 and 5 (normal) and samples 9 and 12 (reversed) from Lake Ellis till in Hobley Valley. Progressive alternating field demagnetization (afd) at levels of 5, 10, 15, 20, 25, 30, 40, 50 and 60 millitesla (mT) gives ratios of M/MO as shown in (A). Magnetization values are normalized to the highest intensity and are unitless. Orthogonal projections (Zijderveld plots) of the changing resultant magnetization vector are shown on the horizontal plane (B); vertical N-S plane (C); and in the vertical E-W plane (D).

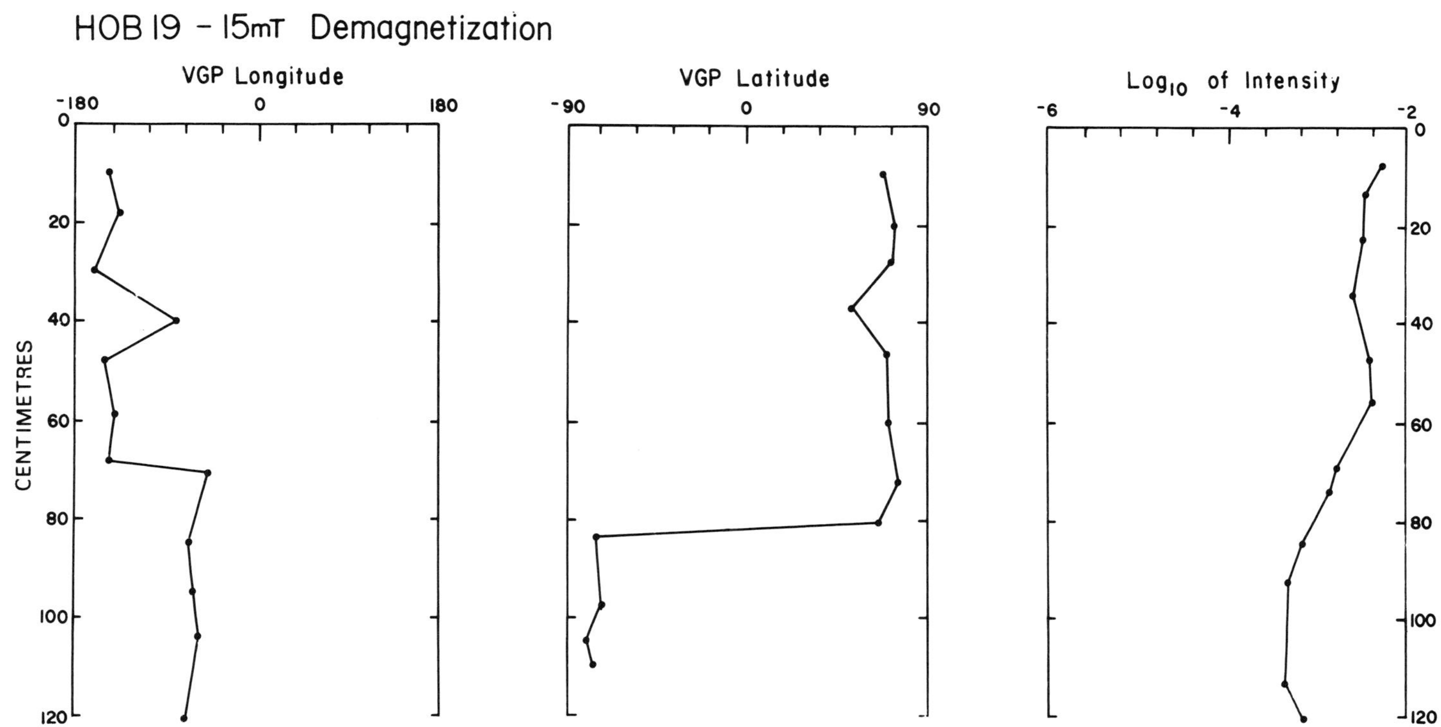

Figure 4.33: Magnetization profiles for HOB19 showing declination, inclination and intensity values for samples at 15 mT demagnetization.

The inferred polarity of the Hobley Valley sediments shows a normally magnetized upper zone (loess) and a reversely magnetized lower zone (till). The change from normal to reversely magnetized sediments is associated with an abrupt shift in declination of nearly 180°, no major change in inclination, and a significant weakening of magnetic moment (Figure 4.31). These changes are typical of magnetic field behavior during reversals at equatorial latitudes. Transition polarities between the upper and lower units are not seen, probably indicating a hiatus between the deposition of the reversely magnetized till and the normally magnetized loess.

To estimate the timing and age of these two polarity events, it is necessary to take into consideration the Quaternary stratigraphic and morphologic evidence of multiple glacial events on Mount Kenya, and the sense of the polarity reversal (with decreasing age, the sequence changes from reversed to normal polarity). The till found at Hobley Valley site HOB19 is at an elevation of approximately 3150 m.a.s.l. and marks the lowermost advance of ice down the southeast slope of the mountain. These oldest glacial deposits named pre-Teleki by Mahaney, (1979a, 1983), have been renamed Lake Ellis (Mahaney, Barendregt and Vortisch, 1989). It is assumed on Mount Kenya, as elsewhere, that the strongest glaciations were of early Pleistocene age, and that these glaciers would have deposited the moraines and associated tills found at lower elevations. Correlation of these deposits with adjacent volcanic outcrops, imply an early Pleistocene age for this till. On the basis of this information, the reversely magnetized till found in Hobley Valley has an age of approximately ~730,000 yr or greater. The magnetic record for HOB19 may mark the Brunhes/Matuyama border (0.73 my) or one of the magnetic events within the Matuyama Chron. Ages of 0.9 and 0.97 my for the Jaramillo, 1.87 my for the Olduvai and 2.04 and 2.12 my for the Réunion are also possible candidates, in that they mark polarity transitions from a reversed to a normal field, with decreasing time during the Pleistocene. There can be no doubt that the Hobley Valley record marks one of these early Pleistocene magnetic boundaries. Because the older Gorges deposits have been placed largely in the Olduvai subchron, it is likely these younger deposits were emplaced in the later Matuyama (~1.0 my).

The lower part of profile GOR71 (B1b horizon) gives a reversed polarity indicating it was emplaced during the Matuyama Chron. The upper unit (till and loess) both give a normal polarity suggesting they were emplaced during one of the normal events (e.g. Jaramillo or Brunhes). Given the age and stage of weathering of these sediments it is likely the till was emplaced long before the loess. The loess is considerably less weathered and was probably emplaced during the Brunhes Chron. Hiatuses between till and loess are common on Mount Kenya making it reasonable to expect a difference of several hundred thousand years between the two units.

At site LE4 a sequence of till, loess, alluvium and outwash reveals reversed magnetization for the till and lower loess and normal magnetization for the upper unit (outwash and loess). The till and lower loess were likely emplaced during the upper Matuyama and weathered prior to the emplacement of the younger alluvial and loess units. The lower unit appears more aggressively leached owing to the presence of increased gibbsite; the upper unit appears less weathered and hence is probably younger.

4.10 Conclusions

The oldest documented early Quaternary glacial deposits were emplaced during the Gorges Glaciation. Three advances are known during the Olduvai subchron and middle Matuyama Chron, and it is possible the oldest tills are pre-Olduvai in age. Certainly the presence of glacially-crushed clasts in Mount Kenya lahar deposits that predate the Gorges Glaciation suggest that an earlier glacial episode occurred prior to 2.0 my. Presumably the moraines of this earliest glaciation were destroyed by episodic volcanism and mass wasting that led to emplacement of the lahars.

The geomorphologic evidence for early Quaternary glaciation is found only in Gorges Valley. Most probably the early Quaternary ice cap was asymmetrical, as during subsequent glacial episodes, producing the largest accumulation of ice on the moist trade wind slopes (southeast) of the mountain. Older glacial deposits on the northern, western, and southern slopes were probably removed by younger glacial advances and/or by episodic mass wasting events. However, the sequence of Gorges-Lake Ellis moraines and associated sediments in Gorges and Hobley Valleys and in the Lake Ellis areas, provides an important amount of information on early Quaternary paleoenvironments. There is sufficient evidence to show that pre-Gorges paleoclimate was wetter (and presumably warmer) than today and probably of equal magnitude with the Gorges-Lake Ellis and Lake Ellis-Naro Moru Interglaciations. From the available iron and clay mineral data it appears that paleoclimate became less intense during the subsequent Naro Moru-Teleki Interglacial period.

Paleosols are abundant in all early Quaternary deposits and buried paleosols provide an important corpus of information on the paleoclimate that existed during the time of their formation. Even though many of these paleosols failed to provide recoverable pollen, they did provide important clay mineral and Fe ratios that assist in interpreting age and paleoclimate. The greatest information on age, however, comes from the paleomagnetic investigations which show that all of these old glacial deposits have immense antiquity and were emplaced in the Matuyama Chron starting around ~2.0 my ago. Because there is not one exposure that contains a "stacked" sequence of superposed tills, outwash deposits and loesses, it is necessary to look at several sections in an effort to piece the story together from scattered deposits at several different sites.

CHAPTER 5

NARO MORU AND TELEKI GLACIATIONS

5.1 Introduction

The Naro Moru and Teleki glaciations were the two major glacial geologic events of the middle Pleistocene on Mount Kenya. Both glaciations were extensive, reaching below 3000 m.a.s.l. on the western flank of the mountain. In many drainages, moraines were presumably overrun by younger ice or destroyed by combinations of meltwater release (Charsley, 1989) and/or mass wasting. In most cases these deposits are associated with thicker loess covers and colluvial-slope wash sequences. In nearly all cases, middle Pleistocene deposits are weathered producing very complex associations of surface and buried paleosols (see Vortisch, Mahaney and Fecher, 1987; and Mahaney, Barendregt, and Vortisch, 1989). Only on the eastern flank of the mountain is it possible to find end moraines *in situ* without extensive erosional effects. On the western side an "older" moraine, first recognized by Gregory (1894b), provided most of the information contained in this chapter. Paleomagnetic dating places all of these deposits in the Brunhes Chron (Mahaney, Barendregt and Vortisch, 1989).

Most middle Pleistocene sites in lower Teleki Valley are in dense forests of mixed bamboo (*Arundinaria alpina*) and *Podocarpus* (*P. gracilior*) that have a documented downward altitudinal shift of ~700 m. during glacial periods (Mahaney, Harmsen and Spence, 1989). An example of the denseness of the mixed forest is seen in Figure 5.1 at site TV23A, one of many sites in the "Met Station Catena" (Figures 5.2 and 5.3).

Understory shrubs consisting of *Cyperus rotundifolius* and species of *Carex* alternate with clusters of *Lycopodium saururus* and *Dryopteris inaequalis* (Mahaney and Boyer, 1986b). The site shown in Figure 5.1 had been opened by cape buffalo (*Syncerus caffer caffer*). Here Naro Moru till underlies a thick mass of cover sediments consisting of loess of mixed Naro Moru-Teleki and Teleki-Liki ages. The Teleki ice terminus was about 30 m away on top of the ridge (to the right in Figure 5.1).

5.2 Moraine Sequences

Teleki and Naro Moru moraine sequences are found only in Gorges and Teleki valleys (see Chapter 6 and Plate 2). In general, these are flat-topped moraines with gentle slopes of 5° to 8° and moderate relief features of only a few meters. In Gorges Valley remnants are spread out at different elevations above and below "The Gates", between 3150 and 3300 m.a.s.l., the higher remnant escaping erosion during younger glaciations (*e.g.* Teleki and Liki glaciations) by virtue of its elevation. The lower moraines, protected by a bedrock bar at "The Gates" lie just beyond the terminal Liki position.

In Teleki Valley, a moraine of Teleki age below the Met Station (originally the older Moraine of Gregory, 1894b) marks the terminal position of Teleki ice of middle Brunhes age (~300,000 yr). The end portion of this moraine is missing, presumably eroded by meltwater release during the Teleki and Liki glaciations (Figure 5.4). The left lateral portion of this moraine is referred to as the Met Station Catena (Figures 5.2 and 5.3). In the Burguret Valley, to the north of Teleki Valley (Fig. 5.2), large boulders of porphyritic phonolite and syenite, lying beyond the Liki limit, may be remnants of large clasts left behind by Teleki ice. Certainly in Teleki Valley, release of meltwater deepened the valley by ~130 m over the last ~300,000 yr, or since the end of the Teleki Glaciation. A similar model for erosion of Liki end moraines in Hausberg Valley was discussed by Charsley (1989) on aerial photographic evidence.

Figure 5.1: Bamboo forest (*Arundinaria alpina*) at site TV23A (2980 m.a.s.l.) in lower Teleki Valley.

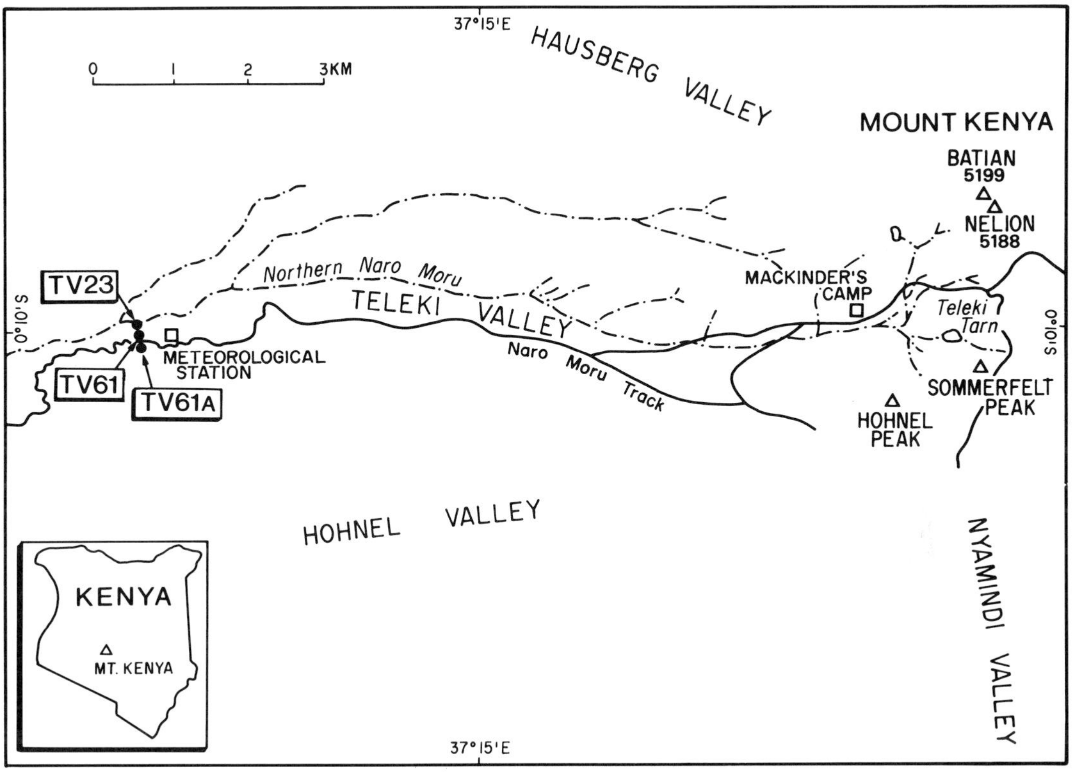

Figure 5.2: Location of sites TV23, TV61, TV61A in lower Teleki Valley, Mount Kenya.

5.3 Surface Weathering Features

Gorges Valley is the only drainage with Naro Moru and Teleki age moraines exposed at the surface. Attempts at using weathering ratios to distinguish between these deposits were frustrated by boulder spalling, which is very pronounced on the eastern (and hence drier) side of the mountain. Charcoal is frequently found in the surface horizons of ground soils, and Ab horizons of buried paleosols, indicating that fires occur frequently in eastern drainages. A high frequency of fires may cause spalling of large boulders such as at site GOR48 (Figure 5.5) making it impossible to use rock-weathering characteristics in differentiating deposits (see Mahaney 1978 for a discussion).

Weathering rinds were also measured in an attempt to determine if they could be used to separate deposits in the geologic succession. Samples collected on five different moraine surfaces showed considerable overlap in maximum rind development (Table 5.1) among the different sites studied.

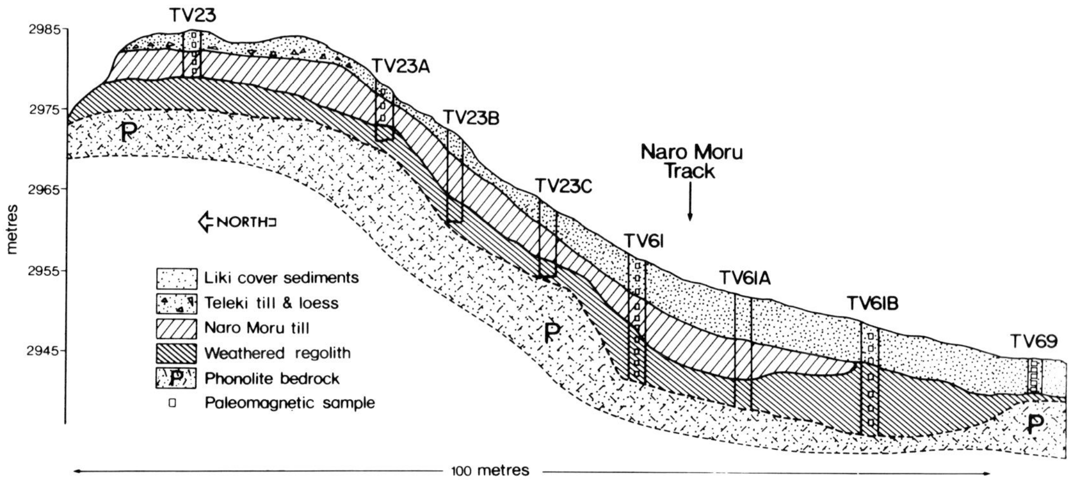

Figure 5.3: Met Station catena at ~3000 m, western flank of Mount Kenya.

Table 5.1
Weathering Characteristics for Deposits of Naro Moru and Teleki age, Gorges Valley, Mount Kenya

Site[a]	Age	Maximum Rind (mm)	Minimum Rind (mm)	n
GOR36		4.51	0.74	50
GOR37	Teleki	4.98	0.46	50
GOR38		4.60	0.72	50
GOR39	Naro Moru	6.20	1.20	50
GOR48		4.76	0.98	50

a. For location see Plate 2

Insofar as the minimum rinds are concerned it is possible to differentiate the two units. As in other mountain areas, such as the Rocky Mountains of the western U.S., it is possible to achieve differentiation of units of glacial rank, by using *minimum* weathering rind development. For all practical purposes, minimum rinds are less than ~1.0 mm on Teleki age deposits and greater than ~1.0 mm on older deposits of Naro Moru age.

5.4 TV23 Section

Deposits belonging to the "older glaciation" (see Figure 2.27) were first described nearly a century ago by Gregory (1894) and Nilsson (1935). These deposits are now regarded as pre-dating the last (Liki) glaciation (Mahaney, 1982b) and are considered to consist of more than one till

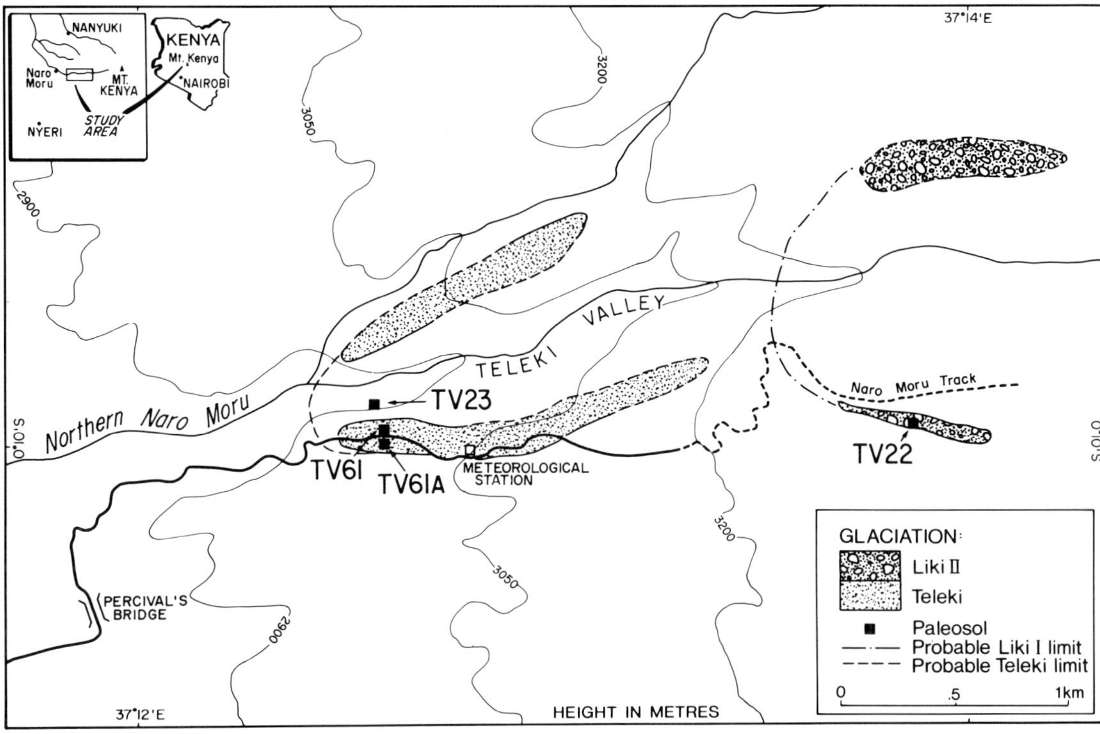

Figure 5.4: Teleki and Liki moraine limits in lower Teleki Valley.

(Mahaney, 1986). At site TV23 (Figure 5.2) (first described by Mahaney, 1979a) the lithostratigraphy and paleosol sequence (Figure 5.6) provide important insight into the sequence of tills, loesses, and tephras. Using the mineralogy and chemistry of the fine and very fine sand fractions (63-250 μm), it becomes possible to differentiate deposits and reconstruct the paleoclimatic setting in which they attained their secondary weathering characteristics.

Formed in the upper 4.6 m of transported regolith (mainly till and loess with minor tephra deposits), the paleosols (Figure 5.6) described herein were produced largely during the middle Pleistocene (lower and middle Brunhes Chron). Throughout the TV23 sequence, the primary mineralogy consists of feldspar (mainly sanidine and K-analbite), quartz, magnetite, and amphibole, as determined by polarizing microscope and SEM-EDS. Metahalloysite and gibbsite were identified by XRD and SEM-EDS. In many samples, some grains of nepheline were identified by petrographic microscope but the amount was too small to appear on X-ray diffractograms. Large amounts of volcanic glass in various alteration stages were identified (Figure 5.6) with the aid of the petrographic microscope (Vortisch, Mahaney and Fecher, 1987).

The particle size distributions of the two paleosols in the TV23 section are shown in Figures 5.6 and 5.7. A clear distinction emerges between the post-Naro Moru and post-Teleki paleosols. The older post-Naro Moru paleosol contains more clay and silt, some of which is well-weathered tephra (Figure 5.6). The post-Teleki paleosol contains less clay & silt and overall it has a considerably "lighter" texture compared to the older unit.

Throughout the section feldspar of the anorthoclase (sanidine and K-analbite) series dominates over significantly fewer plagioclase grains. Feldspar etching follows cleavage planes, but most feldspar grains are fresh internally and feldspar corrosion was not of sufficient intensity

Figure 5.5: Spalled boulder on surface of "older" moraine (Baker, 1967) in Gorges Valley at ~3350 m.a.s.l.

to alter the grain percentages shown in Figure 5.6. In the post-Naro Moru paleosol, feldspar corrosion seems more intensive (Vortisch, Mahaney and Fecher, 1987). It is therefore possible in the older paleosol that feldspar grain frequency might be lower due to dissolution effects. Most volcanic feldspar surfaces are clean, indicating that feldspar corrosion did not lead to *in situ* clay mineral formation.

Loessic sediments in the sequence are rich in quartz grains which often show inclusions of one type or another. Some samples showed a high relative intensity of the 100 reflection which suggests a high amount of low-temperature quartz (chalcedony) (Eslinger *et al.*, 1973). In most cases chalcedonic quartz occurs in devitrified volcanic glass.

Volcanic glass in profile TV23 is differentiated as follows: a) transparent to translucent glass (inclusion-rich) with colors ranging from near colorless to brown; and b) opaque grains or "droplets" having dark brown to black colors with the outer portions faintly translucent, *e.g.* sideromelane (Heiken, 1972). All volcanic glass grains are of basic composition-leucitic-tephritic to basaltic with refractive indices above 1.54 (Bambauer, *et al.*, 1971). Nearly every transitional stage between weathered/altered and unweathered volcanic glass was observed throughout the section (Vortisch, Mahaney and Fecher, 1987). Transparent to translucent glass grains were predominantly transformed to goethite, becoming progressively more opaque with increased weathering. Opaque grains are referred to as altered glass in Figure 5.6 and tend to provide sites for metahalloysite formation (Figure 5.8 A, B). Other opaque glass particles have been weathered to hematitic grains which are also opaque. Both goethitic and hematitic grains are combined in some cases making it difficult to distinguish them with the light microscope (Vortisch, Mahaney & Fecher, 1987). Observations with the SEM showed that even transparent grains which appear fresh under the light microscope have surficial coatings of clay minerals and oxides of one type or another.

The tills in the TV23 sequence are composed predominantly of magnetite which dominates as the chief ferromagnetic mineral. Most magnetite is Ti-rich and is generally fresh and unweathered (Vortisch, Mahaney and Fecher, 1987). Amphibole occurs less frequently and may be represented as microliths in volcanic glass. Often biotite and amphibole appear together with the former being very rare and confined mainly to the loessic units.

The distribution of metahalloysite correlates with the altered glass (Figure 5.6). A detailed SEM study of the TV23 section showed fibrous crystallites of metahalloysite on volcanic glass grains (Figure 5.8A, B; 5.9). These grains probably formed originally as halloysite (Vortisch, Mahaney & Fecher, 1987). The weathering of volcanic glass to halloysite and metahalloysite is a well studied process (Holler *et al.*, 1976 and Kirkman, 1975). The weathering sequence put forward by Kirkman is: feldspar + glass $\rightarrow$ allophane $\rightarrow$ poorly ordered halloysite $\rightarrow$ well ordered halloysite $\rightarrow$ metahalloysite. The amount of metahalloysite formation depends on the weathering intensity and the amount of volcanic glass present (see A11 *vs.* IIIAb horizons in Figure 5.6).

The presence of gibbsite is indicative of more aggressive leaching (Figure 5.10) and more intensive weathering. The distribution of gibbsite is particularly revealing in profile TV23 because it is present mainly in the post-Teleki paleosol. The frequency with which gibbsite occurs does not correlate with altered glass or hematitic grains (Figure 5.6). Often gibbsite was observed on altered glass grains with the SEM, and in several cases it occurred on the same grain with metahalloysite.

The major lithostratigraphic units can be differentiated with respect to their primary mineralogy which varies between tills, loesses and tephras. Horizons within the paleosols can be differentiated by their weathering mineralogy. The important difference between the two paleosols is the presence of gibbsite in the post-Teleki paleosol and its relative absence in the post-Naro Moru paleosol (Vortisch, Mahaney and Fecher, 1987).

Although gibbsite varies to a certain degree in the post-Teleki paleosol (Figure 5.6), its distribution is too uniform for a detrital origin. If gibbsite in the younger paleosol had a detrital

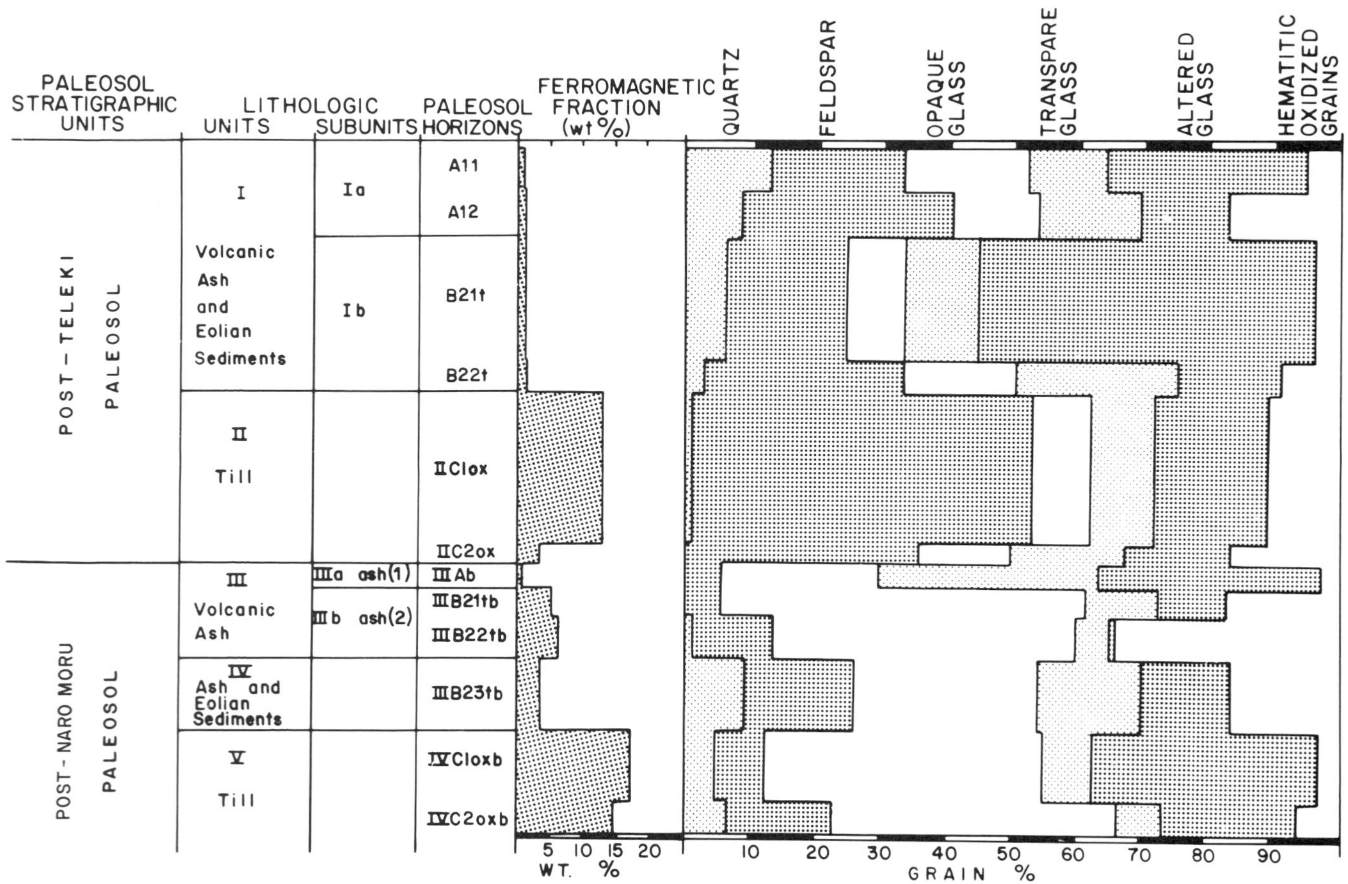

Figure 5.6: A, Lithostratigraphy, paleosol sequence and mineralogy of the fine and very fine sand (63-250μm) fractions in Profile TV23 (modified from Figure 2 in Vortisch, Mahaney and Fecher, 1987).

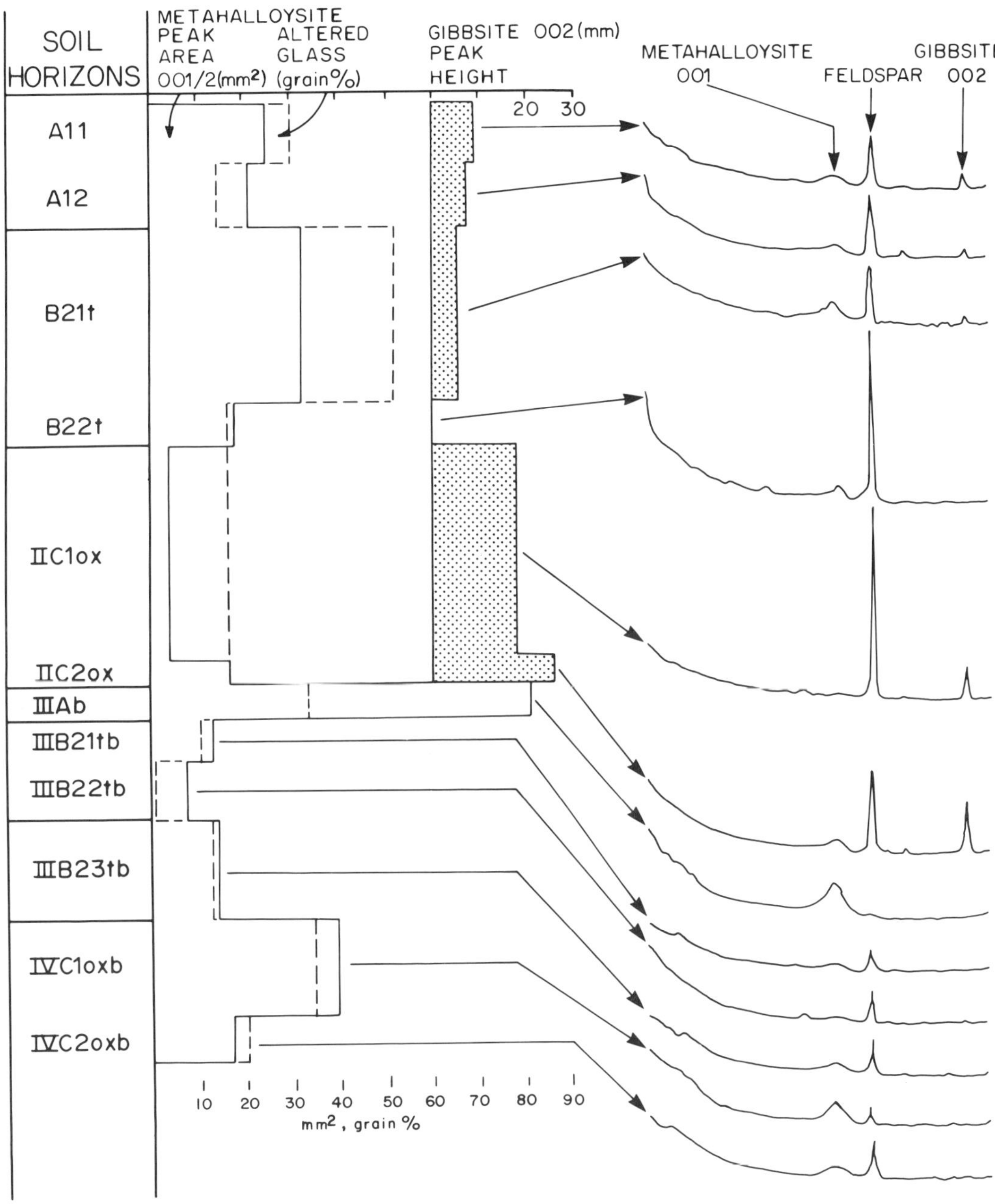

Figure 5.6: B, Clay mineralogy in profile TV23 showing relationships between percentage of altered glass grains and metahalloysite peak areas.

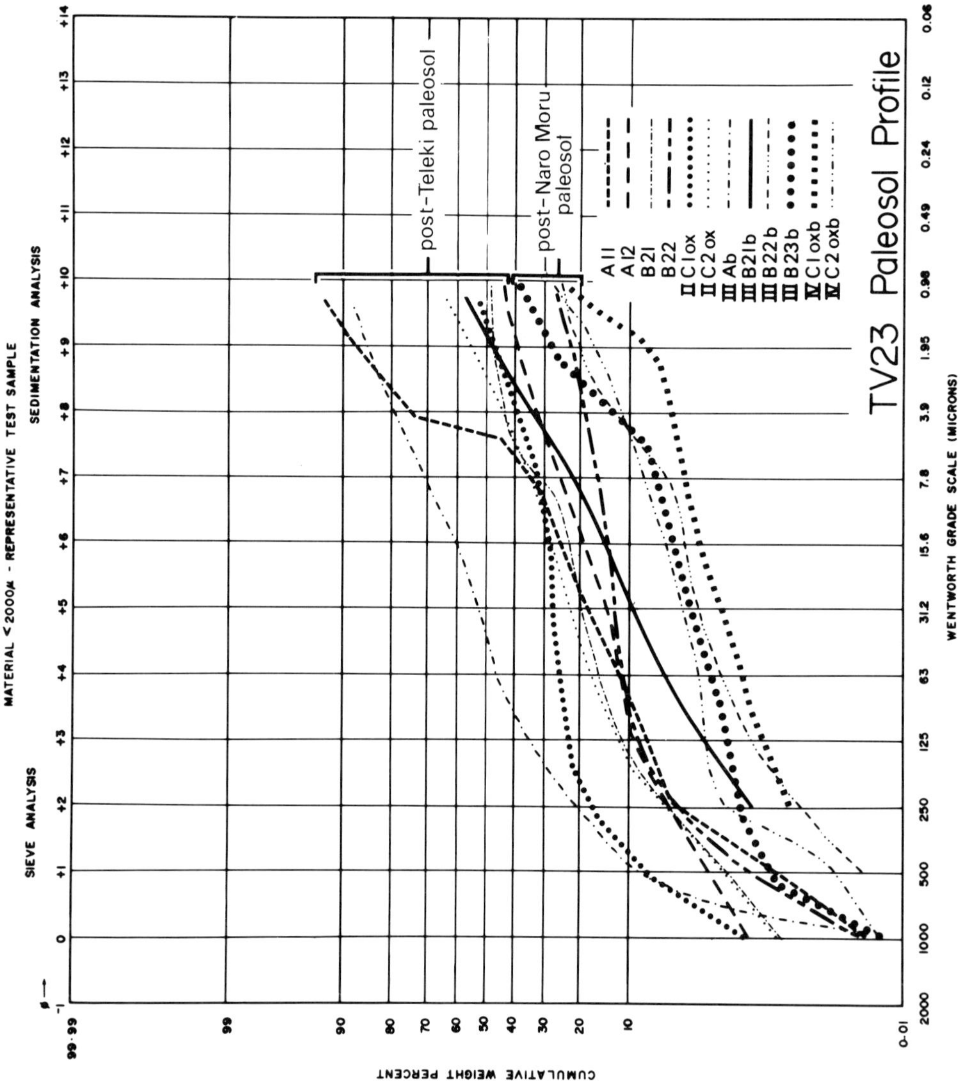

Figure 5.7: Particle size distributions in the TV23 paleosol sequence.

A.

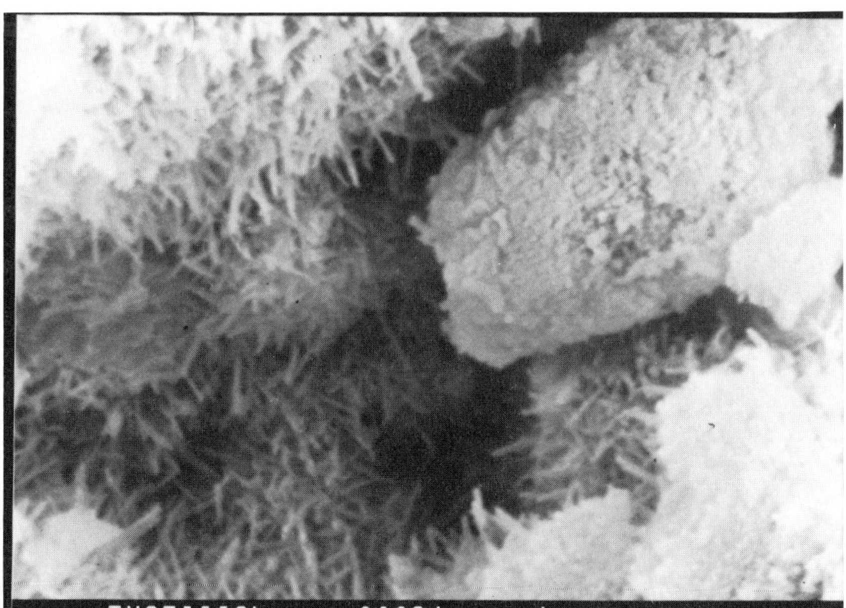

B.

Figure 5.8: A, Neoformed clay minerals on a volcanic glass grain; B, enlargement of A center showing fibrous form of metahalloysite crystals.

Figure 5.9: Formation of tubular metahalloysite on volcanic glass grain in the IIIAb horizon. The inclusion is a well etched alkali feldspar.

origin, the post-Naro Moru paleosol horizons should also contain gibbsite. The relative absence of gibbsite in the older paleosol argues for neoformation of gibbsite in the younger unit. Alternatively the transformation of gibbsite to halloysite-metahalloysite, with increasing concentrations of dissolved Si in the pore solutions, may account for the absence of gibbsite in the older paleosol. Because the chemistry of pore solutions and drainage are the important factors in gibbsite formation (Dejou, *et al.*, 1977), its distribution in profile TV23 is considered to be closely linked with paleoclimate and percolation of soil water. All horizons in the post-Naro Moru paleosol (Figure 5.6), with the exception of the IIIAb horizon, have high clay contents, and less permeability than in the horizons of the younger paleosol. As a result drainage conditions might intervene to control the occurrence of gibbsite.

On average, the higher volcanic glass content in the fine and very fine sand fractions of the post-Naro Moru paleosol reflects on the high content of volcanic ash throughout the older unit. Because Si is released to the pore solutions during the transformation of volcanic tuffs, a high content of tephra combined with low permeability might inhibit the genesis of gibbsite, or even assist in the transformation of gibbsite to halloysite-metahalloysite (Curtis and Spears, 1971; Vortisch, Mahaney and Fecher, 1987).

The above discussion concerning the silicification of gibbsite involves a time span of $<700{,}000$ yr. Is this sufficient time for such complex chemical changes to occur? Dejou *et al.* (1977) have reported gibbsite neoformation in weathering profiles on acid plutonic rocks in France. They argued that gibbsite formed in the Upper Tertiary and occurs today as an "inherited" clay mineral in Quaternary paleosols. Perhaps combinations of time and paleoclimate worked together to produce the gibbsite distribution in the two paleosols of profile TV23. Because the

A.

B.

Figure 5.10: A, Nodular form of gibbsite in IVC1oxb horizon of the post-Naro Moru paleosol; B, fibrous form of gibbsite in the IVC1oxb horizon.

IIIAb horizon has a lower mean clay content than the post-Teleki paleosol, it may be that paleoclimate was less intense during the Naro Moru-Teleki Interglaciation.

5.5 TV61A Section

Section TV61A (Figure 5.11) situated on the margin of glaciation in lower Teleki Valley, was selected for study because it contains a sequence of saprolite, till, and other associated sediments that are all products of landform instability associated with younger glaciations, and with weathering that presumably occurred during interstadial-interglacial periods. This moraine complex, identified by Gregory (1894b), was excavated by Mahaney in 1976 and described, in part, in several subsequent papers (1979a, 1982b). The complexity of the sequence was discussed by Mahaney, Barendregt and Vortisch (1989) and paleomagnetic analysis on a nearby section (TV61) was discussed by Barendregt and Mahaney (1988). Because section TV61A is located precisely on the margin of the Naro Moru Glaciation, its older till and saprolite units are well-preserved and covered by a thick sequence of loess, colluvium and slope wash.

Five paleosols are present in section TV61A beginning with the older saprolite cover (Figure 5.11) over fresh unweathered phonolite. It is most probable that the saprolite weathered first under an interglacial paleoclimate (Lake Ellis-Naro Moru Interglaciation) (warmer and wetter) and was truncated at a later time when the till (Figures 5.12 and 5.13A and B) comprising unit IV (Naro Moru Glaciation) was emplaced (in the lower Brunhes Chron; Barendregt and Mahaney, 1988). This till was later covered by a loess (lower unit III) which weathered to produce a relict paleosol open to influences from the subaerial weathering paleoenvironment. The overlying paleosol in upper unit III (Figure 5.14) is weathered to somewhat greater intensity than the lower one giving the impression that, of the two interglacial paleoclimates (Naro Moru-Teleki and Teleki-Liki), the younger one had the higher weathering potential and was no doubt wetter (and perhaps warmer) than either the Liki interstadial or postglacial paleoclimates. The ^{14}C dates in the IIIAb (Naro Moru-Teleki) and IIIA12b (Teleki-Liki) horizons give ages between ~27,000 and ~34,000 yr BP which are probably too young. Both paleosols are subject to perched ground water discharges, which are known to allow the ingress of organic carbon (Mahaney and Boyer, 1986a) affecting the validity of ^{14}C ages.

The overlying paleosols (units II and I; Figure 5.11) are both Inceptisols (Tropepts) that formed in colluvium and slope wash. The paleosol in unit II contains very little clay compared to the three older paleosols, and considerably less clay than the overlying ground paleosol (Unit I). This indicates differences in weathering intensity between interstadial/stadial paleosols formed during or slightly before the last glaciation, and surface paleosols formed also toward the end of the last glaciation and during the postglacial period. Because a geomagnetic excursion has been located (Barendregt and Mahaney, 1988) in the B2 and upper Cox horizons of unit I in a nearby section (TV61), it is possible that most of the upper unit has been *in situ* since ~25,000 yr BP, but weathered mainly in the postglacial period.

The Ab horizon sequence in unit II comprises a representative buried moorland-like Ab horizon complex that dates from the Liki II stadial on Mount Kenya (Mahaney, Harmsen and Spence, 1990). Because this buried moorland-paleosol complex can be traced ~700 m below timberline on Mount Kenya, it is considered to represent a forest depression of considerable magnitude in response to colder paleoclimate during the Liki II stadial (Mahaney, 1982b). Pollen preserved in these buried horizons contain Afroalpine elements including taxa of Erica and Poa that are considerably different from pollen recovered in the overlying ground paleosol which consists of forest elements including *Hagenia* and *Podocarpus*.

The particle size ratios for silt and clay show rather extreme weathering in the three older paleosols that are representative of the three last interglacials – Lake Ellis-Naro Moru; Naro-Moru-Teleki; and Teleki-Liki from youngest to oldest (Figure 5.11). Because the clay

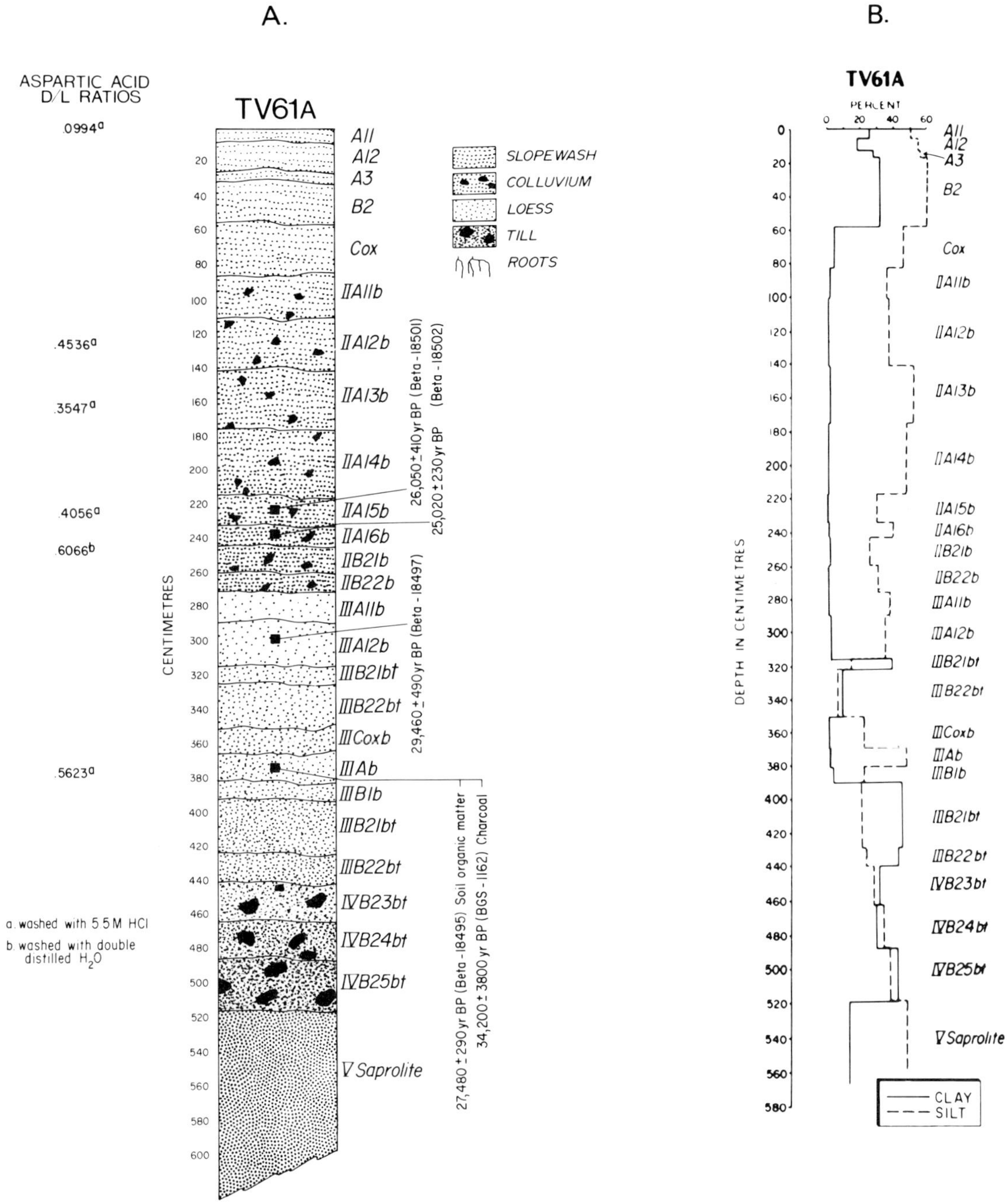

Figure 5.11: Section TV61A in the Met Station Catena, western flank of Mount Kenya: A, stratigraphy; B, particle size.

Figure 5.12: W. B. Vortisch collecting a micromorphological sample in upper part of Naro Moru till (horizon IVB23bt) in profile TV61A (Figure 5.11). Large boulder protrudes to lower right of the micromorph box; overlying sediment is loess in lower unit III.

A.

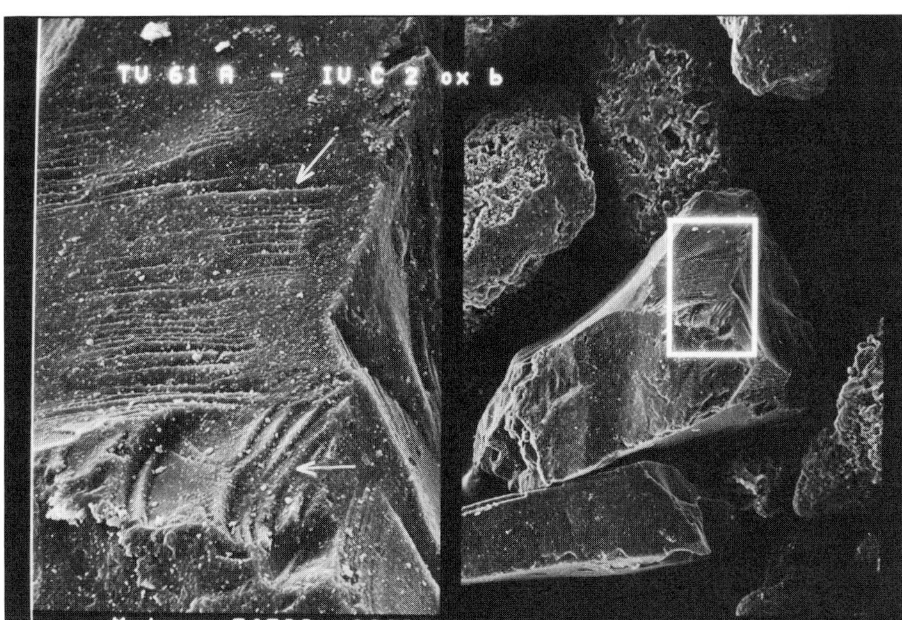

B.

Figure 5.13: A, General view of fine and very fine sand in Naro Moru till (TV61A section); B, glacially crushed quartz grain showing conchoidal and linear (arrows) crushing microtextures.

A.

Figure 5.14: General view of unit III (upper) B21bt and B22bt horizons in profile TV61A, A and B.

percentages are smaller in the saprolite (unit V) relative to the overlying paleosol, it is possible that the saprolite represents somewhat less intense weathering in a drier interglacial or soil formation over a shorter interval of time. The two middle paleosols (units III and IV) contain about the same amounts of clay which indicate similar weathering intensities (differing somewhat with the texture profile in profile TV23). In both profiles, silt tends to increase upward in response presumably to increasing airfall input; first, at the end of each glaciation or during the Naro Moru-Teleki (unit III lower and unit IV), and later during the Teleki-Liki (unit III upper) interglaciations.

In general, the contrast in particle size distributions between units III -IV and unit II, especially with respect to increasing sand and decreasing clay in the upper unit, is striking. This serves to illustrate the change in weathering intensity between Pleistocene interstadial (low intensity) and interglacial paleoclimates (high intensity) (see Chapter 11 for discussion).

The overlying surface paleosol, formed in slope wash cover sediments, provides a good example of Liki II deposits emplaced for the most part after ~25,000 yr BP, and weathered mainly in the postglacial period. While there is some clay movement in the ground paleosol it is not of sufficient magnitude to meet the definition of a Tropudalf (Soil Survey Staff, 1975).

Clay minerals and other weathering products were studied primarily to determine if they (along with other weathering products) could be used to separate paleosols in the sequence. In general, halloysite/metahalloysite is found only in the three interglacial paleosols (units III, IV and V), and it increases in the buried B horizons (Figure 5.15). X-ray reflections are more pronounced in units III and V and somewhat lower in the middle unit suggesting less weathering there. This trend is not unexpected and follows trends in a nearby section (TV23) described by Mahaney (1982b) and Vortisch, Mahaney and Fecher (1987). The presence of halloysite/metahalloysite signals a pre-Pleniglacial age ($>$100,000 yr) (Mahaney, 1989) and is usually indicative of interglacial paleosols.

In units III and IV, the presence of gibbsite in the buried B horizons indicates aggressive leaching (Vortish, Mahaney & Fecher, 1987). Lower amounts of gibbsite in the saprolite and its absence in the I and II units suggest much lower to nil leaching intensities and lower weathering potentials. Hematite and goethite, are for the most part, absent in units I and II, but increase markedly in the buried B horizons of units III and IV (Figure 5.15B), decreasing in the saprolite. The clay mineral, gibbsite, Fe-oxide and hydroxide data suggest a maximum weathering potential in the paleosols in units III (upper), IV and V, somewhat lower potential in the unit III (lower), and the lowest potential in the I and II units.

5.6 Relative Weathering of Middle Pleistocene Deposits

The amounts of extractable Fe and Al were analyzed to determine if age-dependent trends could be used to separate the paleosols. In general, values of Fe_d (Table 5.2) give some idea of the weathering that occurred during the time each paleosol was exposed to subaerial processes. The Fed values are approximately the same in each of the three lower paleosols, somewhat lower in the Liki interstadial paleosol (although some preweathered sediment might have been reworked by aeolian processes into this younger paleosol), and lowest in the surface paleosol. After analyzing the activity ratio (Fe_o/Fe_d) (Table 5.3) the saprolite appears to be less weathered than the till (unit IV) above, mainly because more amorphous Fe has been converted to Fe_d in unit IV than in unit V. The overlying loess in unit III becomes progressively less weathered up to the IIIAb (at 370-381 cm). These values remain the same or fall off in upper unit III in response to slightly increasing weathering in the upper interglacial paleosol (Figures 5.14A and B). The Liki stadial/interstadial paleosol (unit II) yields lower values in the buried B horizon complex (which could be due to incorporation of preweathered sediments) steadily increasing upward in the thick

A.

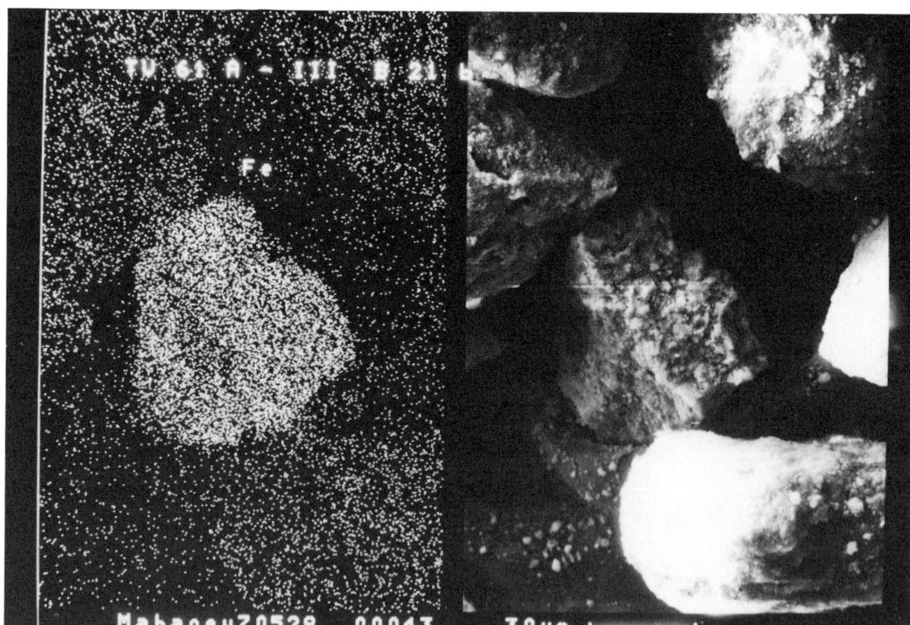

B.

Figure 5.15: A, coatings of 1:1 clay minerals (mainly metahalloysite) on plagioclase grains (note subrounded and subangular grains mixed together in loess of lower unit III, Profile TV61A); B, Fe distribution shown by luminescence on SEM photomicrograph.

Table 5.2

Extractable Fe and Al in a Quaternary Paleosol Sequence on Mount Kenya, East Africa.

			Extractable Fe and Al					
Site	**Horizon**	**Age**[a]	**Fe_p**	**Fe_o**	**Fe_d**	**Al_p**	**Al_o**	**Al_d**
TV61A	A11		1.70	2.16	3.24	1.33	1.89	1.55
	A12	F	1.67	2.02	3.20	1.33	1.68	1.54
	A3		1.63	2.12	2.72	1.21	1.48	1.22
	B2		2.47	2.67	3.58	1.89	2.08	1.71
	Cox		1.98	2.93	4.73	1.70	4.03	3.37
	IIA11b		1.03	2.38	5.03	1.43	4.52	3.44
	IIA12b		0.94	2.50	5.61	1.56	4.44	3.82
	IIA13b		1.11	2.50	4.99	1.61	4.56	3.64
	IIA14b	E	0.19	0.91	4.50	1.08	4.18	3.57
	IIA15b		0.26	1.30	4.84	1.12	4.97	3.04
	IIA16b		0.16	2.28	5.17	0.79	4.93	2.80
	IIB21b		0.11	1.68	6.50	0.60	4.88	3.09
	IIB22b		0.12	1.82	7.02	0.70	4.97	3.11
	IIIA11b		1.26	2.45	4.57	1.67	4.24	2.97
	IIIA12b		0.81	2.06	4.87	1.24	4.42	2.86
	IIIB21bt	D	1.57	2.49	5.61	2.69	1.30	1.36
	IIIB22bt		0.44	2.96	6.21	0.83	2.65	2.73
	IIICoxb		1.51	3.90	7.56	1.34	1.46	3.72
	IIIAb		0.18	2.89	5.56	0.70	3.77	2.56
	IIIB1b		0.11	2.54	5.64	0.53	2.90	2.54
	IIIB21bt	C	0.48	1.26	4.25	0.70	0.71	0.82
	IIIB22bt		0.27	1.24	4.88	0.60	0.62	0.92
	IVB23bt		0.07	1.29	6.16	0.19	0.52	0.59
	IVB24bt	B	0.05	0.88	3.62	0.16	0.41	0.43
	IVB25bt		0.14	0.52	3.15	0.36	0.47	0.51
	VSaprolite	A	0.87	2.29	6.88	0.87	0.70	0.63

a. A = Lake Ellis-Naro Moru paleosol; B = Naro Moru Till; C = Naro Moru-Teleki loess; D = Teleki-Liki loess; E = Liki interstadial paleosol; F = Liki II paleosol.

Table 5.3

Ratios of Fe_o/Fe_d and $Fe_p + Al_p/Fe_d + Al_d$

in section TV61A, Mount Kenya

Site	Horizon	Age[a]	Depth (cm)	Fe_o/Fe_d	$\frac{Al_p + Fe_p}{Al_d + Fe_d}$
TV61A	A11		0-6	0.66	0.63
	A12	F	6-23	0.63	0.63
	A3		23-30	0.78	0.73
	B2		30-58	0.75	0.82
	Cox		58-83	0.62	0.45
	IIA11b		83-110	0.47	0.29
	IIA12b		110-141	0.45	0.27
	IIA13b		141-175	0.50	0.25
	IIA14b	E	175-217	0.20	0.16
	IIA15b		217-234	0.27	0.18
	IIA16b		234-243	0.44	0.12
	IIB21bt		243-260	0.26	0.06
	IIB22bt		260-276	0.26	0.08
	IIIA11b		276-290	0.54	0.39
	IIIA12B		290-316	0.42	1.09
	IIIB21bt	D	316-322	0.44	0.61
	IIIB22bt		322-351	0.48	0.14
	IIICoxb		351-370	0.52	0.25
	IIIAb		370-381	0.52	0.11
	IIIB1b		381-394	0.45	0.08
	IIIB21bt	C	394-429	0.30	0.23
	IIIB22bt		429-440	0.25	0.15
	IVB23bt		440-463	0.21	0.04
	IVB24bt	B	463-488	0.24	0.05
	IVB25bt		488-519	0.17	0.14
	Saprolite	A	519-610	0.33	0.23

a. A = Lake Ellis-Naro Moru paleosol; B = Naro Moru Till; C = Naro Moru-Teleki loess; D = Teleki-Liki loess; E = Liki interstadial paleosol; F = Liki II paleosol.

moorland-like unit (IIA11b-IIA16b). An abrupt transition occurs in unit I where Fe_o/Fe_d ratios rise to >0.6 signalling less weathering or slower conversion to Fe_d.

The movement of organically-bound Fe can be determined from the distribution of Fe_p (Table 5.2). Because the saprolite (unit V) was truncated during the emplacement of till (unit IV), we cannot determine if movement of Fe occurred there. However, because it has higher concentrations than in the unit above, it is possible either that some movement occurred in the older paleosol or even that some Fe moved down from unit IV. The first hypothesis seems more likely because the distributions of Fe_p in unit IV and lower unit III reflect the genesis of the entire paleosol up to IIIAb (370-381 cm), *i.e.* the highest values are in the IIIB21bt and IIIB22bt horizons. In upper unit III (Age D) again the higher values are in the B2 horizon complex and indicate somewhat stronger downward movement. The distributions of Fe_p in unit II show no evidence of movement downward into the B2 horizon complex and overall lower values of organically complexed Fe throughout the moorland-like Ab complex. In the surface paleosol, there is some evidence for downward movement of Fe_p into the B2 horizon.

The ratio of $Fe_p + Al_p/Fe_d + Al_d$ (Soil Survey Staff, 1975) is used to identify spodic horizons in the U.S. Soil Taxonomy and to provide information concerning present and past soil (paleosol) forming environments. In this study, analysis of the ratios shows that downward movement can be demonstrated in the three lower paleosols. Assuming that the saprolite had an A horizon complex on it before truncation and emplacement of the till (unit IV), it has approximately the same ratio as the overlying paleosol. The paleosol in upper unit III (Age D) yields higher values, and hence, greater leaching power than in the two lower paleosols, suggesting a more intense paleoclimate. No movement occurred in unit II even within the A horizon complex indicating substantially drier (and perhaps cooler paleoclimate as indicated by pollen content). Enough movement had occurred in the surface paleosol to support the notion of downward translocation of Fe and Al. However, because there is no downward movement of organic carbon into the B2 horizon, it does not meet the definition of a spodic horizon (Soil Survey Staff, 1975).

The geochemistry of the deposits in profile TV61A (Table 5.4) shows some interesting differences with depth. Sodium is lower in the intensively leached units (upper III and V), somewhat higher in less intensively leached zones (unit IV, lower unit III) and higher in low intensity paleosols (units I and II). Potassium, although less mobile than sodium, shows relatively low variability throughout the sequence. Aluminum shows slight variability in the lower, and hence older units decreasing in the buried A horizons. A significant reduction in aluminum occurs in the ground paleosol, along with potassium, titanium, iron, and sodium, which may reflect higher SiO2 and/or additional organic matter. Because many surface and buried A horizons show reduced levels of Al, it is highly possible that aggressive leaching, induced by higher intensity paleoclimates during interglacial periods, may have led to removal of silica and build up of gibbsitic-rich clays in buried B and C horizons.

Titanium, manganese and iron show relationships of variable importance in separating weathering units. The distribution of Ti is related to the magnetite content of the Naro Moru till (unit IV). Generally the highest values occur in the till and in the succession of younger loessic deposits (illustrating the reworking of till materials by aeolian processes). The distribution of Mn reflects the leaching histories, indicating the greatest translocation in lower unit III, upper unit III, and somewhat less in units II and I. Iron, an important indicator of weathering and age, is higher in the saprolite, lower in unit IV and lower unit III, higher again in upper unit III, and thereafter generally decreasing in units II and I. The slight increase in the buried B horizon complex in unit II may represent the incorporation of preweathered sediments from higher up in the catena.

Amino acid analysis was completed for profile TV61A using the methods described in Mahaney, Boyer and Rutter (1986), with the exceptions that the amount of soil prepared was

Table 5.4

Geochemical analyses of whole-paleosol (< 2mm) samples of profile TV61A, Met Station Catena

Site	Horizon	Depth (cm)	Na %	Mg %	Al %	K %	Ca %	Ti (ppm)	Mn (ppm)	Fe %
TV61A	A11	0-6	4.46	≤0.5	7.10	0.66	0.31	3870	2390	4.77
	A12	6-23	4.03	≤0.4	6.24	0.78	≤0.18	3750	2800	4.32
	A3	23-28	3.43	≤0.4	5.66	0.59	≤0.21	3500	2050	4.08
	B2	28-58	5.10	0.6	6.97	0.74	≤0.08	3540	1520	4.95
	Cox	58-83	5.69	≤0.5	9.74	0.87	≤0.20	4960	1720	6.36
	IIA11b	83-110	5.70	0.7	11.6	1.09	≤0.09	5090	1540	7.36
	IIA12b	110-141	6.33	≤0.5	11.1	1.00	≤0.21	4400	1630	6.94
	IIA13b	141-175	7.05	0.7	11.1	0.84	0.17	4660	1040	6.32
	IIA14b	175-217	2.92	≤0.5	12.9	0.55	≤0.24	2580	2340	5.95
	IIA15b	217-234	5.11	0.5	11.7	0.66	≤0.13	3400	3140	7.33
	IIA16b	234-243	7.48	≤1.0	15.0	1.07	0.30	3950	4230	7.18
	IIB21b	243-260	6.27	≤0.8	15.0	1.03	≤0.30	5190	3920	8.46
	IIB22b	260-276	5.35	≤0.3	14.0	0.81	≤0.11	4980	2410	8.00
	IIIA11b	276-290	5.81	≤0.4	11.4	1.04	≤0.19	5340	878	6.53
	IIIA12b	290-316	5.94	0.4	12.4	0.92	0.09	5110	611	6.91
	IIIB21bt	316-322	3.55	≤0.4	13.3	0.68	≤0.17	6720	1570	8.56
	IIIB22bt	322-351	0.58	≤0.5	12.7	1.14	0.26	4080	4370	8.92
	IIICoxb	351-370	3.05	≤0.6	12.3	0.67	≤0.19	7400	5900	10.50
	IIIAb	370-381	7.42	≤0.5	13.1	0.99	≤0.25	4520	2030	8.04
	IIIB1b	381-394	7.00	≤0.5	16.5	0.79	0.17	4180	3860	7.31
	IIIB21bt	394-429	2.88	≤0.5	15.2	0.52	≤0.18	4890	1800	4.86
	IIIB22bt	429-440	2.24	≤0.4	16.2	0.52	≤0.08	3680	3800	7.44
	IVB23bt	440-463	1.35	≤0.8	16.0	0.24	≤0.38	3820	5350	8.24
	IVB24bt	463-488	1.48	≤0.4	16.8	0.21	≤0.14	4570	1800	5.79
	IVB25bt	488-519	2.13	≤0.4	17.1	0.30	≤0.10	5070	702	5.32
	VSaprolite	519 +	1.26	≤0.7	15.8	1.20	≤0.27	5250	2135	8.07

increased to 3.5 g, and the soil was thoroughly rinsed with purified water to remove carbohydrates and salts. The amount of 5.5 M HCl used for hydrolysis was increased to 10 ml and samples were mixed by shaking and sonication twice during the 24 hour heating period. The acid solution was separated from residual soil by filtration rather than centrifugation and decanting. These minor changes in the method provided a greater concentration of more highly purified amino acids for analysis resulting in better chromatographic separation.

The amino acids were extracted from soil organic matter of A and Ab horizons in selected horizons. With only one exception, the D/L ratios of aspartic acid showed consistently higher ratios with increasing age (Fig. 5.11). Ratios vary from .09 for surface samples to .75 from material that is approximately 34,000 years[a] based on a ^{14}C date on charcoal. No aspartic acid could be extracted from the oldest sample taken from the saprolitic horizon, indicating extreme leaching and antiquity.

In general, the D/L ratios obtained are relatively high. As it would be difficult to derive reliable racemization rate data from such a complex substance as soil organic matter, all that can be safely concluded is that the high ratios are due to a relatively high temperature history (which is unlikely given the pollen data which support a cooler Liki (Würm) stadial paleoclimate) over a short period of time or alternatively a lower temperature history over a longer period of time.

5.7 Loess Cover

Loess deposits associated with Naro Moru and Teleki tills range from 0.3 to 1.0 m in thickness. They are often buried by younger colluvial and slope wash deposits following strong weathering episodes (interglacial periods). Investigations of the fine sand fractions by scanning electron microscopy often shows subrounded grains which indicate an aeolian character (Figure 5.14A). In these samples there are also subangular and angular grains that were obviously derived from till and mass wasted deposits following short-distance aeolian transport. Many grains in aeolian deposits show the effects of glacial crushing (Vortisch, Mahaney and Fecher, 1987; Mahaney, Vortisch and Julig, 1988) suggesting that reworking of aeolian grains is widespread on the mountain.

Many Fe-coated grains occur in thick argillic horizons formed in loessic sediments (Figure 5.15B). The grains shown in Figure 5.14B are mainly subangular in shape and contain Fe coatings – the thickness of the Fe coat can be determined by luminescence using the SEM (left frame in Figure 5.15B). The amount of dithionite-extractable Fe is discussed in section 5.5.

5.8 Paleomagnetic Dating

Several Brunhes-age sections have been investigated on Mount Kenya (Barendregt and Mahaney, 1988; Mahaney, Barendregt and Vortisch, 1989). Notable among these sections is TV61 (Figure 5.16) in Teleki Valley at 2950 m, which is located just below the Met Station. It is therefore referred to as one of the sections in the Met Station Catena, an important sequence of Naro Moru and Teleki tills plus associated loess, colluvial slope wash deposits and saprolite beds (Vortisch, Mahaney and Fecher, 1987). Profile TV61 (Figures 5.16 and 5.17) consists of a multistory sequence of paleosols in till (IV), in loess (III) (Tropudalf), in colluvium (II) (Entisol), and in slope wash (ground soil; Tropept).

Paleomagnetic samples were collected from the Teleki Valley (TV61) site composed of a complex of three buried paleosols and ground paleosol (Figure 5.18). After extensive excavation and removal of overburden, undeformed sediments were found which proved suitable for

a. This material is probably contaminated based on data obtained by Mahaney and Boyer, 1986a.

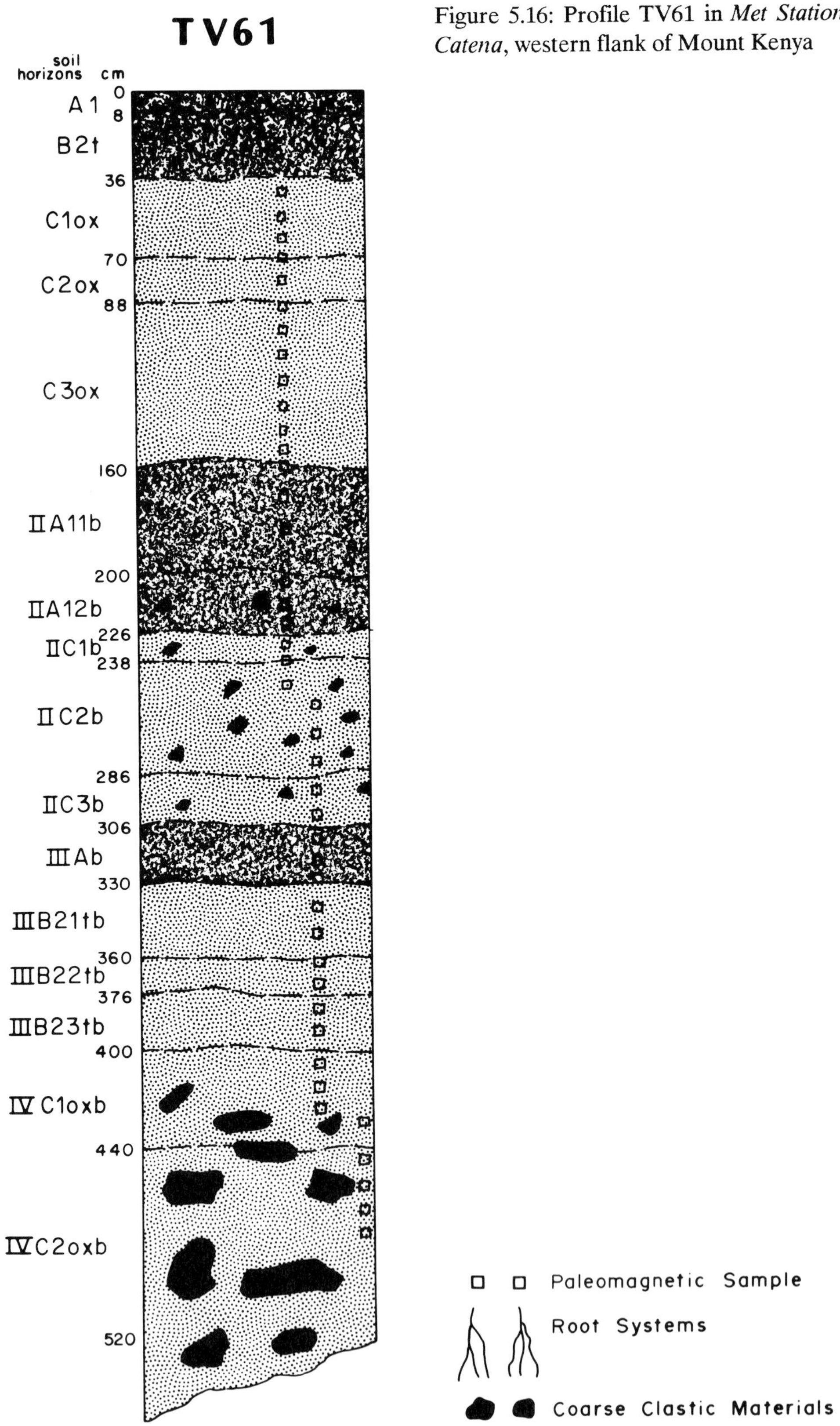

Figure 5.16: Profile TV61 in *Met Station Catena*, western flank of Mount Kenya

paleomagnetic sampling. Azimuthal orientation was made in the field by magnetic compass. The open end of the plastic cube was aligned parallel to the vertical face of the outcrop and samples were taken at 10 cm intervals. Five pilot specimens from the Teleki Valley site were selected and these were subjected to stepwise alternating field (a.f.) demagnetization at levels of 5, 10, 15, 20, 25, 30, 40, 50, and 60 millitesla (mT). After each demagnetization step, the remanence was measured by means of a Schonstedt DSM-1 magnetometer. Qualitative assessment of changes in magnetic vector directions and magnetization values following a.f. demagnetization led to a choice of 15, 20 and 25 mT as the standard demagnetizing fields for the remaining samples. Typical demagnetization plots are shown in Figure 5.19 and most specimens are characterized by a single component remanent magnetization. On the basis of the low median destructive field of these specimens, it appears that detrital magnetite is the predominant remanence carrier [magnetite is one of the principal components of most tills on Mount Kenya (Vortisch, Mahaney and Fecher, 1987)].

Profile TV61 consists of a multistorey sequence of paleosols in till (IV), loess (III) (Tropudalf), colluvium (II) (Entisol), and slope wash (ground soil; Tropept). Horizon boundaries were drawn primarily on the basis of soil color and particle size, there being little variation in soil structure (typical of soils in the Afroalpine and tropical subalpine environments). This paleosol sequence is important because it contains several disconformities, the lowest (III and IV) of these may delineate the upper boundary of the oldest till and loess of Brunhes age ($< 0.7 \times 10^6$ yr BP) on Mount Kenya.

The declination, inclination and intensity profiles for all samples are shown in Figure 5.20 and the summary of remanent magnetization directions is shown in Table 5.5.

Table 5.5

Summary of remanent magnetization directions and mean "A" pole positions

(for definition of "A" pole, see Irving, 1964)

							"A" Pole	
Symbol	**N**	**M**	**D**	**I**	**K**	**α95**	**Lat.**	**Long.**
TV61 (Normal)	35	0.16	1	-5	23	9	88°N	199°
TV61 (Excursion)	9	0.15	158	-1	3	35	68°S	130°

N = number of samples per site.
M = average intensity of magnetization (AM^{-1}).
D = average declination (degrees)
I = average inclination (degrees).
K = precision parameter
α95 = radius of the 95% confidence circle about the mean direction (in degrees).

The variability of NRM intensity is probably related to changes in the composition and quantity of magnetic carriers incorporated in the sediment. In general, the declination and inclination values reflect a normal dipole configuration of the earth's magnetic field. A notable exception is found in the top 60 cm of undisturbed sediments. Samples 2, 3 and 5 near the top of the Teleki Valley profile show persistent intermediate polarity with progressive stepwise

Figure 5.17: Initial single string of paleomagnetic cubes in section TV61 (1984). Paleomag cubes are 10 cm apart; the Mono Lake excursion appeared in surface slope wash (2nd and 3rd cubes from the top). R. W. Barendregt collected the samples.

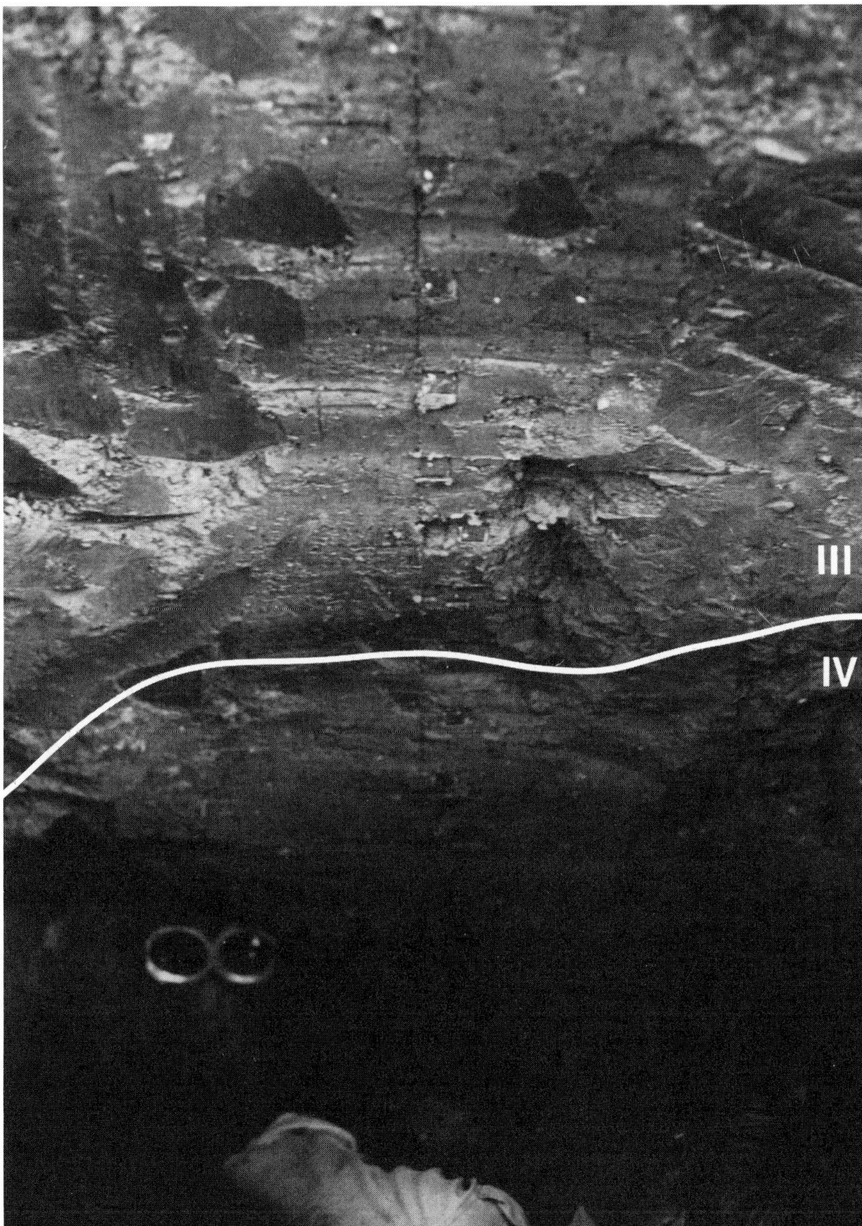

Figure 5.18: Paleomagnetic sampling cubes in loess of unit III and till of unit IV; see Figure 5.16 (profile TV61). Samples collected by R. W. Barendregt, 1984.

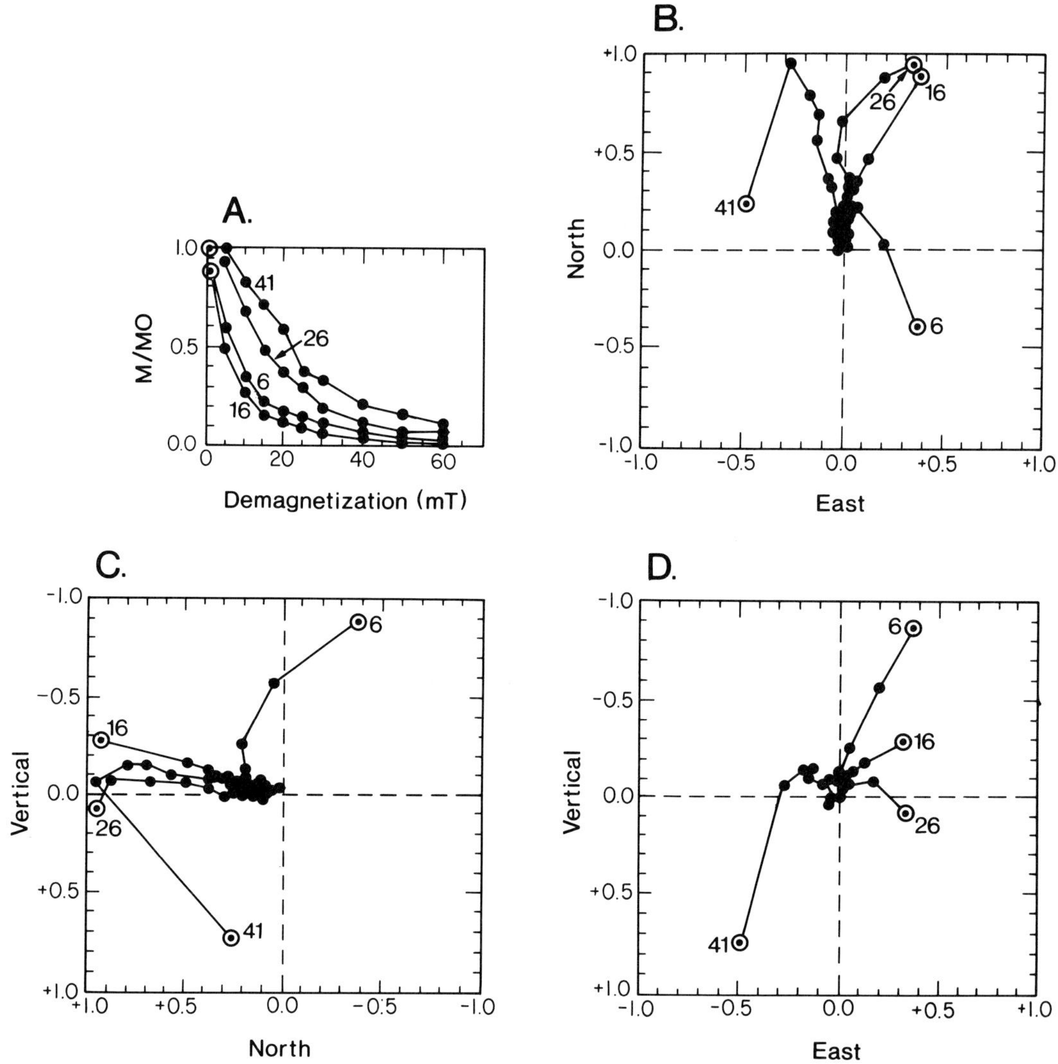

Figure 5.19: Typical demagnetization curves of samples 6, 16, 26 and 41 from Teleki Valley. **[A]** Ratio (M/MO) of natural remanent magnetization (N.R.M.) to residual moment after progressive alternating field demagnetization (a.f.d.) at levels of 5, 10, 15, 20, 25, 30, 40, 50 and 60 millitesla (mT). Magnetization values have been normalized to the highest intensity and are thus unitless. Zijderveld plots showing orthogonal projections of the changing resultant magnetization vector on the horizontal plane **[B]**, on the vertical N-S plane **[C]**, and on the vertical E-W plane **[D]**. The samples are normally magnetized.

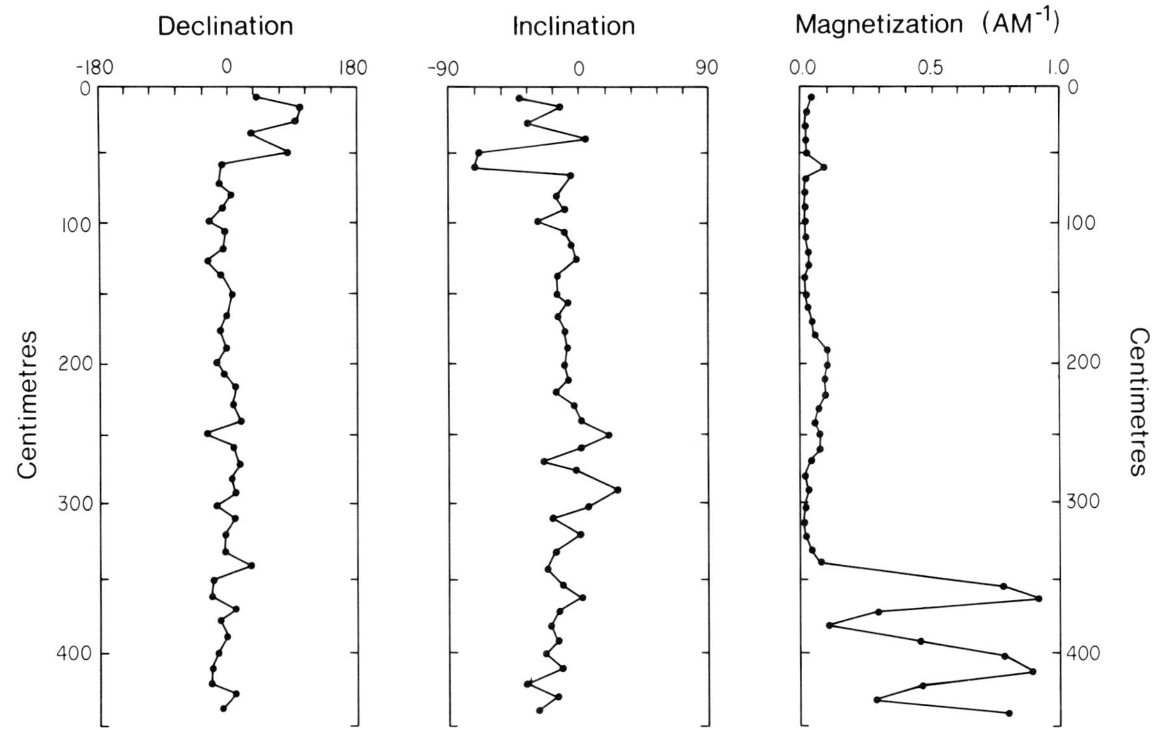

Figure 5.20: Magnetization profiles for site TV61 showing declination, inclination and intensity values for all samples at 15 mT demagnetization

demagnetization. Another six samples below sample 5 show a reversed polarity before demagnetization. It is thought that samples 2, 3 and 5 record a short-lived excursion that affected sediments laid down during this time. Samples 6 to 11 may represent sediments deposited previously but which nevertheless received a weak reversed overprint that is completely removed at 15 to 25 mT demagnetization (see sample 6, Figure 5.19). Perhaps this overprint occurred as a result of active pedogenic or hydrologic conditions which were present at the time and were conducive to remineralization of the remanence carriers. Radiocarbon dates obtained from whole-soil organic materials contained in a zone (IIA11b and IIA12b horizons) just below the area from which paleomagnetic samples 1-11 were collected gives dates of 21,500 ± 430 (Beta-11371) and 24,920 ± 450 yr BP (Beta-11372). This suggests that the younger date may be from sediments which were contaminated by younger ^{14}C moving downward from the surface soil profile.

Short-lived excursions between 20,000 and 25,000 yr BP have been proposed for deposits in France, Denmark, Australia, Iceland, Canada and U.S.A. (Tarling, 1983; Verosub, 1982 and 1986). Whether or not this excursion correlates with any of those found in other countries is difficult to ascertain at present. Further suitable sediments of this age range will have to be found to determine its areal extent in East Africa.

Profile TV61 consists of a complex paleosol sequence formed in till and outwash that predates the last glaciation. Because the lower half of the profile gives a normal polarity, it is considered to fall within the early Brunhes Chron. A high ratio of D to L aspartic acid, very fine soil textures and higher degree of weathering indicate that the lower part of the profile is considerably older than the radiocarbon ages suggested for the upper half of the profile. Sediments in the uppermost part of the profile record a possible magnetic excursion and have yielded radiocarbon dates of around 25,000 yr BP.

5.9 Conclusions

Compound paleosol sequences, such as TV23, TV61, and TV61A, provide important information on present and past soil (paleosol) forming environments and relative weathering effects. The relative weathering differences between the present-day, Liki interstadial, and older interglacial paleoclimates, in terms of their relative abilities to move soluble Fe and Al downward in each system, show the importance of using detailed mineralogical and chemical data to differentiate units in these sections on Mount Kenya. The data also show that, of the three lower paleosols, the middle one seemed to have weathered either in a less intense paleoclimate or over a longer time frame. The striking differences between units II and III-IV-V serve to illustrate the important low energy of a stadial/interstadial *vs.* an interglacial paleoclimate. Interpretation of Fe_o/Fe_d and $Fe_p + Al_p/Fe_d + Al_d$ ratios proved useful in assessing the relative time of subaerial exposure and in downward translocation of soluble Fe + Al. The D/L ratios for aspartic acid increase down-section in profile TV61A, showing reasonable increases with age, and providing a reasonable relative age dating method. The rates of racemization for aspartic acid were influenced by paleoclimate through the successive swings of three glacial periods, two interglacials and the postglacial. Detailed analyses of individual units in section TV61A might yield important data on racemization rates during these glacial-interglacial periods.

CHAPTER 6

THE LIKI GLACIATION

6.1 Moraine Sequence

The time of the onset of the Liki Glaciation on Mount Kenya is not known with precision but is believed to be ~100,000 yr or soon after the end of the Teleki-Liki Interglaciation (isotope stage 5) (Mahaney, 1988a). In most drainages moraine sequences have been subdivided into glacial stages on the basis of topographic position. Thus, Baker (1967) referred to stage I, Ia, Ib etc. as a means of subdividing deposits from the maximum of the last glaciation (late Würm), and stage V moraines (late glacial), to the classical Little Ice Age (stage VI). Because some of these older stage I moraines have buried paleosols in them (*i.e.* Figure 9.2), they must have witnessed Liki interstadial paleoclimates and are older than the last glacial maximum. Most of the stage I moraines described by Baker (1967) in Gorges, Thuchi, and Ruguti drainages are, in fact, considerably older and probably belong to isotope stage 4. Because Baker's chronology was based entirely on topographic position, Mahaney (1979a, 1982b, 1988a) utilized a sequence of Liki I, II and III, where unit I moraines refers to the outermost ridge or Liki terminal positions that are most prevalent on the eastern side of Mount Kenya in the Lake Ellis and Gorges Valley areas. Liki II refers to the next youngest series of moraines behind the Liki I positions. Liki II moraines may be subdivided using small case letters depending on the number of recessional phases that occurred. On the western and northern side of the mountain all Liki moraines present are considered to have been emplaced in Liki II time, while on the eastern side the record is longer and more complex. Liki III moraines are equivalent to Baker's stage V moraines and were deposited during recessional stillstand during the retreat of most valley glaciers.

6.2 Western Drainages

An important sequence of Liki deposits is known from major drainages on the western flank of the mountain including Teleki Valley, Teleki-Burguret Divide (Two Tarn Area), Mackinder Valley and Liki North Valley (Figure 6.1).

Upper Teleki Valley is one of the most imposing and beautiful glaciated areas on Mount Kenya (Figure 6.2).

Surrounded by many high buttresses the U-shaped cross-valley profile obtained its present shape during the last glacial maximum about 20,000 yr BP (Mahaney, 1987a). Ice from the Central Peaks area (Figure 6.2) flowed down Teleki Valley but did not cross the Teleki-Burguret Divide. Evidence for this consists of a well weathered divide surface where pitted boulders and rock outcrops give the appearance of considerable weathering over at least one interglacial period. A core (TTH3) (Figure 6.3) from the Two Tarn area showed a recovered sequence of clay and silt beds over till to a depth of *c.* 100 cm. A ^{14}C date of 33,840 ± 2810 (GaK-11270) at 74 cm depth gives a sedimentation rate of .022 mm/yr, which is slow when compared with late glacial and Holocene rates (Chapter 7). Most probably, the position of this site on the Teleki-Burguret Divide, and the absence of ice bodies that could drain into Two Tarn, provided a setting in which sediment had to be derived from airborne influx or from nearby snowbanks. It is very probable that the silt layer between 30 and 51 cm in this core may represent "glacial flour" being reworked by wind during the height of the Liki II Glaciation (< 25,000 yr BP). The underlying till in this core section may have been deposited during the Teleki Glaciation when the ice descended to < 3000 m.a.s.l. (> 100,000 yr). Clearly Two Tarn is an area that warrants further research especially with respect to recoverable pollen which might provide new insight into the paleoclimate that existed during the Liki Interstade (*i.e.* Kalambo Interstadial).

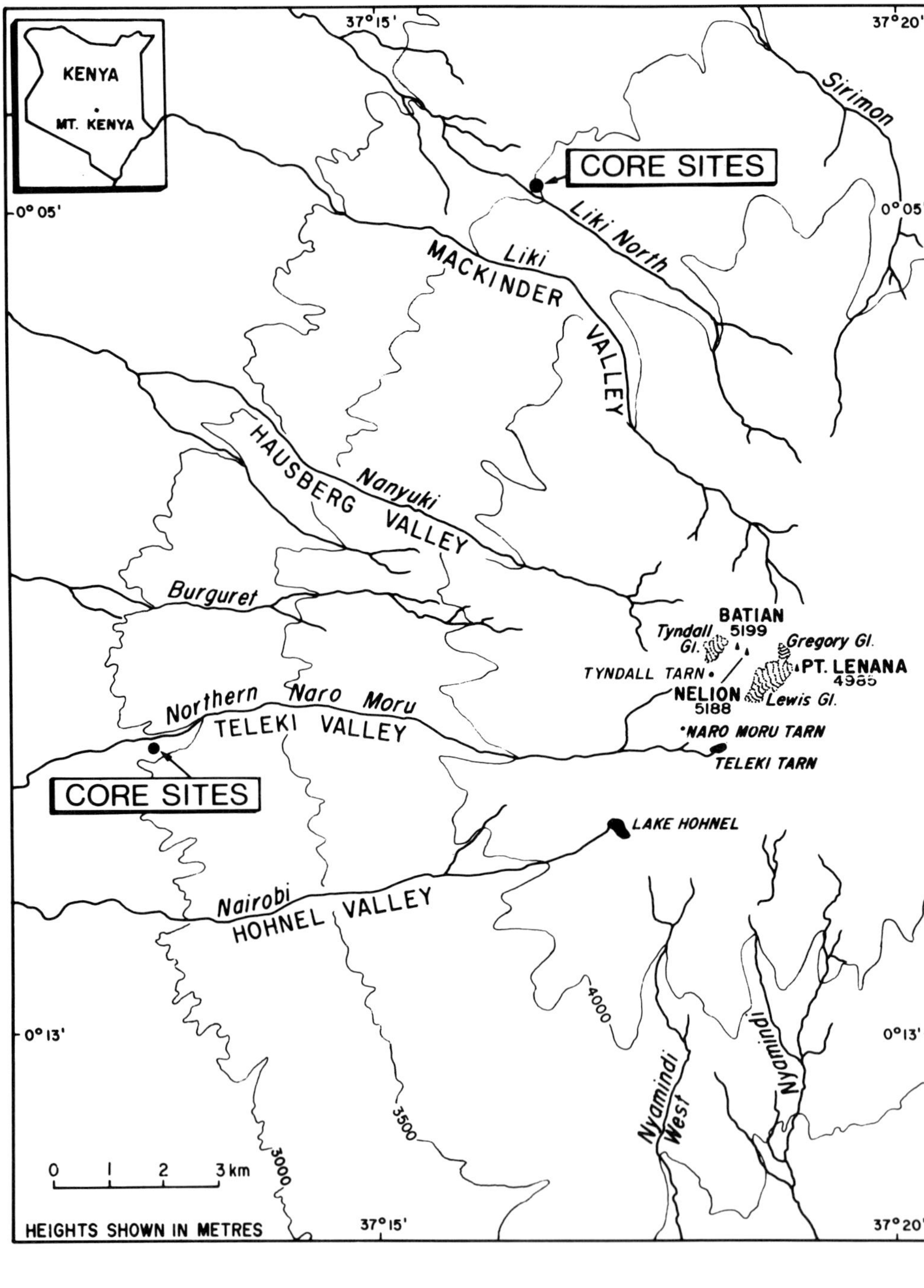

Figure 6.1: Location of core sites in Teleki and Liki North valleys, Mount Kenya.

Figure 6.2: Upper Teleki Valley showing the Tyndall Glacier, Central Peaks and Two Tarn.

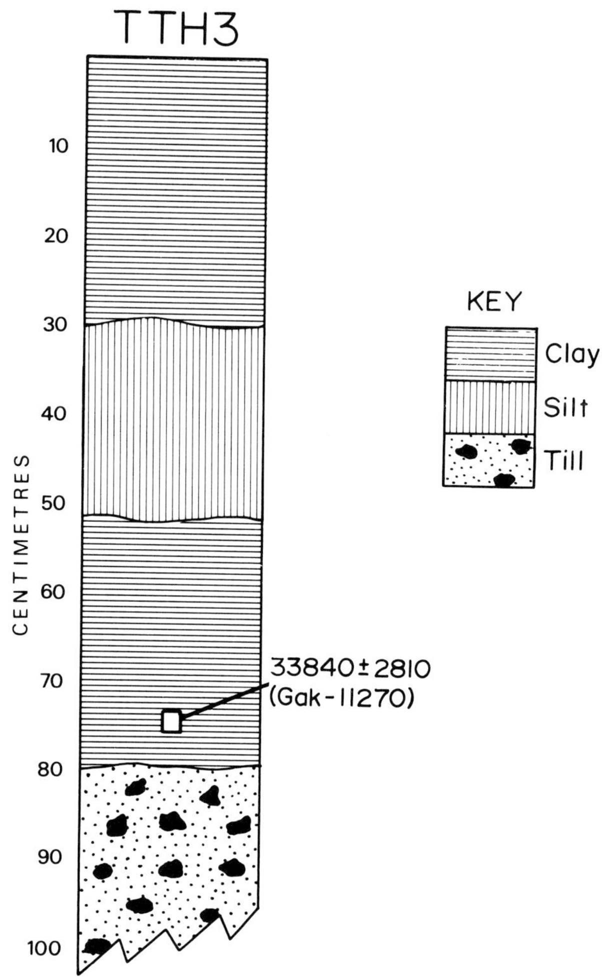

Figure 6.3: TTH3 core recovered from the Two Tarn area on the Teleki-Burguret Divide.

Evidence for the presence of an ice cap on Mount Kenya comes primarily from Lenana (3rd highest summit on the mountain) (Figure 6.4) where polished and striated bedrock indicate an ice flow direction to the east and southeast down Gorges and Hobley valleys, respectively. Because the earlier pre-Pleniglacial glaciers were larger during the first four glaciations, it is probable that an ice cap existed on the mountain during most Pleistocene glaciations.

Upper and lower Teleki Valley (Figure 6.5) was filled with an impressive valley glacier that descended to ~3200 m.a.s.l. in Liki II time. Evidence for an earlier Liki glaciation is lacking in Teleki Valley; however, basal dates from bog cores on Liki II substrates give ages shown in Figure 6.6 and sedimentation rates shown in Table 6.1. Extrapolated ages of the moraine substrates approximate 15,000 yr BP (Table 6.1). This age estimate assumes a near uniform sedimentation rate below the ^{14}C dated beds in all three cores. Given the fine grain size characteristics of the lower core sections it is probable that the extrapolated ages are close approximations of the time

Figure 6.4: Polished and striated glacial pavement on top of Lenana (4985 m).

Figure 6.5: Upper Teleki Valley from close to the present position of the Lewis Glacier terminus. The summit of Mt. Shipton towers above the Lewis moraine. Liki III rock glaciers can be seen in the lower valley with Mackinder's Camp on the crest of the middle unit.

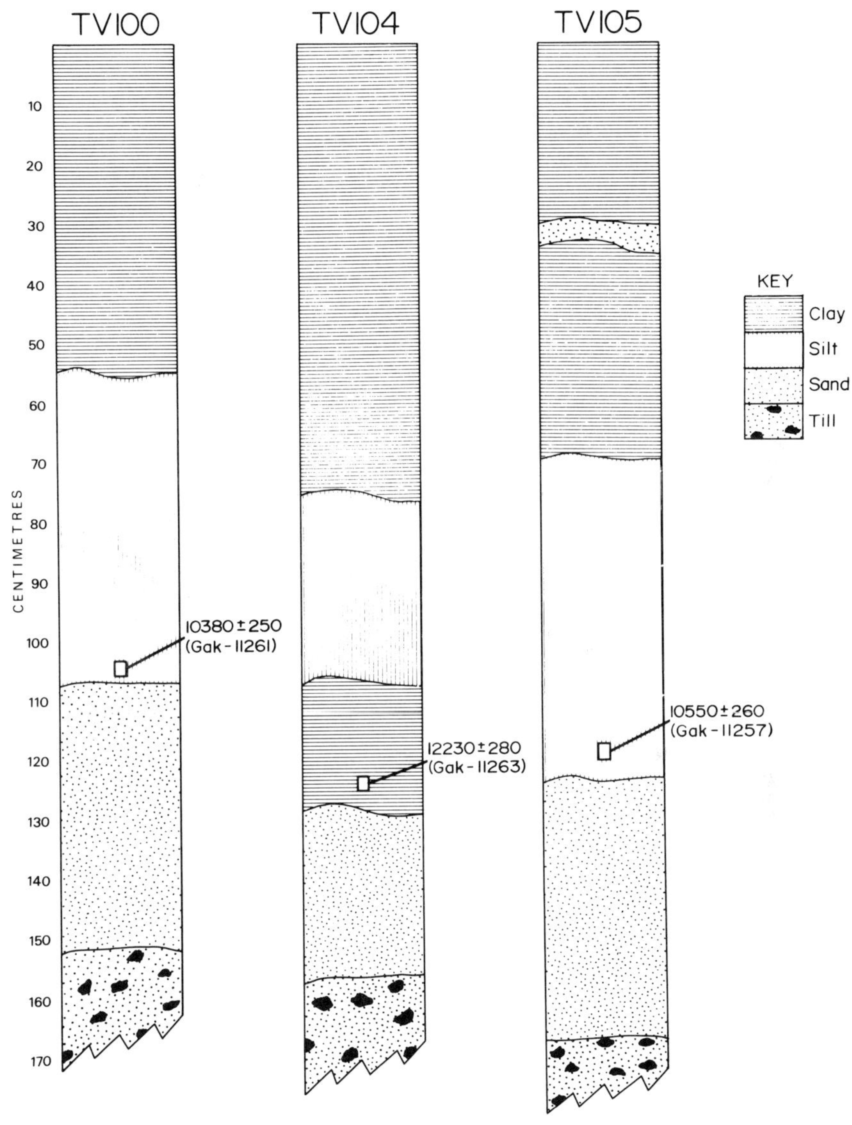

Figure. 6.6: Cores from end moraine complex of Liki II age in lower Teleki Valley.

of deglaciation in lower Teleki Valley. Thus ~ 15,000 yr BP is a probable time for the termination of the Liki II substage (Figure 6.7).

In Liki North Valley on the northwestern flank of the mountain (Figure 6.8), the outer sequence of Liki II end moraines is located at an elevation of 3950 m. Liki I and older pre-Liki moraines are not present in this drainage and have presumably been lost by erosion. An impressive sequence of lacustrine beds are located behind the Liki II end moraines in Liki North Valley.

Cores recovered from bogs in Liki II end moraines in Liki North Valley gave ^{14}C dates of 10,430 ± 370 and 12,590 ± 300 yr BP in the lower beds (Figure 6.9). Extrapolation of these dates to the base of the cores yielded basal substrate dates of 14.2 and 12.8 x 10^3 yr BP. While these ages are somewhat younger than extrapolated substrate ages in lower Teleki Valley, the age difference may be related to lower bog succession rates on the drier northwestern flanks of the mountain (Mahaney, 1987a).

Table 6.1

Sedimentation rates and estimated ages of substrates in Liki North and Teleki valleys, Mount Kenya.

Core	Sedimentation rate (mm/yr)	Depth (cm)	Required time for sedimentatation of lower core (in yr)	14C date	Substrate Extrapolated Age (yr BP)
LN5	.09	(0-112)	1667	12,590 ± 300	14257
LN26	.16	(0-18)		1140 ± 120	
	.14	(18-28)	2444	1880 ± 110	12874
	.11	(28-45)		3490 ± 190	
	.09	(45-105)		10,430 ± 320	
TV100	.10	(0-104)	4900	10,380 ± 250	15280
TV104	.10	(0-126)	3300	12,230 ± 280	15530
TV105	.12	(0-123)	3917	10,550 ± 260	14467

From close to the moraine systems in similar topographic positions at nearly the same elevation, in Liki North Valley (Figure 6.9), cores LN5 and LN26 yielded sediment accumulation rates that compared closely with those from Hausberg Valley, but contrasted somewhat with those in Teleki Valley (Chapter 7). This trend may reflect the general tendency for increased precipitation to occur from the northwestern to western drainages (Mahaney, 1981, 1982b, 1985), which may have influenced weathering and sedimentation rates. Basal dates in both bog cores [plant fragments in LN5; brown (10YR 4/4) sediment in LN26] yielded later Pleistocene ages and the estimated substrate ages fall within late glacial to glacial time limits (Mahaney, 1987a). The discussion above illustrates the problems of interpreting bog-bottom dates as discussed by workers in other localities (Davis and Osborn, 1988; Mahaney, 1988b).

Paleosols in Liki II moraines in Liki North Valley are somewhat thinner and contain less humus than in Teleki Valley (Figure 6.10; Table 6.2). When compared with paleosols of the same age in lower Teleki Valley (Chapter 9) it is apparent that wetter paleoclimate in the western drainages gives rise to somewhat deeper soil and higher amounts of peat in the surface horizons. While B horizons have about the same thickness, color expression is stronger in the wetter areas, usually by 2 or 3 chromas (see Oyama and Takehara, 1970).

Table 6.2.

Paleosol profile LN2

Horizon	Depth (cm)	Description[1]
01	1-0	Black (10YR 2/1m)
A11	0-6	Black (10YR 2/1m,d) loam, granular structure, friable consistence, plastic and nonsticky
A12	6-14	Black (10YR 2/1m) and brownish black (10YR 3/2d) clay, granular structure, friable consistence, plastic and slightly sticky
IIB2	14-25	Dark brown (10YR 3/3m, 3/4m), brown (10YR 4/5m and 7.5YR 4/4m) and dull yellow orange (10YR 6/3d) sandy loam, weak blocky structure, firm consistence, plastic and sticky
IIC1ox	25-36	Dark brown (10YR 3/3m) and greyish yellow brown (10YR 6/2d) pebbly sandy loam, massive, friable consistence, nonplastic and nonsticky
IIC2ox	36-61	Brownish black (10YR 3/2m) and greyish yellow brown (10YR 6/2d) pebbly loamy sand, massive, friable consistence, slightly plastic and nonsticky
IICu	61+	Grey (5Y 4/1m) and yellowish grey (2.5Y 6/1d) pebbly sand, massive, loose consistence, nonplastic and nonsticky

6.3 Northern Drainages

The Sirimon and Kazita valleys on the northern and northwestern flanks spawned small valley glaciers that descended to ~3200 and 3500 m.a.s.l., respectively. Of these two glaciers the Sirimon was the largest, being fed by ice from the Central Peaks, that spilled over and around Tereri and Sendeyo (highest elevations on the Mackinder-Sirimon Divide).

A few glacially-scoured lakes in bedrock attest to the erosive action of Liki II ice in the Sirimon Valley (Figure 6.11). Because the cross-valley profiles (Figure 6.12) show only minor U-shape characteristics, it is highly probable that these glaciers were thin (~300 m). No moraines other than Liki II have been found in the Sirimon, which suggests either that pre-Liki ice did not

1. Soil descriptions follow Birkeland (1984) and Soil Survey Staff (1975). Colors were taken from Oyama and Takehara (1970) in the moist (m) and dry (d) states. Consistence is given in the moist state.

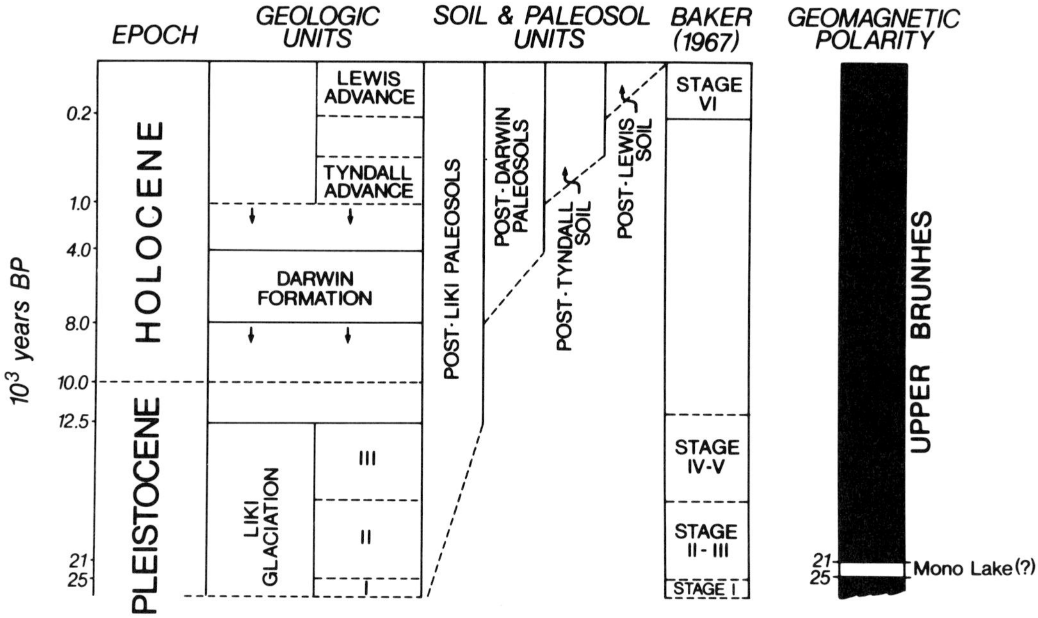

Figure 6.7: Later Pleistocene and Holocene magnetostratigraphy of Mount Kenya.

reach the Sirimon from the Central Peaks or that meltwater release removed all older moraine material from the lower Sirimon Valley. A deep gorge in the lower Sirimon Valley, below ~3000 m.a.s.l., suggests enormous down-cutting in fairly recent time lending support to the meltwater erosion hypothesis. In addition, Charsley (1989) argued for meltwater release leading to emplacement of large bodies of outwash at ~2000 m.a.s.l. on the northwestern flank of the mountain. Catastrophic meltwater release would also explain the absence of Liki and older end moraines in the Hausberg and Teleki valleys, as well as large boulders lacking matrix material in the lower Burguret Valley (see Chapter 5).

6.4 Eastern Drainages

Valleys on the eastern flank of Mount Kenya (Figure 6.13) contain well-preserved moraine sequences. Because glaciers in Gorges and Hinde valleys were fed by moisture-laden winds from the Indian Ocean, it is likely they were protected by more frequent cloud cover produced by rapid adiabatic cooling in air masses forced over the steep eastern slopes of the mountain. Some Liki moraines in Gorges Valley approach ~20 m in thickness indicating a rather large ice body capable of emplacing large end moraines. Indeed some glaciers were large enough to spill into adjoining drainages such as from Hinde Valley to Lake Ellis (Arrow, Figure 6.13). In Gorges Valley, Liki ice descended to "the Gates" (Figure 6.14), emplacing massive moraines on agglomerate (lahar) bedrock of early Quaternary age (see Chapter 2). The expression of Liki moraines here is the most pronounced of any drainage on the mountain. Inside the massive Liki I end moraines (Figure 6.15), the drift thickness becomes apparent along the Nithi River which drains the valley. Just below "The Gates" (Figure 6.14), moraine successions of Teleki, Naro Moru, Lake Ellis and Gorges age can be traced some 7 km to the east (and below to 2850 m). Above and to the south, the massive remnant of a Naro Moru moraine (recognized by Baker 1967 as "older moraine") has

Figure 6.8: Liki I/II moraine and lacustrine landform complex in Liki North Valley.

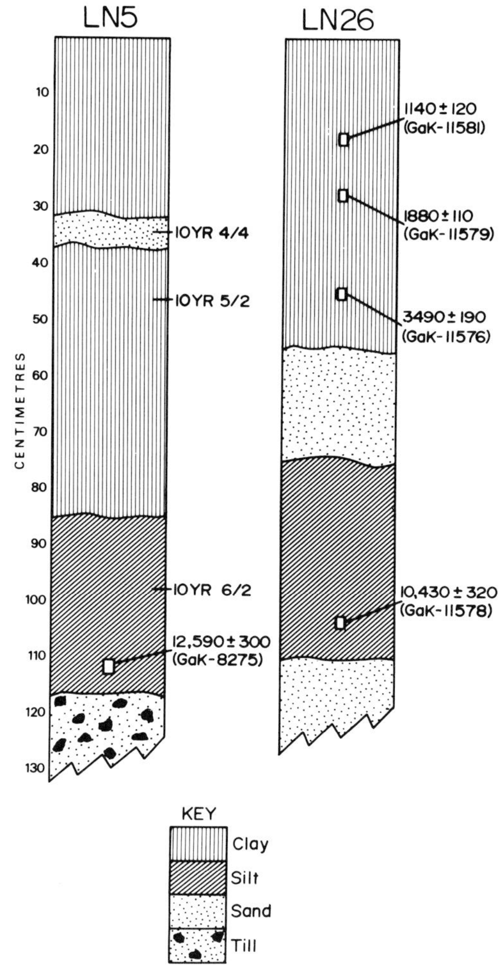

Figure 6.9: Liki North Valley bog cores from Liki II moraine crest.

scattered weathered boulders of porphyritic phonolite on a thin body of till over agglomerate. The paleosol in loess plus till contains a tephra that has a probable source at Ithanguni volcano, some 10 km to the northeast.

Liki moraines farther up Gorges Valley can be subdivided into various substages of the Liki II advance (Figure 6.16) on the basis of topographic position and soil profile development differences [see Chapter 9, profiles GOR1 and 2 (Liki I) and GOR8 (Liki II)].

6.5 Ithanguni

To the northeast of Mount Kenya a small parasitic cone, Ithanguni, rises to 3891 m. During the Liki Glaciation, Ithanguni was high enough to support the growth and development of small valley glaciers which radiated outward and downward to elevations of 3350 m.a.s.l. on the northeast and southeastern flanks (Plate 4). Somewhat smaller glaciers descended to 3600 m.a.s.l. on the northwestern and southwestern sides. Two of these small glaciers descended into Lake Alice on the western flanks (Figure 6.17). Ithanguni is the largest of the parasitic cones on the eastern flank of the Mountain. Because it is younger than ~300,000 yr, it has been glaciated only during the

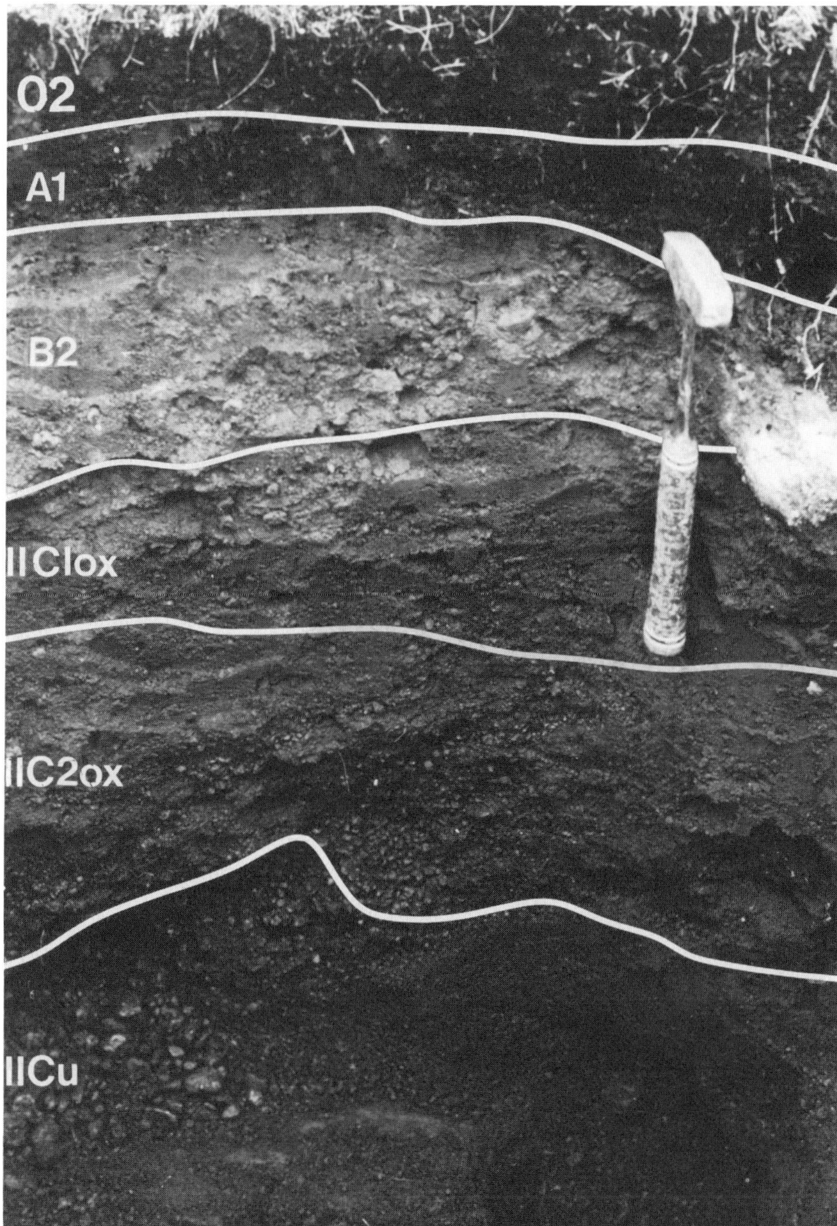

Figure 6.10: Post-Liki II paleosol at site LN14, Liki North Valley.

Figure 6.11: Tarns in bedrock basins, Sirimon Valley, north flank of Mount Kenya.

Figure 6.12: Upper Sirimon Valley showing edge of young lava flow eroded by Liki ice. The flow in the background lies beyond the margin of Liki glacial movement.

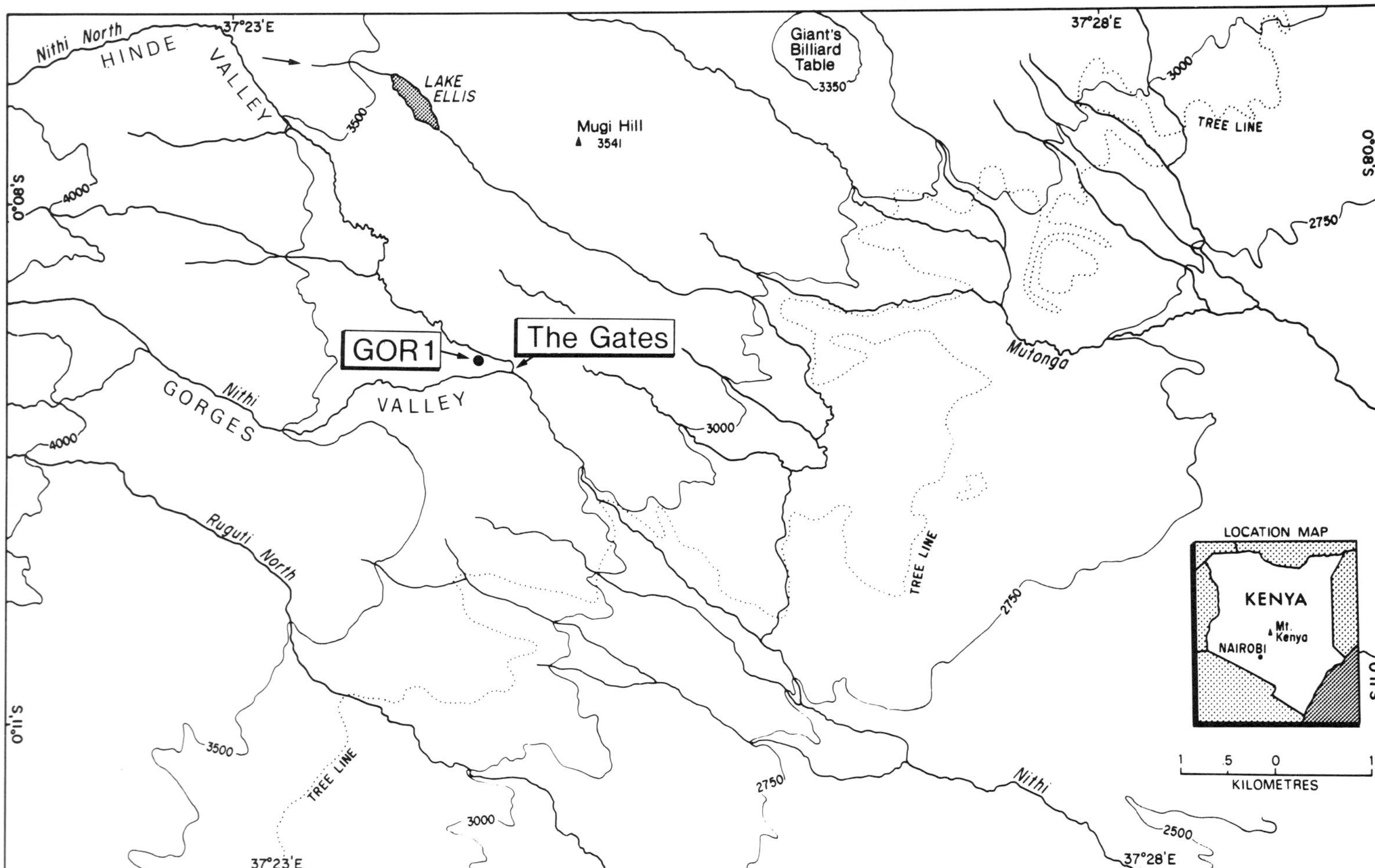

Figure 6.13: Location of sites on the eastern flank of Mount Kenya

Figure 6.14: The Gates from the Lake Ellis moraine complex in Gorges Valley. The sharp crests and steeper flanks of the Liki moraines contrast with the gentle slopes of older Teleki and pre-Teleki moraines (fore and middle ground).

Figure 6.15: Liki I drift capping lahar in Gorges Valley above "The Gates". GOR1 site is at the crest of the end moraine to the right; GOR2 is at the crest of the recessional moraine to the left.

Figure 6.16: Liki II moraine complex (including site GOR8) from the crest of the Naro Moru age moraine (Baker's, 1967 "older moraine").

Figure 6.17: Lake Alice in a young explosion crater on the western flank of Ithanguni.

Liki and possibly during the Teleki Glaciation (Derbyshire and Mahaney, 1989). Dissection of the cone is in its early stages which means that rock exposures are still fresh and relatively easy to examine in the field. The summit which supported a small ice cap during the Liki Glaciation, is a flat plateau area about 1 1/2 km in diameter; to the south and north the summit area tapers off at angles of 20°. To the east, steep slopes plummit down ~300 m to the base of the cone, and towards the west a rounded ridge of pyroclastic rock can be traced to Mount Kenya itself (Baker, 1967). To the northwest of the Summit Plateau a deep explosion crater is occupied by Lake Alice (Figure 6.17). To the east of Lake Alice the neck of the volcano forms a cliff ~300 m high. Below the volcanic neck two small lobate-shaped rock glaciers, that probably formed after the retreat of Liki II ice provide the only evidence for any late glacial weathering and landform activity (see Mahaney, 1980, for a discussion of rock glaciers on Mount Kenya).

Above Lake Alice the upper ~200 m of bedrock contains mainly brown and yellow tephra with beds of lahar and trachytic lavas. Trachytic outcrops are more common in the lower 100 m of cliff above Lake Alice and to the west of the lake they are interbedded with tephra and lahar.

The Ithanguni volcanic neck was probably surrounded originally by a crater wall, mainly consisting of pyroclastic rock. The prism of lava that froze in the crater and adjacent lahars represents the remains of the volcanic plug that withstood ice action during the Liki Glaciation.

Paleosols formed in tills on Ithanguni appear to belong to the Liki II stage (Plate 4). Profile (ITH11) on the northwestern slope is closely related to profile LN14 on the northwestern flank of Mount Kenya (Table 9.5), but is considerably different from paleosols on the western and southeastern flanks (in the wetter areas). Paleosol ITH11 is an Inceptisol with a pebbly texture that fines upwards in response to the incorporation of aeolian material (Table 6.3).

Table 6.3

Profile ITH11, northwestern flank of Ithanguni volcano, northeast of Mount Kenya.

Horizon	Depth (cm)	Description[a]
A11	0-8	Brownish black (10YR 3/2m) pebbly silt loam, granular structure, very friable consistence, nonsticky and nonplastic
A12	8-17	Brownish black (10YR 2/3m) pebbly silt loam, granular structure, friable consistence, slightly sticky, nonplastic
B2	17-31	Dark brown (10YR 3/4m) pebbly silt loam, blocky structure, friable consistence, slightly sticky, nonplastic
IICox	31-57	Dull yellowish brown (10YR 5/4m) and yellowish brown (10YR 5/6m) pebbly sandy loam, massive, friable consistence, nonsticky and nonplastic
IICu	57+	Olive brown (2.5Y 4/3m) pebbly loamy sand, loose consistence, nonsticky and nonplastic

a. Profile nomenclature follows Birkeland (1984), Soil Survey Staff (1951, 1975), and Hodgson (1976).

There is no evidence for movement of fine-grained material downward into the B2 horizon. Rather the B2 horizon appears representative of the cambic (color) B horizons that have formed between ~3200 and ~4000 m.a.s.l. in till and aeolian materials since the last glacial maximum (see Figure 9.4). The other soil properties appear representative of Liki II paleosols in general. Structure, consistence, stickiness and plasticity characteristics are all within the range of postglacial weathering.

Weathering rinds on the ITH11 surface give maximum values of 2.4 and minimum values of 0.56 that are consistent with postglacial weathering on the drier northern and northwestern flanks of Mount Kenya (see Tables 6.4 and 6.5).

6.6 Relationship to Loess Cover

Most Liki tills are covered with thin (0.2-0.4 m) aeolian deposits that together have weathered to form post-Liki paleosols. These aeolian deposits are identified by a fining upwards sequence where silt increases toward the surface (Mahaney, 1979b). These deposits are so young that silt has not yet been translocated downward in most of them to form a mix of till and loess (see Mahaney, 1978, for a discussion of silt movement in similar soils of the Rocky Mountains). However, in some sections, a mix of till and loess tends to occur in the A and B horizons of post-Liki paleosols. This results either from mixing of aeolian and till sediments during deposition, or possibly from movement of larger pebble size clasts by burrowing animals, such as the grove-toothed rat (*Otomys orestes orestes*) and hyrax (*Procavia Johnstoni Mackinderi*) (see Mahaney, 1986). It may also be that mice (*Lophuromys a. aquilus,.*) and moles (*Tachyoryetes rex*) are capable of moving larger clasts to the surface, especially on the eastern flanks of the mountain (personal observations).

Scanning electron microscopy of the loess cover sediments associated with Liki-age tills shows that many grains have complex histories that can probably be traced to the lower Brunhes Chron. Many grains have overprinted crushing features or microtextures that show as many as two or three weathering episodes. The examples shown in Figures 6.18 and 6.19 provide some idea of the complex histories of many grains. Glacially crushed (= arrows) quartz and feldspar (sanidine) grains in the B2 horizon of the LN14 profile in Liki II loess in Liki North Valley provide excellent examples of reworked clasts. A number of V-shaped percussion cracks with various degrees of weathering are also visible on the quartz grain.

6.7 Relationship to Buried Paleosols

Buried paleosols within Liki drifts are rather uncommon on Mount Kenya as elsewhere in other mountain areas (see Mahaney, 1978; and Mahaney and Spence, 1984). Occasionally buried soils have been found in young drifts (see profile GOR8 in Chapter 9), but on the whole it appears the Liki Glaciation was sufficiently vigorous enough to destroy many pre-existing paleosols, and in some drainages where catastrophic release of meltwater occurred, pre-existing paleosols were apparently swept away (section 6.3).

In embayments where relatively thin masses of Liki ice spilled over drainage divides (*e.g.* Liki North Valley and Lake Ellis areas) it might be possible to find buried paleosols that could be sampled for age determination by ^{14}C and TL. Indeed, the LE3 profile in early Liki drift near Lake Ellis (Figure 6.20) provided considerable insight into the nature of Liki I and II paleoclimate and weathering histories. This profile (see Chapter 9, section 9.6) yielded a buried A horizon in loess over Liki I till. The pollen in this Ab horizon contained less *Podocarpus* and *Compositae* and higher *Ericaceae* and*Alchemilla* compared with the A11 horizon (E.M. van Zinderen Bakker, personal communication, 1986). This indicates cooler and drier conditions during interstadial time when timberline was depressed several hundred meters.

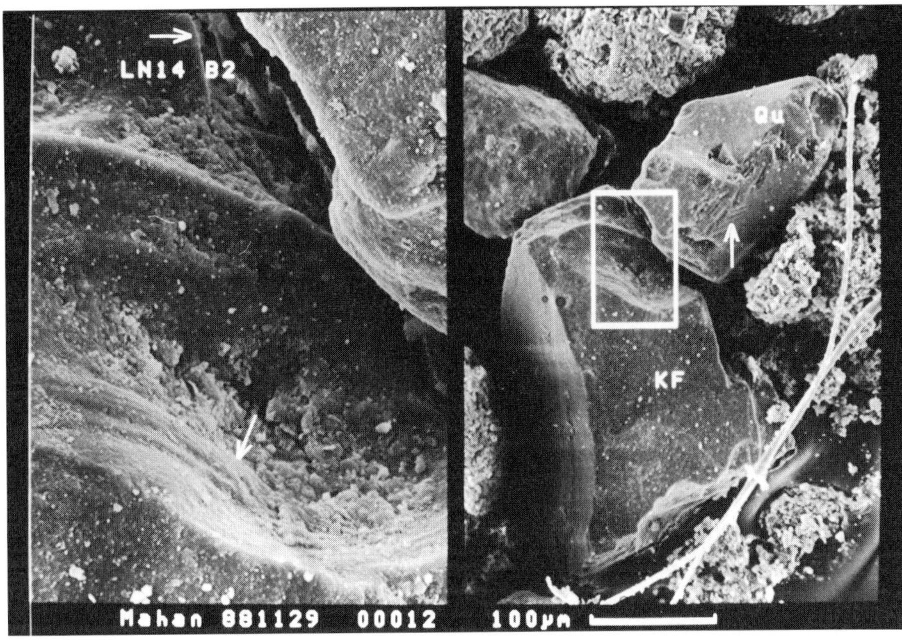

Figure 6.18: Glacially-crushed (= arrows) quartz and feldspar (sanidine) grains in the B2 horizon of the LN14 profile (Liki II loess) in Liki North Valley. A number of V-shaped percussion cracks with various degrees of weathering are visible on the quartz grain.

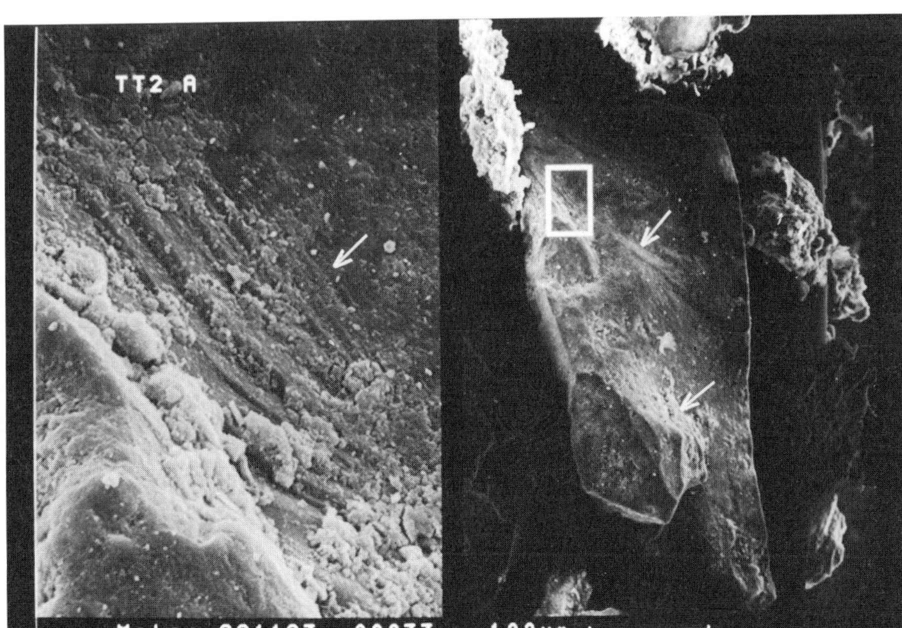

Figure 6.19: A number of glacial crushing features (= arrows) on a quartz grain in the A horizon of the TT2 profile in Liki III loess near Tyndall Tarn. The enlargement shows crushing planes that might date from the last glaciation. The arrows indicate older crushing planes from pre-Liki glaciations.

Figure 6.20: Right and left lateral end moraines of Liki I age at Lake Ellis.

6.8 Relative Weathering Intensities

Attempts at dating Liki moraines by the use of the acoustics produced after striking boulders with a rock hammer (Mahaney *et al.*, 1984a, 1984b) were confounded by the presence of so much porphyritic phonolite on drift surfaces (e.g. phonolite means "to ring" and gives a sharp sound even when weathered). Nevertheless Liki drifts can be distinguished from older Teleki moraines by the use of boulder weathering (see Sharp, 1969, for a discussion of other "boulder banging" methods).

In Gorges Valley boulder weathering measurements on the Liki sequence of deposits gave the results shown in Table 6.4 (sites are located on Plate 2).

Table 6.4

Relative weathering characteristics on cobbles and pebbles in the Liki moraine sequence, Gorges Valley, eastern slope of Mount Kenya

Site	Age	% Fresh	% Weathered	Maximum Weathering Rind (mm)	Minimum Weathering Rind (mm)
GOR10		98	2	2.8	0.20
GOR9	Liki II	97	3	2.6	0.18
GOR8		96	4	3.8	0.58
GOR7		98	2	3.3	0.42
GOR6		90	10	5.5	0.92
GOR5	Liki I	91	9	5.0	1.2
GOR3		87	13	5.3	1.7
GOR2		91	9	5.7	1.1
GOR1		90	10	5.3	1.2

Clearly there are two groups of clasts – those on Liki I and II moraines. The percentage of weathered boulders increases from between 2-4 on younger Liki II surfaces and to 9-13 on the older Liki I deposits. One must use caution in interpreting results of this kind primarily because the frequent occurrence of grass fires (Coe, 1967) may spall rocks on deposit surfaces and leave fresh clasts behind. However, the ratios of fresh to weathered stones follows a pattern seen in other areas (Mahaney, 1978), and the data appear useful in assessing the degree of post-depositional weathering that has occurred.

The weathering rind data in Table 6.4 shows that the mean maximum rind increases with age from 2.6-3.8 mm in the Liki II group, to 5.0-5.7 mm in the Liki I data set. The minimum rind also shows an increase from .18-.58 mm (Liki II) to .92-1.7 mm (Liki I), demonstrating that the *minimum rind* is an important age indicator (Mahaney *et al.*, 1984a). The gap between these two data sets is not great primarily because the Liki I set obtained its greater development by weathering through cool and dry interstadial paleoclimates (Coetzee, 1967).

Weathering rind data from the Liki moraine sequence in Kazita Valley (Plate 3) (northeastern slope) show some overlap among the deposits studied. The data in Table 6.5 show a slight increase in the maximum rind from Liki II to Liki I moraines. All Liki II stations give values <3.0 mm; Liki I moraines show some overlap with 4 stations >3.0 mm. Minimum weathering rind thicknesses are about the same in the Liki I and II groups (only one station is higher in the Liki I samples studied). Compared with Gorges Valley these maximum and minimum rinds are smaller, which is probably due to relatively less moisture on the northeastern slopes. Overall, the minimum rind is not as useful in differentiating Liki I and II deposits when compared to the Gorges Valley (Table 6.4). While one site in the Kazita (e.g. KAZ9) shows a marginal increase in the minimum rind, the overall values are very similar among the sites studied. The minimum rind values, which are often sensitive indicators of substages within a glaciation, have obvious limitations in the Kazita sequence (Table 6.5).

Table 6.5 Weathering rind data from stations on Liki moraines in Kazita Valley, northeastern flank of Mount Kenya.

Site[a]	Age	Maximum Rind (mm)	Minimum Rind (mm)	n
KAZ5		2.6	0.3	50
KAZ6	Liki II	2.8	0.3	50
KAZ8		2.2	0.2	50
KAZ11		2.8	0.4	50
KAZ12		2.6	0.3	50
KAZ1		4.8	0.3	50
KAZ2	Liki I	2.2	0.1	50
KAZ3		3.0	0.3	50
KAZ4		2.5	0.2	50
KAZ7		2.7	0.4	50
KAZ9		3.4	0.6	50
KAZ10		3.2	0.3	50

a. Site locations on Plate 3

6.9 Paleomagnetic Dating

In the upper belt (2850-3000 m) of the *Podocarpus* forest (Figure 6.21) on Mount Kenya, a number of sites have yielded sequences of till, loess, colluvium, and slope wash deposits that have been weathered to form complex sequences of paleosols over saprolite. Some sequences span the whole of the Brunhes Chron ($<0.73 \times 10^6$ yr). Generally, these deposits yield normal magnetic remanence (R.W. Barendregt, personal communication, 1986, 1987, 1988).

In the ground paleosol (Figure 6.22), at site TV61, a geomagnetic excursion (Barendregt and Mahaney, 1988) at ~50 cm depth yields a paleomagnetic signal that appears to correlate with the Mono Lake excursion of western North America. As documented by Barnedregt and Mahaney (1988) samples in the upper part of the Liki II slope wash unit show variability of magnetic intensity that is undoubtedly related to magnetic remanence carriers (Figures 5.19 and 5.20). Throughout the upper end of this section declination and inclination values reflect a normal dipole configuration of the earth's magnetic field, a notable exception occurring at 30-60 cm depths where a number of samples showed persistent intermediate polarities after progressive stepwise demagnetization. These samples record a short-lived geomagnetic excursion that would

Figure 6.21: *Podocarpus* forest and slope wash deposits on 15-25° slopes at 2750 m near Percival's Bridge. (From Mahaney, 1986.)

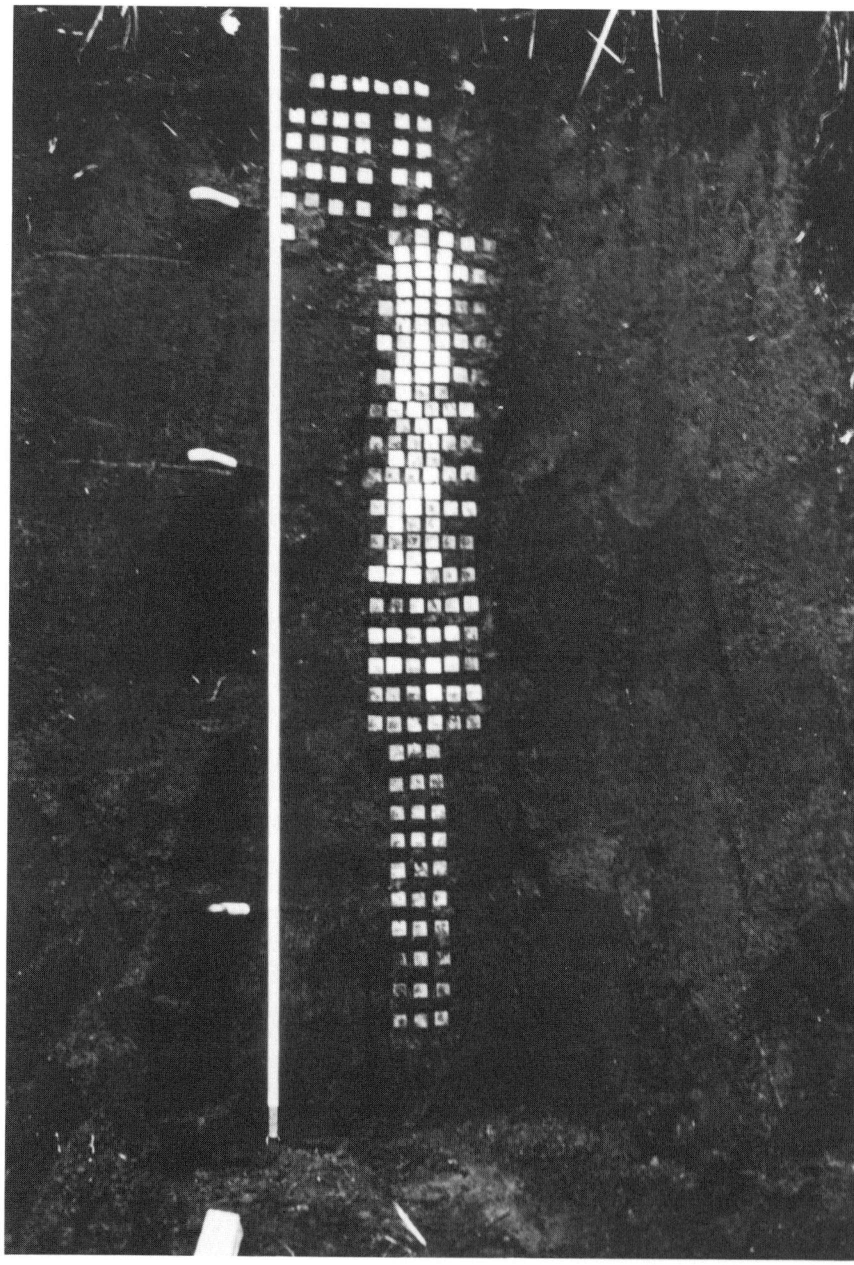

Figure 6.22: The upper beds (I and II) in section TV61, sampled for paleomagnetic dating in 1986. The geomagnetic excursion located between 30-55 cm depth may correlate with the Mono Lake excursion of western North America. Compare this sampling network with the single sample string shown in Figure 5.17.

have affected all sediments emplaced during this time (21-25 x 10^3 yr). Samples just below this zone carry reversed overprints that are removed at 15 to 25 mT demagnetization. This overprint may have occurred as a result of hydrologic changes due to soil water or ground water fluctuations present at the time of emplacement that were conducive to remineralization (principally remobilization of Fe).

6.10 Conclusions

The Liki Glaciation was the least extensive of the Pleistocene glaciations on Mount Kenya. Nevertheless, it followed the usual pattern whereby the largest glaciers were on the eastern and southeastern flanks, and the smallest ice bodies were on the northern flanks. In places there was considerable catastrophic meltwater release which removed pre-existing deposits and even washed away remnants of early Liki deposits. Active volcanism on Mount Kenya had all but ceased prior to the development of Liki glaciers making it impossible to use tephra to date deposits. The absence of buried paleosols of widespread occurrence makes it difficult to obtain ^{14}C or TL age controls.

The termination of the Liki II Glaciation occurred ~15,000 yr BP. The dates obtained are on bog-bottom sediments as close to the underlying substrates as possible, which are subject to some interpretation (Davis and Osborn, 1987; Zielinski and Davis, 1987; and Mahaney, 1988c). In general, the available ^{14}C dates are not in the bottom bog beds and as a result these dates have to be extrapolated to the underlying substrates.

Loess covers on Liki tills are thinner than on older deposits of pre-Pleniglacial age and generally weather to produce Inceptisols of ~80 cm depth that usually lack crystalline clay minerals. It is commonly accepted that halloysite-metahalloysite and other weathering products, such as gibbsite, are indicative of pre-Pleniglacial paleosols.

A geomagnetic excursion in Upper Liki deposits appears to correlate with the Mono Lake excursion of western North America.

CHAPTER 7

HOLOCENE (PRE-NEOGLACIATION) AND LATE GLACIAL RECORD

7.1 Introduction

The Holocene and late glacial records in the East African Mountains have been the subject of considerable study and controversy over the last century as different workers have arrived at conflicting interpretations of the style of deglaciation and ages of moraine systems. During the last two decades of the 19th century most investigators concentrated on Mounts Kenya and Kilimanjaro (Gregory, 1894a, b; Meyer, 1900). Later, during the first decades of the 20th century, Nilsson (1931, 1935, 1940) produced several papers which touched on the problems of interpreting the late glacial and Holocene moraines of Mounts Elgon, Kenya, and Kilimanjaro, as well as the Ruwenzori Mountains. During the 1950's and 1960's additional information on Holocene glacial and nonglacial records emerged from the work of Baker (1967; on Mount Kenya), Osmaston (1965; the Ruwenzori), and Downie (1964; on Kilimanjaro).

On Mount Kenya, as on the other mountain massifs a sequence of prominent end moraines at ~4000 m, situated below Neoglacial limits, have been assigned various ages by different researchers. Some workers regarded these deposits as recessional moraines, formed as a result of recessional stillstand of late Pleistocene ice (Baker, 1967; Gregory, 1894a; Meyer, 1900; Nilsson, 1931, 1935; and Osmaston, 1967; Mahaney, 1979a). Other workers including Downie (1964), Downie and Wilkinson (1972), and Heinzelin (1951, 1952, and 1953) believed these older moraines were emplaced by ice advances in the early Holocene. The hypothesis of the latter group is that some moraines, located between 3800-4200 m. a.s.l., might have been emplaced by an ice advance in the early to middle Holocene, rather than by a stillstand during late Pleistocene ice recession. This hypothesis has received attention recently from other workers (Johansson and Holmgren, 1985; Karlén, 1985; and Perrott, 1982). They argue that Pleistocene deglaciation was rapid, resulting in ice free valleys some time after about 15,000 yr BP, and that at various times during the early and middle Holocene ice might have advanced and emplaced some or all of these prominent end moraines. Nearly all the supporting data for this hypothesis is derived from a series of *four* bog and lake sediment cores, recovered from high Afroalpine valleys on Mount Kenya, rather than directly from the moraine systems. Mahaney (1987a, b, 1989a) discussed the evidence for middle Holocene ice advances and concluded that these moraines should be correlated with the late glacial.

7.2 Deglaciation of Mount Kenya

All glaciated valleys on Mount Kenya are floored with moraines from the last Pleistocene glaciation. Baker (1967) mapped and correlated moraines in several valleys, as low as ~3200 m, distinguishing six stages in the retreat of the ice cap, which he defined as I-VI. Subsequent work by Mahaney in 1979a, 1981, 1982b, 1983a led to a revision of the distribution, age and correlation of these moraines where Liki I/II stades refer to moraines in the lower valleys (3300-3200 m), ^{14}C dated at *circa* 15,000 yr BP. Liki III refers to moraines in high valleys (about 4000 m) ^{14}C dated at *circa* 12,000 yr BP (Figure 7.1). Because only discontinuous ground moraine exists between the Liki I/II end and Liki III recessional moraine systems, it is clear that deglaciation from the lower Liki I/II moraine limits to the Liki III moraines in the high valleys, was rapid and without apparent recessional stillstand of the ice (Mahaney, 1983a, 1985, 1986, 1987a). Hence, the style of deglaciation is remarkably similar to modes of deglaciation reported for the Rocky Mountains in the U.S. (Mahaney, 1978; Mahaney *et al.*, 1984a, 1984b).

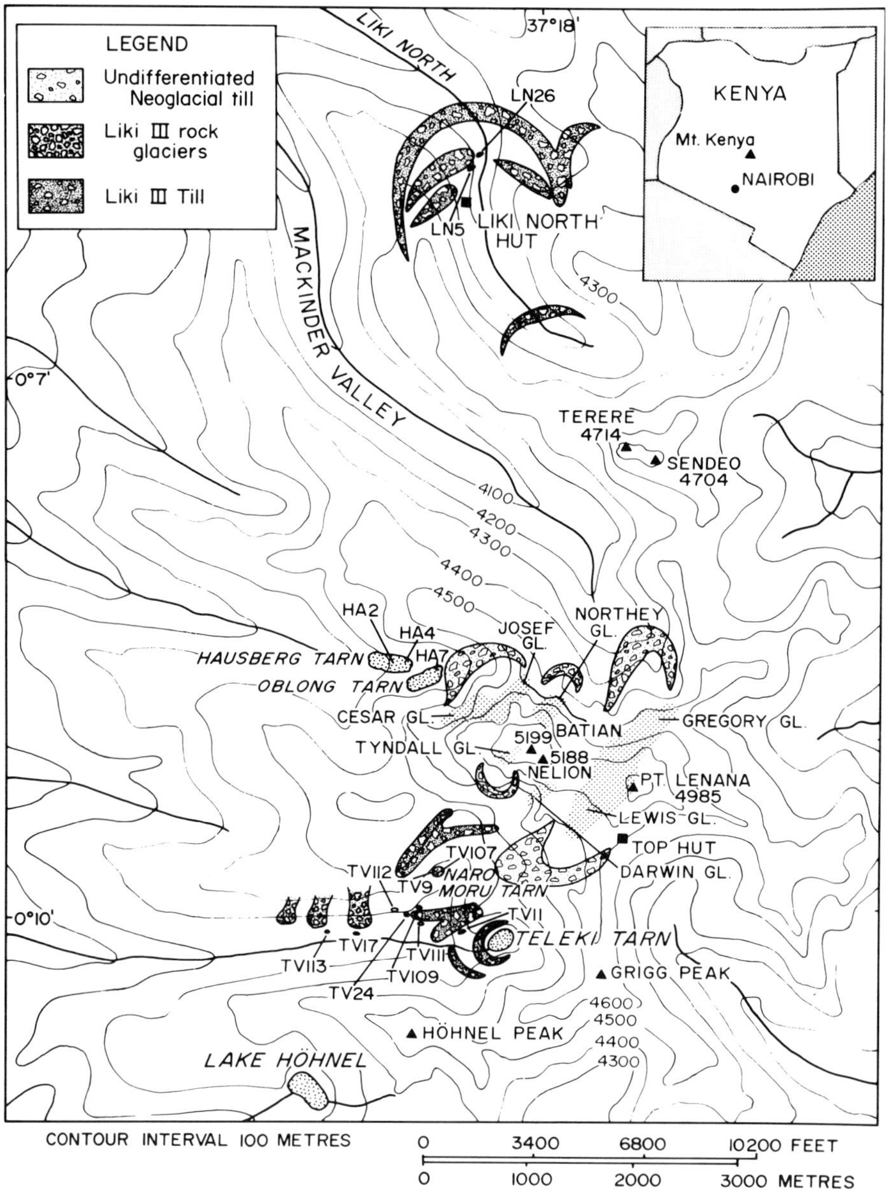

Figure 7.1: Map of Liki III moraine positions at ~4000 m on Mount Kenya.

Moraines at *circa* 4000 m on Mount Kenya, considered by Baker (1967) to represent recessional stillstand during the retreat of late Pleistocene ice are termed Stage V moraines (= Liki III), an interpretation supported by Mahaney (1979a, 1981). However, in the 1980's other workers attempted to core bogs behind Liki III moraines in an effort to recover datable organic material that could be used to provide bog-bottom or minimum ages. Perrott (1982) sampled *one* core from Thompson Tarn situated behind a Liki III moraine in Hobley Valley (4200 m) for radicarbon datable material and pollen. On the basis of radiocarbon dates he concluded that the small moraine ponding Thompson Tarn dates from between 5425 (minimum) and 6277 (maximum) ^{14}C yr BP (no standard deviations published), and represents a mid-Holocene glacial advance. In Teleki Valley a small moraine of very similar appearance in front of Naro Moru Tarn (Figure 7.2) was ^{14}C dated by Johansson and Holmgren (1985) at 6070 ± 225 ^{14}C yr BP (a minimum age). The Naro Moru Tarn, situated in a rock basin scoured by late Pleistocene ice, is not dammed by the small moraine in front of it (Figures 7.1 & 7.2). Contrary to this evidence, extrapolated substrate ages from bog-bottom radiocarbon dates in Teleki Valley provide new minimum ages for the Liki III end moraines that place them in the later Pleistocene. Moreover, the Liki III moraines (Figure 7.3) all have similar weathering features, topographic positions, and degrees of soil development (Mahaney, 1982b, 1983a), suggesting a similar pre-Holocene age. Although variable minimum radiocarbon ages have been obtained for similar moraine systems in different drainages on Mount Kenya, + 10,000 yr ^{14}C extrapolated dates suggest a later Pleistocene age for these features.

7.3 Liki III Moraines and Bog Sites

At the ~ 4000 m elevation Teleki Valley is a wide, U-shaped valley, surrounded by a number of rugged peaks, all of which are at least 450 m higher than the adjacent valley floors. The Liki III moraines (Figure 7.3) located in the upper valley at about 4000 m, have topographic relationships which indicate that after the recessional moraines (Figures 7.1 and 7.3) were constructed, the valley glacier receded first to the position of Naro Moru Tarn, and then upvalley, either disappearing during the postglacial period, or remaining only as isolated cirque glaciers about the size of Lewis and Tyndall glaciers today.

In Hausberg Valley (Figure 7.1) the Liki I/II episode left only lateral moraines down to 3500 m; below that elevation end moraines appear to have been washed away by meltwater (Charlsey, 1989). Only patches of ground moraine exist on valley floors to about 4200 m where Oblong and Hausberg tarns are located (Figure 7.4). Because they are situated in juxtaposition to moraines resulting from late Holocene ice advances (*i.e.* Lewis advance), it is obvious that dates from Oblong and Hausberg tarns are important in establishing the glacial history of Mount Kenya.

7.4 Bog and Lake Cores

Much of the Holocene stratigraphy in the high valleys is known from bog and lake cores recovered with modified Livingstone-type piston corers (Coetzee, 1967; Karlén, 1985; Mahaney, 1987a, b). Because some of the tarns are deep, it proved impossible to sample the deepest parts; however, many investigators on Mount Kenya have recovered long and presumably old bog and lake cores using chest waders and working the margins of the deeper tarns. All workers have described the difficulties of finding bog-bottom organic-rich materials that might yield radiocarbon dates. Most researchers have concentrated on collecting organic-rich sediment, detrital plant fragments, or wood from the bottom (or near the bottom) of each core which could be radiocarbon dated, and hence, provide minimum ages on underlying substrates. In almost every case there is nearly always a lower core section that lacks organic-rich sediment and is impossible to date by radiocarbon.

In an attempt to reconstruct short-term glacial fluctuations Karlén (1985) recovered cores of lacustrine sediments spanning the last 3200 years. Oblong Tarn (*cf* in front of Cesar and Josef glaciers, Figure 7.4) yielded high amounts of silt in certain core segments that Karlén believes correspond to times of glacial advance (on the order of < 100 m) between 3200-2900 yr BP, 2350-1700 yr BP, and 600-50 yr BP. It is important to note that his periods of short-lived ice advances correlate with periods of increased runoff in the Kenyan Rift Valley, but are

Figure 7.2: Upper Teleki Valley showing Naro Moru Tarn in a bedrock basin.

Figure 7.3: Double-crested end moraines of Liki III age in upper Teleki valley.

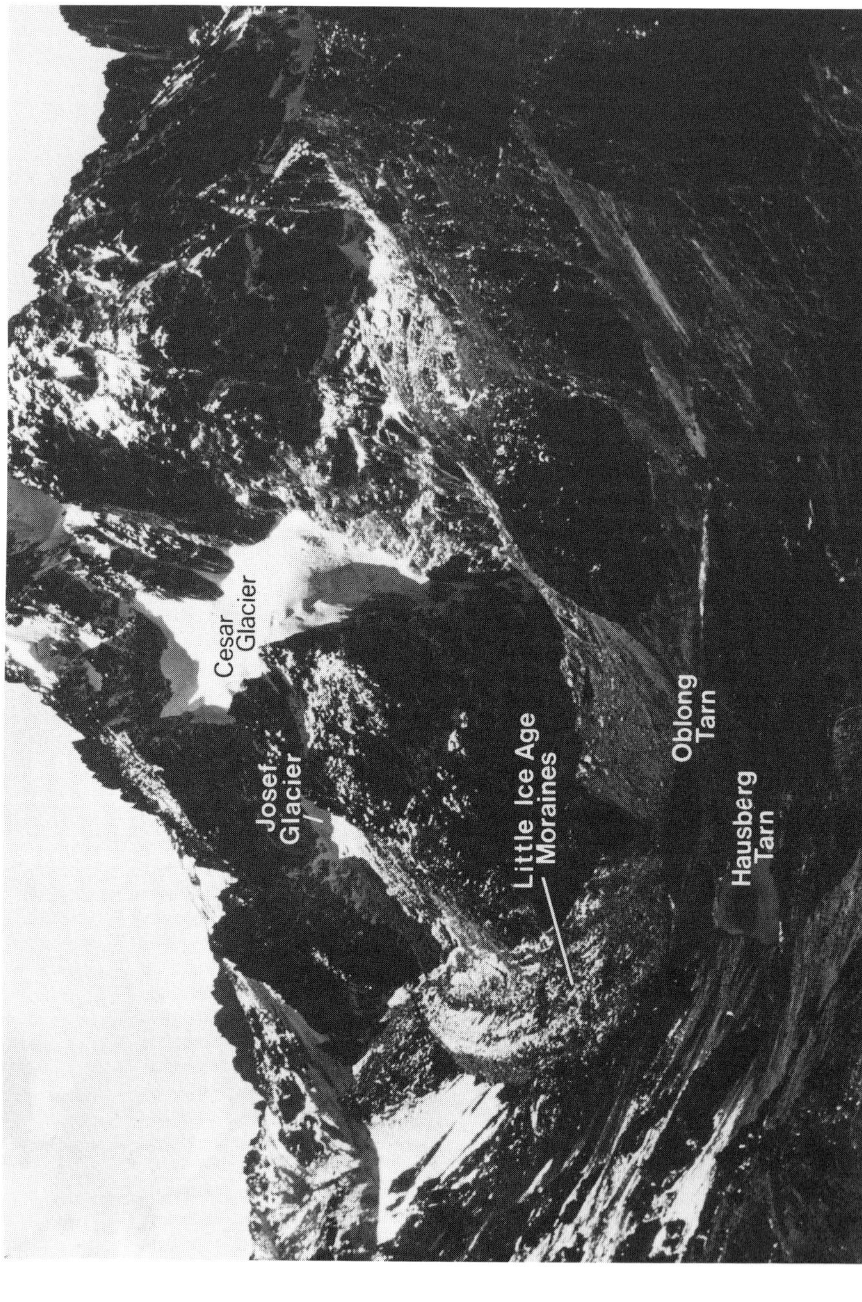

Figure 7.4: Oblong and Hausberg tarns adjacent to Neoglacial moraines on the northwestern flank of the mountain.

out-of-phase with palynologic reconstructions of precipitation from the Mount Kenya area (Coetzee, 1967).

Altogether thirty bog and lake sediment cores have been collected in several valleys with the objective of dating end moraines in the high valleys at about 4000 m (Mahaney, 1987a). These bog and lake sediment cores (Figures 7.5A and B) are of variable length, tend to coarsen downward, and generally lack datable organic materials in their lower sections.

In Teleki Valley eight bog cores taken in front of and behind the Liki III moraines (Figure 7.3) showed that, while core depths (Figure 7.5A) were greater than those reported by Johansson and Holmgren (1985), it was imposible to obtain bog-bottom sediments sufficiently high in organic matter for radiocarbon dating. In some instances tills were encountered at depth in the cores; however, usually the core could not penetrate sand, or pebbly sand, which extended downward to an unknown depth. Mahaney (1987a) used the sedimentation rates in overlying core sections (Table 7.1) to calculate the ages of underlying deposits. Even though these sedimentation rates vary depending on particular sedimentation processes the extrapolations provide a reasonable estimate of substrate age. The time required for sedimentation in the lower core sections varied considerably between cores (Table 7.1), and also with respect to topographic position on the valley floor. For example, sites on the lower valley train deposit in Teleki Valley (Figure 7.5A) are younger, while those adjacent to the Liki III end moraines are older (*e.g.* TV109, TV17, TV24 in Figure 7.5A), and may be somewhat younger or older than the Pleistocene/Holocene boundary. The estimated ages of these various deposits suggest that some were formed in the later Pleistocene or considerably before the time suggested by Perrott (1982) and Johansson and Holmgren (1985).

The inferred substate ages in Table 7.1 should be considered *minimum* ages (Mahaney, 1987a) because the length of time required to form and infill bog depressions following deglaciation is unknown. Because these cores coarsen downward they may have been subjected to sediment transport, presumably by meltwater from Pleistocene ancestors of the Lewis and Tyndall glaciers.

In Hausberg Valley four cores from Hausberg and Oblong tarns were described and analyzed (Figure 7.5B) (Mahaney, 1987a). Core HA2 in Hausberg Tarn yielded an age of 9880 ± 860 yr BP age on wood at a depth of 60 cm, and an estimated substrate age of about 11,600 yr BP. This date suggests that most of Hausberg Valley was ice free prior to the Holocene period. Moreover, with an estimated substrate age *circa* 11,600 yr BP it is highly probable that deglaciation occurred even earlier, perhaps even by 12,500 yr BP.

Two additional cores (HA4 in Hausberg Tarn, and HA7 in Oblong Tarn) provided additional evidence for variable sedimentation rates and time of deglaciation. Core HA4, although shallow (68 cm), yielded an average rate of sedimentation of 0.10 mm/yr. This very closely compares with sedimentation rates in cores HA2 and HA7. Unfortunately core HA4 could not penetrate a layer of sand at 65 cm. Core HA7, which was recovered from considerably greater depth (172 cm), revealed additional sedimentary details with granulometric variations extending to *circa* 15,000 yr BP (assuming a constant rate of sedimentation below 33 cm). Assuming that sedimentation rates were faster in the lower 140 cm of this core, it is quite probable that the lower core dates from late glacial time (*circa* 12,500 yr BP). Core HA5 which produced rather young ^{14}C ages at shallow depth, could not be dated below ~ 60 cm.

7.5 Formation of Recessional Stillstand Moraines

The hypothesis that Liki III moraines at about 4000 m on Mount Kenya were emplaced as a result of renewed glacier activity, and hence *advances* during the mid-Holocene, as proposed by Perrott (1982) and Johansson and Holmgren (1985), clearly requires additional evidence and considerable refinement. The advance hypothesis is based on the analysis of *two cores* taken from

Table 7.1:
Sedimentation rates and estimated ages of substrates in Teleki, Hausberg and Liki North Valley, Mount Kenya (for location and ^{14}C dates see Fig. 7.1).

Core	Sedimentation Rate (mm/yr)[1]	Depth (cm)	Time required for sedimentation of lower core (in yrs.)	^{14}C date	Extrapolated age of Substrate (yr BP)[2]
TV107	.17	(0-46)	3647	2640 ± 200	6287
TV109	.16 .07	(0-30) (30-43)	6714	1850 ± 300 3700 ± 300	10414
TV111	.19 .35	(0-20) (20-45)	2857	1760 ± 130	4617
TV112	.39	(0-38)	1435	1000 ± 100	2435
TV113	.13 .51	(0-25) (25-105)	686	3520 ± 350	4206
TV9	.20	(0-76)	4950	3730 ± 100	8680
TV17	.23	(0-60)	5087	2650 ± 130	7737
TV24	.21	(0-65)	5048	3050 ± 80	8098
HA2	.11 .13	(0-40) (40-60)	1769	5560 ± 390 9880 ± 860	11649
HA4	.10	(0-45)	2300	4710 ± 390	7010
HA7	.11	(0-33)	12636	3030 ± 480	15666

1. Core section in brackets; depth in cm.

2. See Mahaney, 1987a, for discussion.

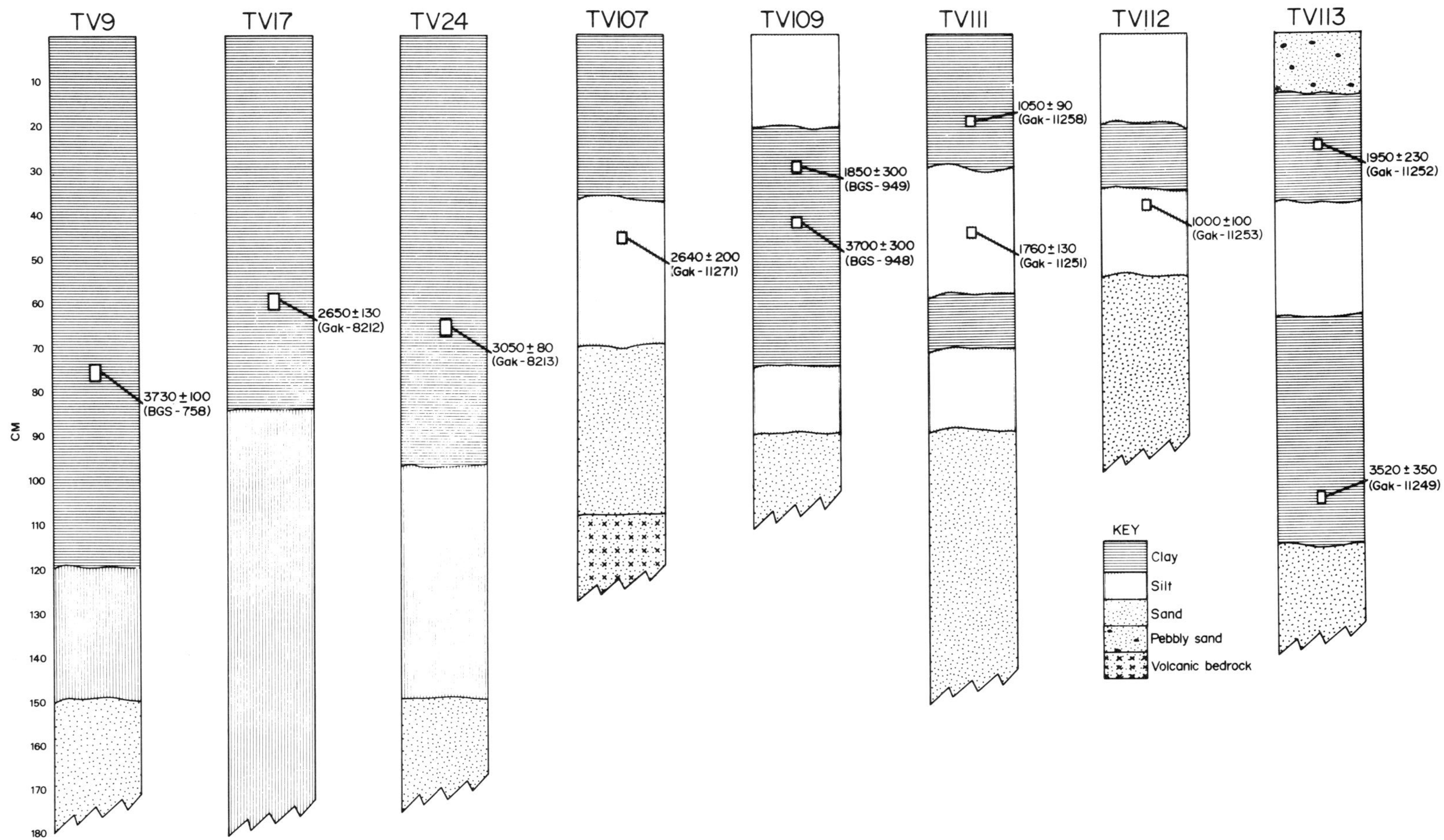

Figure 7.5: A, Bog cores in upper Teleki Valley

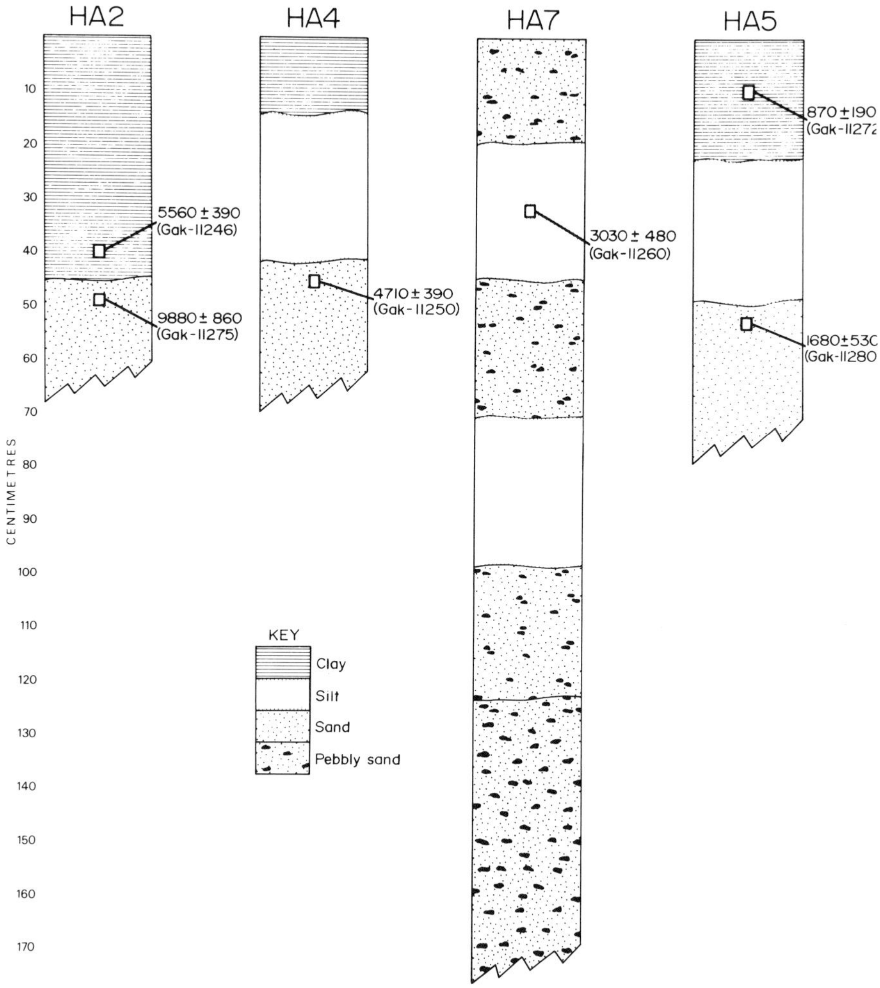

Figure 7.5: B, Bog cores in Hausberg Valley.

the southeastern (Perrott, 1982) and southwestern (Johansson and Holmgren, 1985) drainages on Mount Kenya. They consider the gravelly layers in basal core sections to be till, but make no attempt at correlation with other sections, or with local end moraines. In the case of Johansson and Holmgren (1985) one published SEM micrograph of glacially-crushed quartz was used to establish the glacial origin of basal sediments. The analysis of several paleosols on Mount Kenya (Vortisch, Mahaney and Fecher, 1987), however, showed that quartz is most abundant in loesses and mixed tephra-loess units, and least abundant in tills. As shown in Figure 7.6A and B glacially crushed quartz is often reworked into airfall sediment, so that crushed features on quartz grains do not *conclusively* prove a glacial origin for any particular part of a core section.

In order to test the theory that Liki III moraines were formed by glacial advances as opposed to a recessional stillstand, the distal flanks of seven different moraines at about 4000 m in Teleki, Gorges, Hobley, and Sirimon valleys were trenched (Mahaney, 1987a). The objective of this trenching operation was to locate bog-bottom sediments, weathered bedrock, or buried soil, which would serve to prove conclusively that these valleys had been deglaciated prior to the

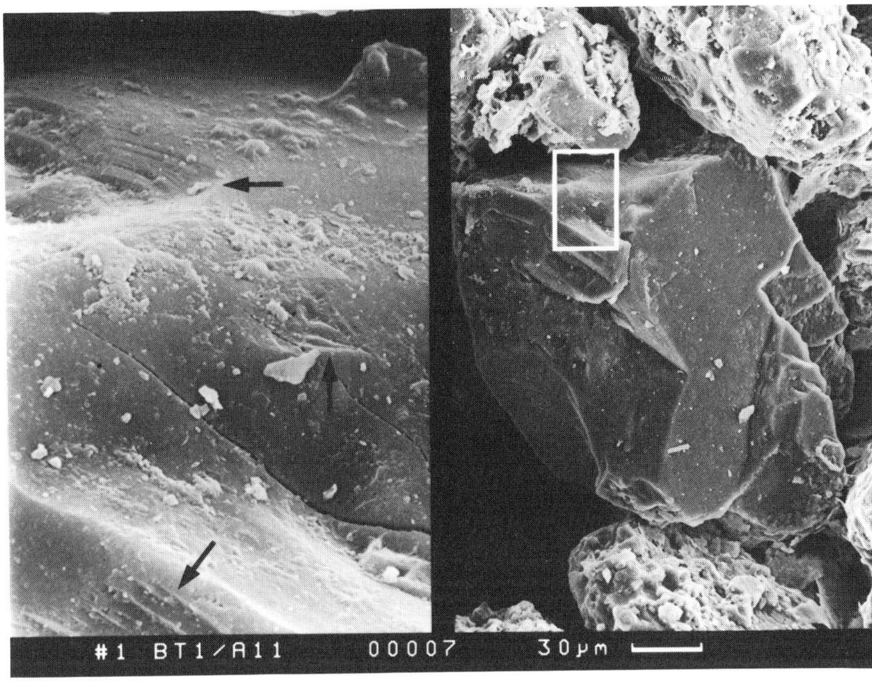

Figure 7.6: A, Glacially-crushed reworked quartz in late Holocene loessic sediments on Mount Kenya; arrows indicate subparallel crushing planes.

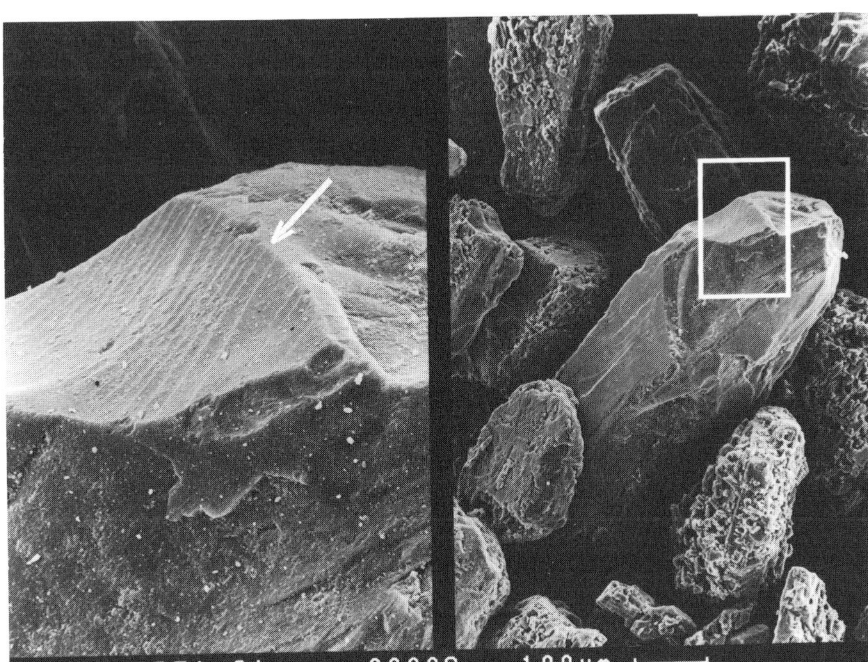

Figure 7.6: B, Glacially-crushed reworked quartz in late Holocene loessic sediments on Mount Kenya; arrows indicate subparallel crushing planes.

advance of Holocene glaciers (Mahaney 1987b). Previous work by Mahaney (1979a, 1981) established that surface soils form A horizons relatively quickly (Mahaney and Boyer, 1986a) and bog core data (Figure 7.5A, B) suggest that some small bogs may form in about 2000-3000 yr BP or less. If these valleys had been deglaciated, and then reinvaded by Holocene ice, one would surely expect to find some evidence of preweathered till beneath post-Liki III paleosols to a depth of about 2-3 m (Figure 7.7A, B) on distal moraine slopes. Therefore, the evidence indicates that these moraines represent a recessional stillstand as originally suggested by Baker (1967) and Mahaney (1979a, 1984b).

7.6 Weathering Characteristics

Weathering rinds on surface stones of Liki III deposits in several cirques are shown in Table 7.2 (see Plate 5 for location of sites). Compared with weathering rinds on younger Neoglacial deposits the data in Table 7.2 show mean maximum values of ~2.0 mm. Overall, the mean maximum rind is thicker than on younger late Holocene clasts. Rinds are also more uniformly developed on individual clasts as the minimum rinds tend to increase as well.

Table 7.2

Weathering Rinds on Surface Stones on

Liki III Moraines in the Central Peaks

Site	Age	Mean Maximum Rind (mm)	Mean Minimum Rind (mm)	Number Counted
TT3		2.36	0.50	25
HA6	Liki	2.01	0.20	50
MK2	III	1.61	0.27	50
MK10		1.56	0.34	50
MK11		2.00	0.12	50
MK13		1.88	0.24	50
MK14		1.76	0.22	50

7.7 Liki III Paleosols

Paleosols in Liki III deposits are nearly always Inceptisols with A/B/C horizon sequences and depths of ~50 to ~80 cm. They have ~10 to ~20 cm thick cambic (color) B horizons and show no translocation of clay within the soil solum (A & B horizons). They usually contain pebbles and cobbles throughout the profile; often the sand-silt-clay matrix shows a fining-upward sequence with silt and sometimes clay increasing towards the surface. This is considered to result from a mix of airfall-influx sediment with till following ice recession. As a result the overlying loessic sediments may be somewhat younger than the underlying till.

Attempts at ^{14}C dating of these paleosols have proved fruitless largely because the till in which these soils are found lacks organic-rich material. It is possible, however, to obtain relative age determinations using amino acid racemization (see Chapter 10). These dates may be obtained by using D/L ratios of aspartic acid and fitting them to an amino acid curve for ~4000 m sites. Because the moraines are impoverished of wood and other datable materials, the only means by

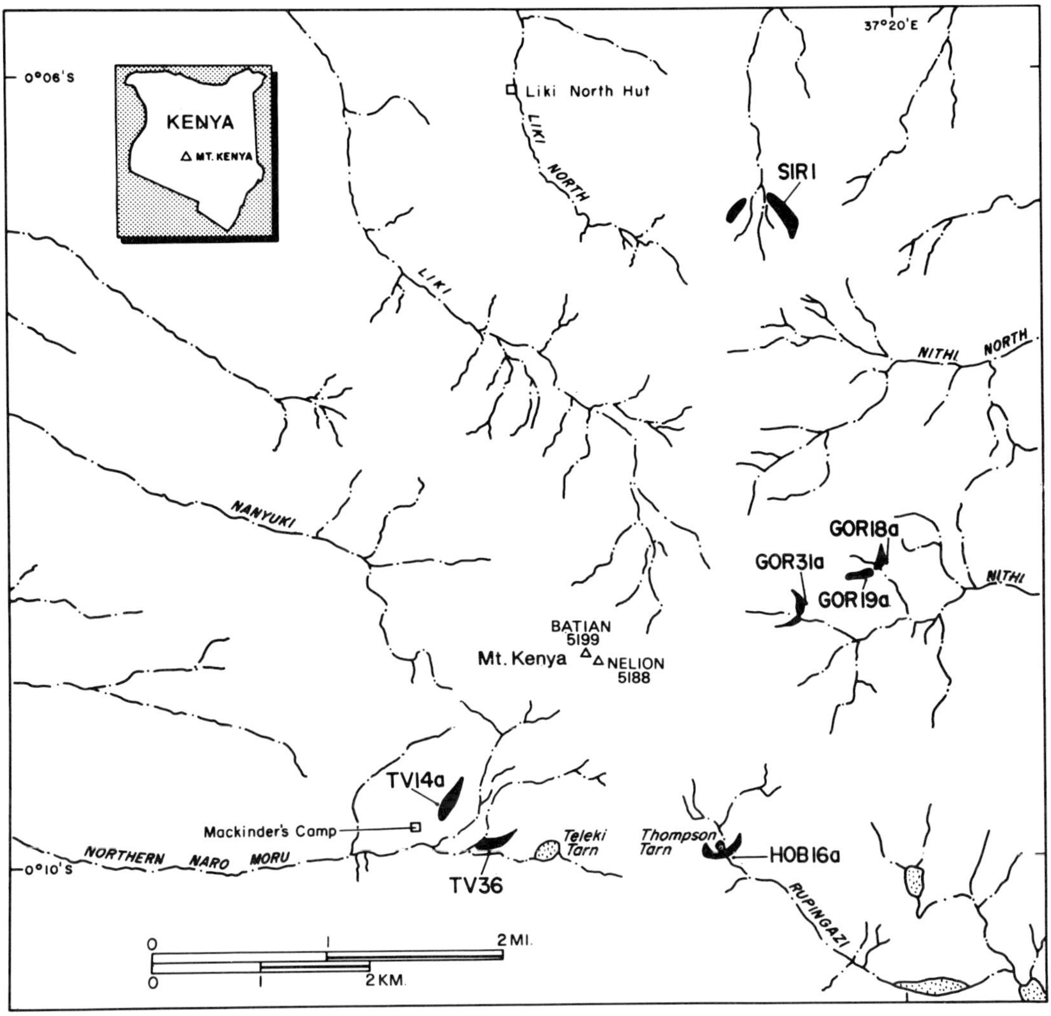

Figure 7.7: A, Location of slit trench sections on distal flanks of Liki III moraines on Mount Kenya.

which ^{14}C dates can be obtained is from bog-bottom sediments recovered in juxtaposition to the moraine.

Careful scrutiny of the Liki III paleosol descriptions shows that they vary in profile characteristics and thickness from the wetter to the drier sides of the mountain. The thickest and best developed profiles are found in Teleki Valley (as shown in Figure 7.8 for location; Figure 7.9 for profiles). These soils have stronger expressions, greater sola depth, and thicker and more deeply developed root systems. In all the A horizons tend to develop into peaty materials with low counts of fungi and bacteria (Mahaney and Boyer, 1986a) that are likely to produce high amounts of organic matter (organic carbon is often +20 percent).

In drier drainages, such as the Sirimon (Figure 7.10) and Gorges (Figure 7.11) valleys, on the northern and eastern flanks of the mountain, post-Liki III profiles are thinner and have weaker expressions in terms of color and soil structure. Representative profiles are shown in Figure 7.12.

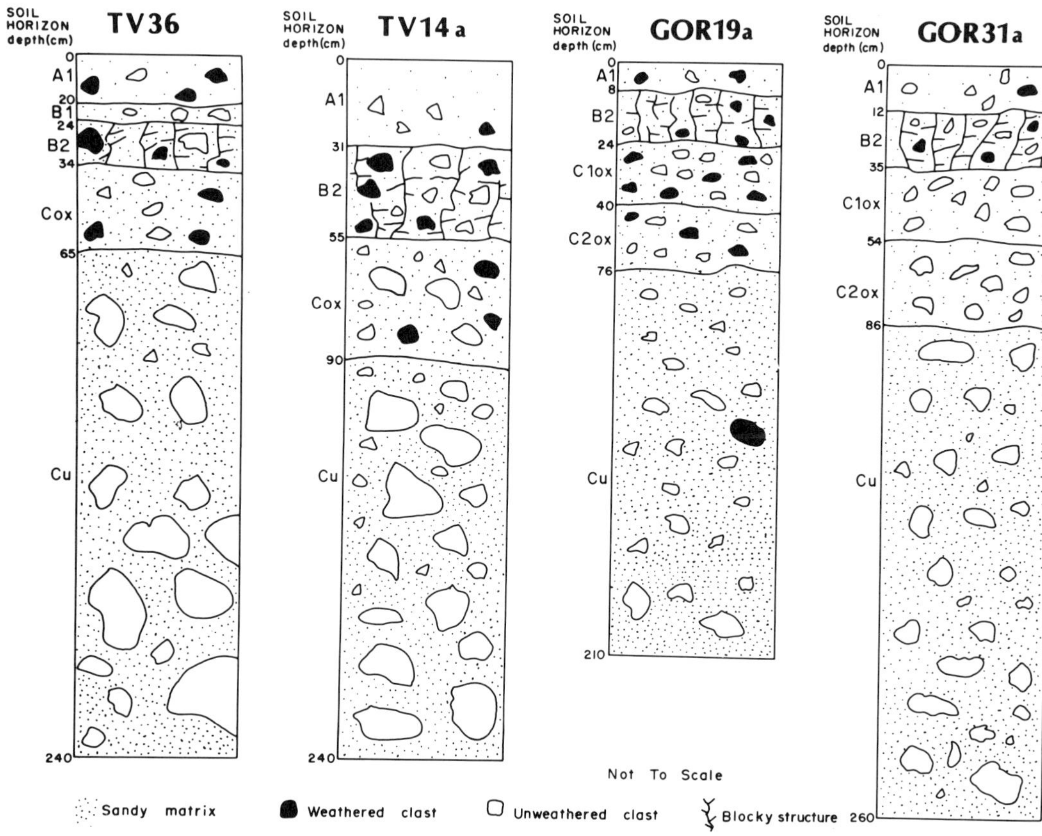

Figure 7.7: B, Location of slit trench sections on distal flanks of Liki III moraines on Mount Kenya: sections TV36, TV14a, GOR19a, and GOR31a.

Even the root systems penetrate to shallow depths in the surfaces of these pedons and the A1 horizons are considerably thinner compared with profiles in wetter drainages.

7.8 Liki III Rock Glaciers

Rock glaciers on Mount Kenya are found only in upper Teleki Valley (Mahaney, 1980) and on Ithanguni near Lake Alice. All are of the lobate-shaped valley side type (Figure 7.13) and no interstitial ice was observed in the three systems (Mahaney, 1980, p. 493). Minimum temperatures taken at the bottom of an ~ 80 cm open soil pit in 1976 gave values of -4.1°C (n = 5), while daytime temperatures range from + 7°C to 15°C. All three rock glaciers have several concave depressions behind their outer lobes, longitudinal furrows, and "drunken stands" of *D. keniodendron* and *L. telekii*, which suggest that relict Pleistocene or late Holocene ice may have melted out.

Three lobes, developed below avalanch talus (Figure 7.14) on the north side of Teleki Valley, have estimated thicknesses of 100 to 150 m (Mahaney, 1980). The frontal lobes of two systems appear to have been reactivated, possibly during the Neoglacial. Surface stones, however, even on the reactivated fronts, carry lichen cover of 80% and all the common species of *Rhizocarpon* section *Rhizocarpon*, *Buellia*, *Lecidea*, and several foliose lichens discussed at length in Chapter

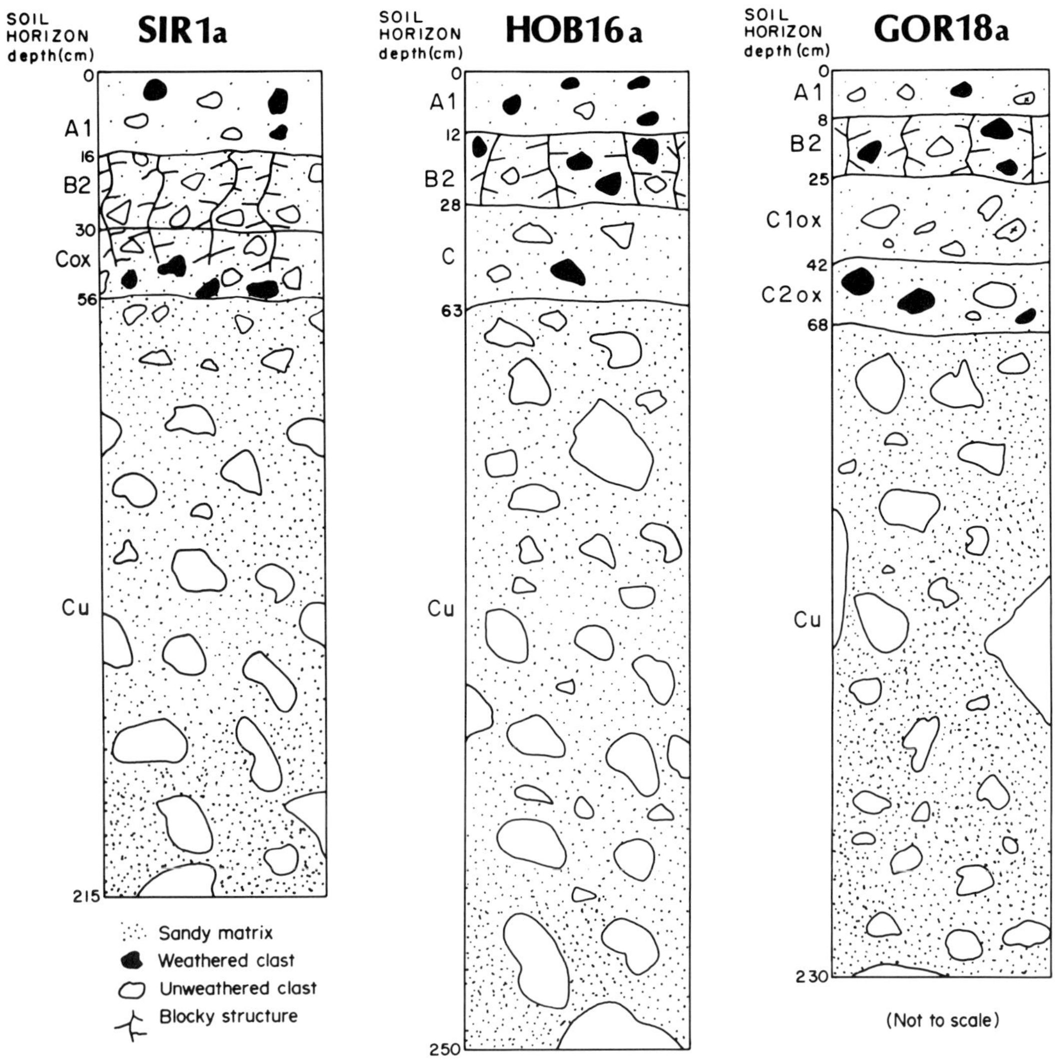

Figure 7.7: C, Location of slit trench sections on distal flanks of Liki III moraines on Mount Kenya: sections SIR1a, HOB16a, and GOR18a.

8. Isolated boulders and very large blocks (some with high lichen cover) were observed in front of the two reactivated fronts.

Weathering rinds on surface stones tend to give mean maximum thicknesses of between 4 and 6 mm; weathering pit depths range from 10 to 11 mm. These data are 2x the rind and pit size normally measured on Liki III end moraines in nearby valleys. The most logical explanation for this is that some of the measurements may have been made on preweathered clasts derived from the valley walls above.

Soils are discontinuously distributed over the surface of the rock glacier complex. Soil profile TV1 (Figure 7.15) has a thick (40 cm) B horizon giving strong dark brown (10YR 3/4) and brown (10YR 4/4) colors as well as weak grades of blocky structure. Overall, the total depth is similar to other paleosols of similar age in upper Teleki Valley (Mahaney, 1980). Particle size

Figure 7.8: Liki III moraines in upper Teleki Valley; site TV3 on inner moraine (site TV3 is located 150 m to the southeast of the Liki III moraine below Teleki Tarn – not on photograph).

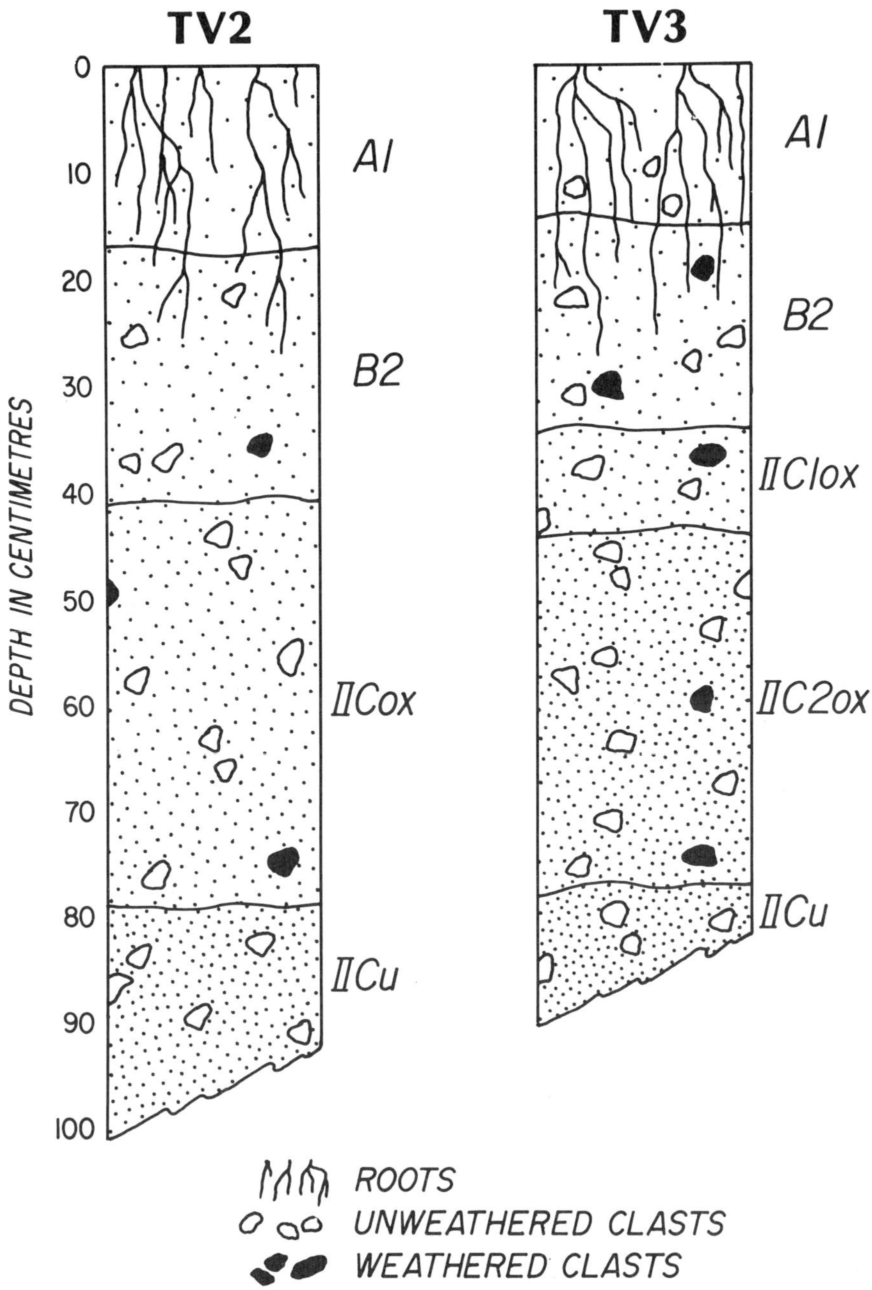

Figure 7.9: Profiles TV2 and TV3

Figure 7.10: Distal side of Liki III moraine in upper Sirimon Valley, northern flank of Mount Kenya.

Figure 7.11: Upper Gorges Valley showing Liki III moraine systems below Simba Col and in front of other small cirques at 4000 m.

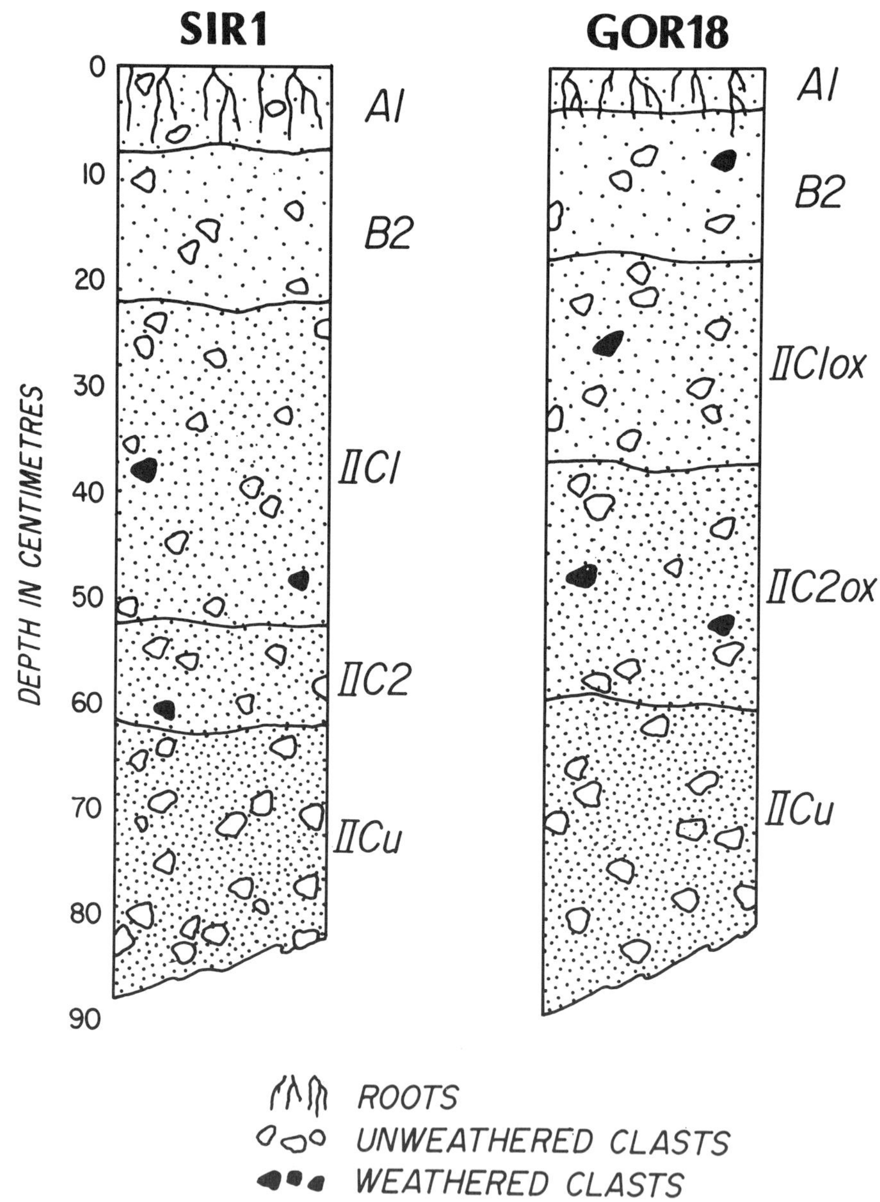

Figure 7.12: A, Profile SIR1 in upper Sirimon Valley; B, Profile GOR18 in upper Gorges Valley.

Figure 7.13: Rock glacier complex in upper Teleki Valley.

Figure 7.14: Talus feeding lobate-shaped valley side rock glaciers in upper Teleki Valley.

distributions show the presence of two parent materials and loess (based on upward fining tendencies into the A11 horizons).

Gregory's (1894a) map shows four "end moraines" on the north side of the valley. Close inspection shows that these lobe-shaped entities do not have counterparts on the south side of the valley. Even closer scrutiny of the rock fabric showed a random orientation. Taken together with the other available data it has been necessary to reassign these deposits as periglacial features.

7.9 Holocene Stratigraphy and Paleosols

The Holocene on Mount Kenya is known primarily from the record of alluvial/colluvial fans, recovered bog and lake cores, outwash deposits, dune deposits, and associated paleosols (both relict and buried). There are no early to middle Holocene tills on Mount Kenya, despite reports to the contrary (Perrott, 1982 and Johansson and Holmgren, 1985), making it imperative to use data from nonglacial deposits in reconstructing Holocene events. Fortunately many nonglacial deposits have an abundance of buried paleosols that can be radiocarbon dated, thus providing significant age controls. In addition, some of these paleosols have been dated by thermoluminiscence and amino acid racemization (Chapter 10).

A summary of radiocarbon-dated sediments is shown in Table 7.3. A range of materials including plant fragments, whole-soil organic material, and buried soil organic carbon have been dated. Seventeen dates are available, ranging from ~8000 to < 1000 yr BP, in a variety of different sediments (Figure 7.16). These deposits consist of alluvial/colluvial fan sediments emplaced over till or older mass-wasted deposits. Profile TV28 (Figure 7.16A, B), is representative of colluvial fan deposits in upper Teleki Valley, emplaced largely in the middle Holocene, and weathered sufficiently to produce an Inceptisol (Table 7.4) with a ~20 cm thick cambic horizon (B21/B22 complex). While the deposit contains mainly fine-grain (< 2 mm) clastic material, there is enough coarse clastic sediment (granules and small pebbles) for it to qualify as a colluvial deposit.

Profiles TV33, TV39 and TV53 (Figure 7.16A) are representative of alluvial fans constructed in upper Teleki Valley starting ~8000 yr BP and continuing intermittently until ~3600 yr BP. These fan deposits consist of alternating beds of fine (< 63μm) and coarse (2mm-63μm) sediment that were emplaced during stages of instability, which either continued uninterrupted for long periods of time (TV53) (Figure 7.16B) or were interrupted periodically for periods of sufficient duration to allow weathering and soil genesis to occur (TV33 and 39). In the case of TV33 (Figure 7.17A, B) the upper radiocarbon date of 3610 ± 180, is considered too young by several hundred years (see Mahaney and Boyer, 1986a). A previous radiocarbon date in this same horizon yielded an age of 4310 ± 180 yr BP (GaK-8214) which may be closer to the true age (Mahaney and Boyer, 1986a). In any case, after comparing dates for this upper paleosol with the middle one dated at 5380 ± 130 yr BP (Figure 7.16) it is clear that considerable time is required to produce a B horizon in the Afroalpine environment (*e.g.* + 1000 yrs).

In upper Teleki Valley at site TV4 a compound paleosol consisting of two units is shown in Figure 7.16A and 7.18. The lower unit consists of fine-grained silty sand and silty clay-rich sediment with an Ab horizon that ^{14}C dates at 1940 ± 120 yr BP (GaK-8273). This buried paleosolum yields a TL date of 1300 ± 400 yr (G. Berger, personal communication, 1985). TL dates are generally 25 to 30% lower than ^{14}C dates, making the true age of this profile approximately ~2000 yr. Because profile TV4 is the youngest dated profile on Mount Kenya with a B horizon, it likely supports the argument that B horizons require at least ~2000 yr to form.

The upper unit in profile TV4 (Figure 7.16 and 7.18) is an Inceptisol with coarse-grained sediment consisting of beds of small pebbles and granules as well as sand and silt lenses. It appears that sometime after ~2000 yr BP a ground soil formed in this coarser outwash material to a depth of ~55 cm. The upward coarsening trend in this profile suggests that higher-flow regime currents

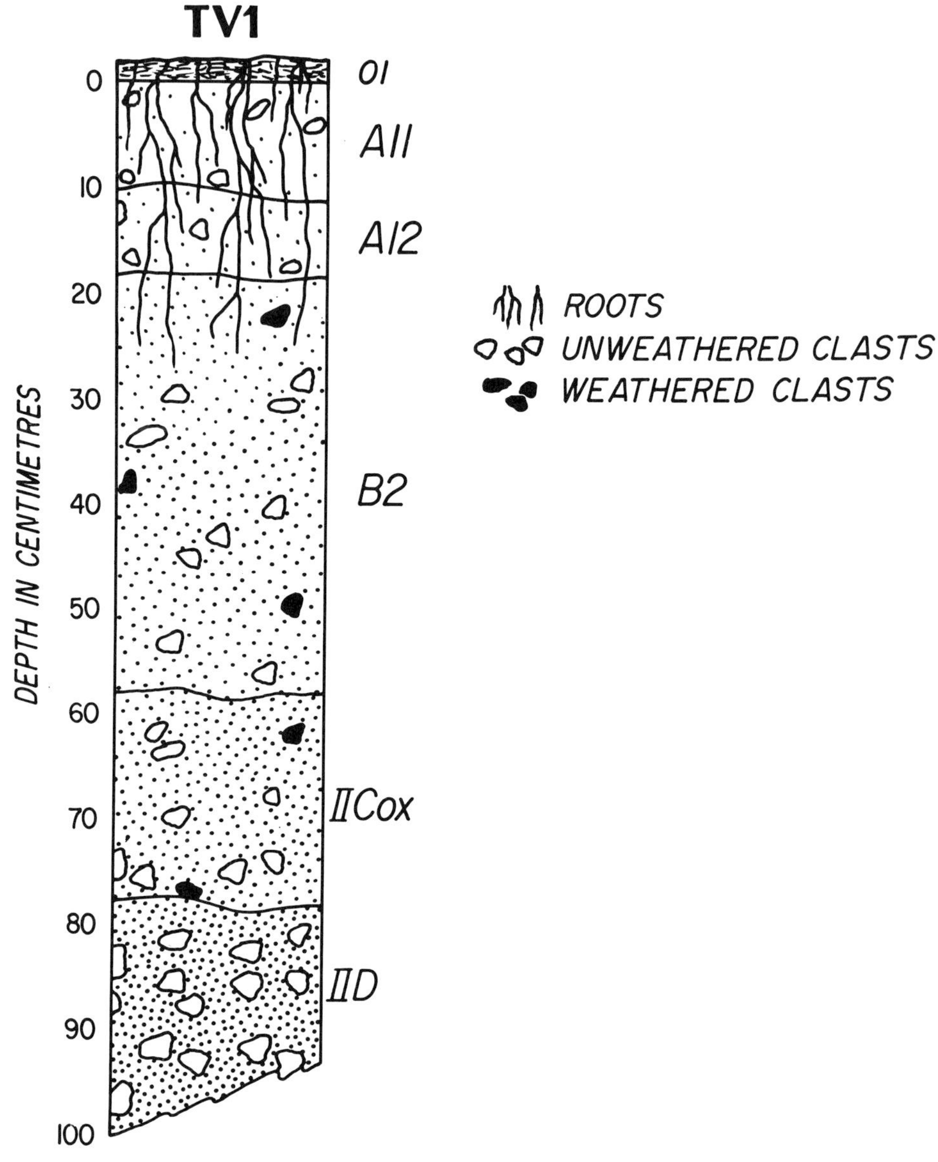

Figure 7.15: Soil profile TV1 in Liki III rock glacier, upper Teleki Valley.

in the Naro Moru River may have been generated by meltout from the Tyndall and Lewis glaciers during the Tyndall advance (see Chapter 8 for discussion). In addition to sites LG5 and TV42 (see Chapter 8) profile TV4 provides the only available evidence of increased runoff in upper Teleki Valley in the later Holocene.

Profile LN9 (Figure 7.16) in Liki North Valley is useful for comparative purposes. It is a compound paleosol (Figure 7.15) with a thin, buried Inceptisol formed in till (unit II) and overlain with colluvium. Burial of the paleosol formed in till occurred approximately 3430 ± 100 yr BP (BGS-756) at a time close to the *Podocarpus* rise of ~ 3700 yr BP (see section 7.10). Profile LN9,

Table 7.3:

Organic samples and radiocarbon dates for Holocene sediments on Mount Kenya

Sample No.	Material Dated	Date (yr BP)	Lab No.
GOR53-A11	whole soil organic matter	0 ± 100	BGS-1217
LG5-Ab	whole soil organic matter	830 ± 100	GaK11239
TV42-Ab	whole soil organic matter	920 ± 100	GaK-11240
BT1-Ab	organic carbon from Ab horizon	1490 ± 80	Beta-10820
TV4-Ab	whole soil organic matter	1940 ± 120	GaK-8273
LN9-Ab	whole soil organic matter	3430 ± 100	BGS-756
TV33-Ab	upper Ab whole-soil organic matter	3610 ± 190	GaK-11320
ONT1	*in situ* plant matter in alluvial fan	4375 ± 100	BGS-753
TV39-Ab	whole soil organic matter	4540 ± 130	GaK-11241
GOR52-Ab	Organic carbon from Ab horizon	5150 ± 80	Beta-10821
GOR8-Ab	whole soil organic matter	5110 ± 120	GaK-11570
LN18	organic-rich bog bottom sediment	5280 ± 110	BGS-757
TV33-Ab	intermediate Ab in middle paleosol	5380 ± 130	GaK-11321
TV33-Ab	whole soil organic matter in lower paleosol[a]	3400 ± 170	GaK-11322
TV28-Cu	*in situ* plant fragments in alluvial fan	6670 ± 170	GaK-8276
TV53-Ab	organic carbon in Ab horizon	6820 ± 160	Beta-9087
TV53-C2b	buried organic matter	8070 ± 100	Beta-9088

a. Contaminated Sample – see Mahaney & Boyer, 1986a, for discussion.

Table 7.4:

Soil Profile TV28

Horizon	Depth (cm)	Description[a]
01	2-0	Black (10YR 2/1m)
All	0-5	Brownish black (10YR 2/3m, 2/3d, 3/2d) clay loam, granular structure, friable consistence, plastic and nonsticky.
A12	5-10	Dark brown (10YR 3/3m, 3/3d) and greyish yellow brown (10YR 5/2d) silty clay, granular structure, friable consistence, plastic and sticky.
B21	10-26	Brown (10YR 4/6m), and dull yellowish brown (10YR 5/3d) pebbly loam, weak blocky structure, firm consistence, plastic and sticky.
B22b	26-34	Brown (7.5YR 4/3m) and greyish yellow brown (10YR 6/2d) pebbly silt loam, blocky structure, firm consistence, plastic and sticky.
Clbox	34-44	Brown (7.5YR 4/4m) and yellowish brown (7.5YR 5/3d) pebbly coarse sandy loam, massive, firm consistence, plastic and slightly sticky.
C2box	44-54	Dull yellowish brown (10YR 5/3m) and greyish yellow brown (10YR 5/2m, 6/2d) pebbly loam, massive, very friable consistence, plastic and slightly sticky.
C3box	54-61	Dull yellowish brown (10YR 5/4m) and dull brown (7.5YR 5/3d) pebbly sandy loam, massive, firm consistence, slightly plastic and slightly sticky.
C4box	61-81	Brown (7.5YR 4/6m) and dull brown (7.5YR 5/3d) pebbly loam, massive, firm consistence, plastic and slightly sticky.
Cub	81 +	Dark greyish yellow (2.5YR 4/2m) and greyish brown (7.5YR 5/2d) loam, massive, very firm consistence, plastic and sticky.

a. Soil descriptions follow Birkeland (1984) and Soil Survey Staff (1975). Colors were taken from Oyama and Takehara (1970) in the moist (m) and dry (d) states. Consistence is given in the moist state.

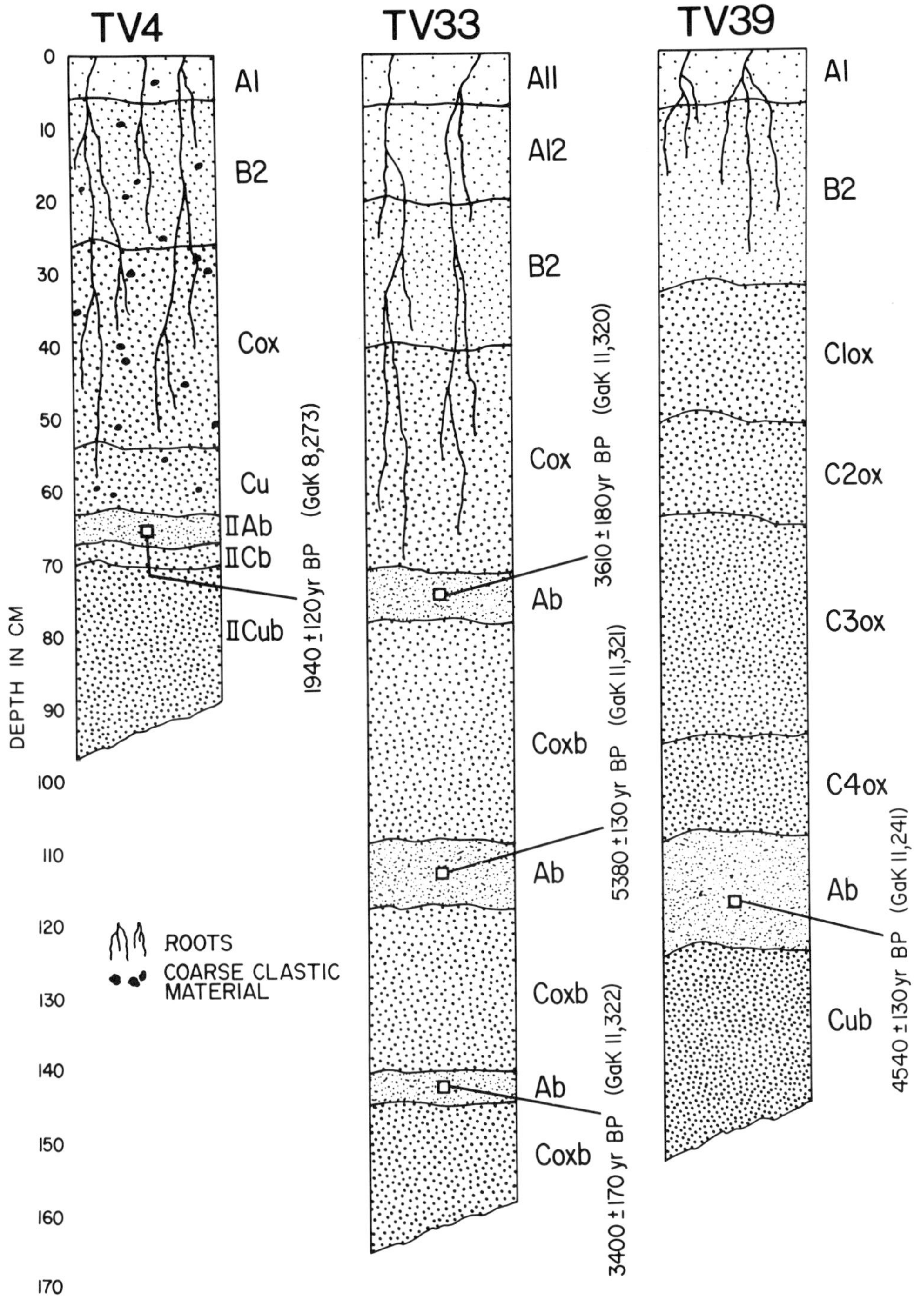

Figure 7.16: A, Sections in alluvial fans and outwash deposits in Teleki and Liki North valleys, Mount Kenya. Profiles TV4, TV33, and TV39.

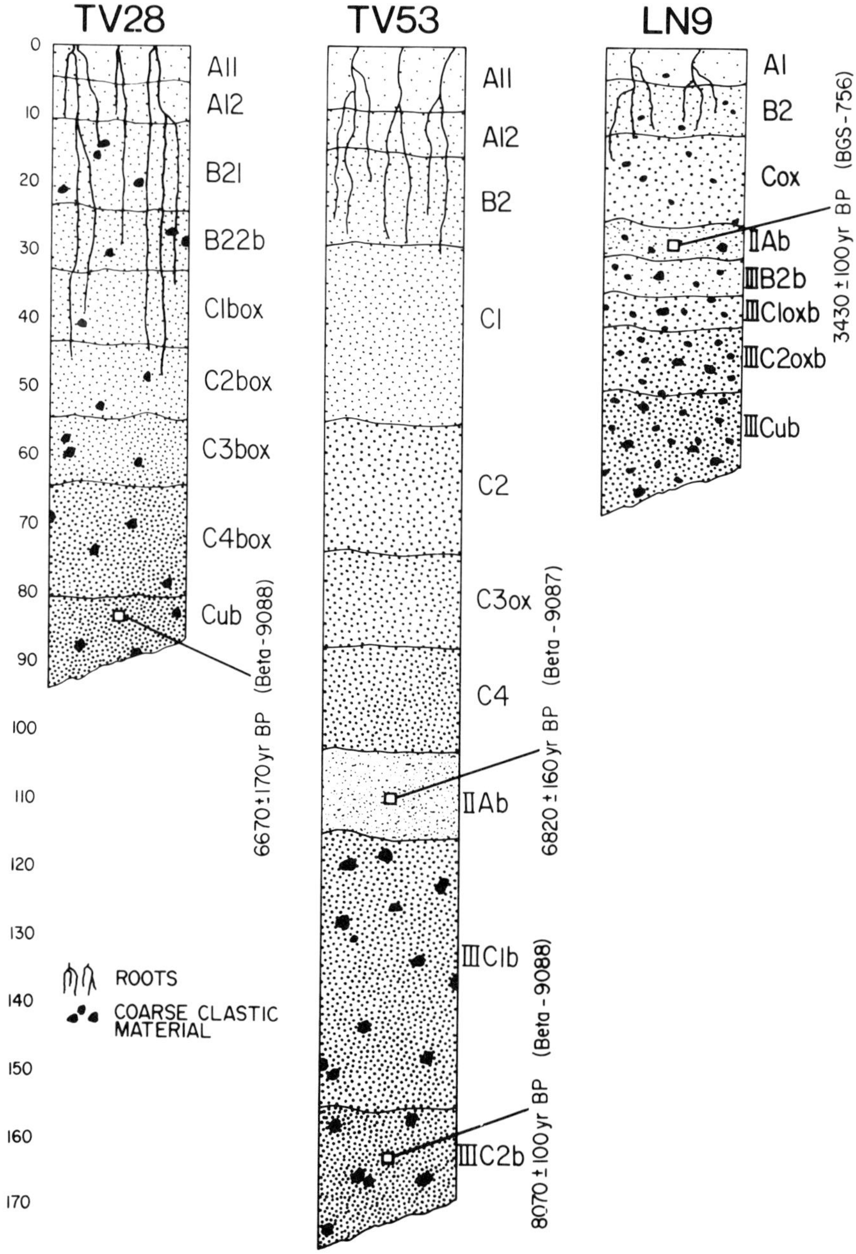

Figure 7.16: B, Sections in alluvial/colluvial fans and outwash deposits in Teleki and Liki North valleys, Mount Kenya: Profiles TV28, TV53, and LN9.

Table 7.5:

Soil Profile LN9

Horizon	Depth (cm)	Description[a]
01	1-0	Black (10YR 1/1m)
A1	0-5	Brownish black (10YR 2/2m, 2/3d, 3/2d) pebbly sandy clay loam, granular structure, firm consistence, plastic and nonsticky.
B2	5-13	Brownish black (10YR 2/3m) pebbly sandy clay, blocky structure, firm consistence, plastic and sticky.
Cox	13-23	Dark brown (10YR 3/3m) and dull yellowish brown (10YR 4/3d) pebbly sandy loam, massive, friable consistence, plastic and sticky.
IIAb	23-30	Black (10YR 2/1m) and greyish yellow brown (10YR 4/2d) pebbly loam, weak granular structure, firm consistence, plastic and very sticky. ^{14}C dated at 3430 ± 100 yrs BP (BGS-756).
IIIB2b	30-35	Dark brown (10YR 3/3 to 3/4m) and dull yellowish brown (10YR 4/3d) pebbly sandy clay loam, weak blocky structure, firm consistence, plastic and sticky.
IIICloxb	35-41	Brown (10YR 4/4m), and yellowish brown (10YR 5/6m) pebbly sandy loam, weak blocky structure, very friable to loose consistence, nonplastic and slightly sticky.
IIIC2oxb	41-51	Dull yellowish brown (10YR 4/3m, 5/4m), pebbly loamy sand, massive, loose consistence, nonplastic and slightly sticky.
IIICub	51-63	Dark greyish yellow (2.5YR 4/2m, 5/2m,d) pebbly loamy sand, massive, loose consistence, nonplastic and nonsticky.

a. For color and horizon designations see page 188.

however, is important also with respect to the thin B horizon in the ground soil when compared against the B horizons in profiles TV28, TV53, TV4, TV33 and TV39. This difference is attributed to its position on the mountain on the drier northwestern side where increased radiation and higher evapotranspiration tend to produce lower available moisture for infiltration into surface soils.

The profiles (Tables 7.6 and 7.7) in alluvial/colluvial fan sediment on the northwestern flank of the mountain – ONT1 and LN18 – are representative of weathering since the middle Holocene. Both profiles are weathered to depths of 75 and 78 cm with thinner B horizons and total depths when compared with profiles of similar age on the southwestern side (Figure 7.16). In addition to thinner profiles and B horizons, these soils have weaker grades of granular structure

Table 7.6:

Soil Profile ONT1

Horizon	Depth (cm)	Description[a]
A1	0-9	Black (10YR 2/1m) and brownish black (10YR 2/2d) loam, weak granular structure, friable consistence, slightly plastic and nonsticky.
B2	9-20	Dull yellowish brown (10YR 4/3m) and dull yellow orange (10YR 6/3d) sandy loam, very weak blocky structure, firm consistence, plastic and sticky.
C1ox	20-31	Dark brown (10YR 3/4m) and dull yellow orange (10YR 6/3d) sandy loam, massive structure, firm consistence, plastic and sticky.
C2ox	31-43	Brown (10YR 4/4m) and dull yellow orange (10YR 6/3d) loam, massive, firm consistence, plastic and sticky.
C3ox	43-60	Brown (10YR 4/4m) and dull yellow orange (10YR 7/3d) sandy loam, massive, firm consistence, slightly plastic and nonsticky.
IIC4ox	60-75	Dull yellow brown (10YR 4/3m), and greyish yellow brown (10YR 6/2d) pebbly sandy loam, massive, loose consistence, nonplastic and nonsticky.
IICu	75+	Dark olive brown (2.5YR 3/3m) and dull yellow (2.5YR 6/4d) pebbly loamy sand, massive, loose consistence, nonplastic and nonsticky.

a. Soil descriptions follow Birkeland (1984) and Soil Survey Staff (1975). Colors were taken from Oyama and Takehara (1970) in the moist (m) and dry (d) states. Consistence is given in the moist state.

in the A horizons and blocky structure in the B horizons when compared with profiles of similar age on the southwestern side.

Towards the end of the Holocene there is substantial geomorphic evidence for cooler and drier conditions following the wetter and presumably warmer climatic optimum (4500-2000 yr BP; Coetzee, 1967). To some degree minor amounts of stream alluviation (profile TV4, Figure 7.16) and wind-blown deposits in the Mutonga Drainage (Mahaney & Spence, 1986; Mahaney, Vortisch and Spence, 1988) provide evidence useful in substantiating this paleoclimatic shift. An area of hummocky terrain around the Mutonga River to the northeast of Mugi Hill has been attributed to periglacial activity (Baker, 1967). More recently, Mahaney and Spence, 1986, have

Table 7.7:

Soil Profile LN18

Horizon	Depth (cm)	Description[a]
01	1-0	Black (10YR 2/1m)
A1	0-10	Brownwish black (10YR 2/3m) and dark brown (10YR 3/3d) sandy loam, weak granular structure, friable consistence, plastic and nonsticky.
B2	10-28	Dark brown (10YR 3/3m) and dull yellowish brown (10YR 4/3d) silt loam, massive structure, firm consistence, plastic and slightly sticky.
Cox	28-78	Brownish black (10YR 3/2m) and dull yellowish brown (10YR 4/3d) clay loam, massive, firm consistence, plastic and sticky.

a. Soil descriptions follow Birkeland (1984) and Soil Survey Staff (1975). Colors were taken from Oyama and Takehara (1970) in the moist (m) and dry (d) states. Consistence is given in the moist state.

argued for an aeolian origin for these deposits. These deposits have lunate shapes with backslopes oriented to the northeast and slip-faces of 30° oriented southwest. Analysis of a compound paleosol (BT1-Ab, Table 7.3) in a representative deposit (Table 7.8) showed a ground soil over a buried paleosol and gave a coarsening trend downward into sandy beds at a depth of 94 cm. The data given in Table 7.8 show a paleosolum with strong 7.5YR hues formed in aeolian sediments with a loam texture over sandy substrates considered to be dune sediments. This lower unit was buried approximately 1490 yr BP (Table 7.3) and covered with aeolian sediment of sand and silty-clay material. The 53 cm thick A horizon complex in the ground soil is typical of peaty surfaces in many deposits within the ericaceous and Afroalpine vegetation belts. X-ray analysis of the BT1 profile revealed a similar mineralogic composition of the $<2\mu m$ fraction in all horizons. Quartz and plagioclase feldspar dominate throughout with lesser amounts of magnetite, hematite and pyroxene. The data suggest local reworking of fine-grained material by aeolian processes. The pH profile is fairly uniform in paleosol BT1, ranging from 6.0 to 6.4, suggesting little if any downward movement of H^+ ions. Measurements of electrical conductivity in each horizon show that total salt content is low (<0.1) throughout the profile. Ratios of Ca/Mg (exchangeable ions) diminish with depth in the profile, indicating little downward movement and low leaching power. The data on soil chemistry for this profile will be discussed later in Chapter 9.

The Holocene stratigraphy is shown in Figure 7.19. It is based on the sequence of deposits and paleosols, ^{14}C dates for deposits and paleosols, pollen sequence (Coetzee, 1967), paleoclimate inferred from landform evidence, and relationship to the European chrono-

Table 7.8:

Soil Profile BT1

Horizon	Depth (cm)	Description[a]
A11	0-12	Brownish black (10YR 3/1m) (10YR 3/2m) (10YR 3/2d) loam, very friable consistence, nonplastic and sticky, and wavy, transitional boundary.
A12	12-43	Brownish black (10YR 3/2m) and dark brown (10YR 3/3d) loam, very friable consistence, nonplastic and nonsticky, and wavy, transitional boundary.
A13	45-53	Brownish black (10YR 2/2m) and dark brown (10YR 3/4d) loam, friable consistence, nonplastic and nonsticky, and clear, abrupt boundary.
C1ox	53-76	Dark brown (10YR 3/3m) brownish black (10YR 2/3m) and dull yellowish brown (10YR 4/3d) sandy loam, friable consistence, nonplastic and nonsticky, and clear abrupt boundary.
Ab	76-94	Brownish black (7.5YR 2/2m) and dark brown (10YR 3/4d) loam, friable consistence, nonplastic and nonsticky, and wavy, transitional boundary. Radiocarbon dated at 1490±80 yr BP (Beta-10820).
Cbox	94+	Brown (7.5YR 4/6m), and yellowish brown (10YR 5/6d) sand, loose consistence, nonplastic and nonsticky.

a. Soil descriptions follow Birkeland (1984) and Soil Survey Staff (1975). Colors were taken from Oyama and Takehara (1970) in the moist (m) and dry (d) states. Consistence is given in the moist state.

stratigraphy. Very little information is available for the period 10,000 to 8000 yr BP but presumably surfaces remained stable under dry and relatively cool paleoclimate. We have no information on glacier positions during this period but presumably they were close to present limits or somewhat smaller than today.

Sometime after 8000 radiocarbon years BP, increased precipitation and warmer temperatures combined to produce unstable slopes that resulted in the formation of alluvial and colluvial fans which lasted until *c.* 4000 yr BP. Deposits constructed during this interval are known collectively as the Darwin Formation (Mahaney, 1985) and are well documented in nearly every valley on the mountain. In all the Darwin Formation is probably equivalent to the Atlantic Period

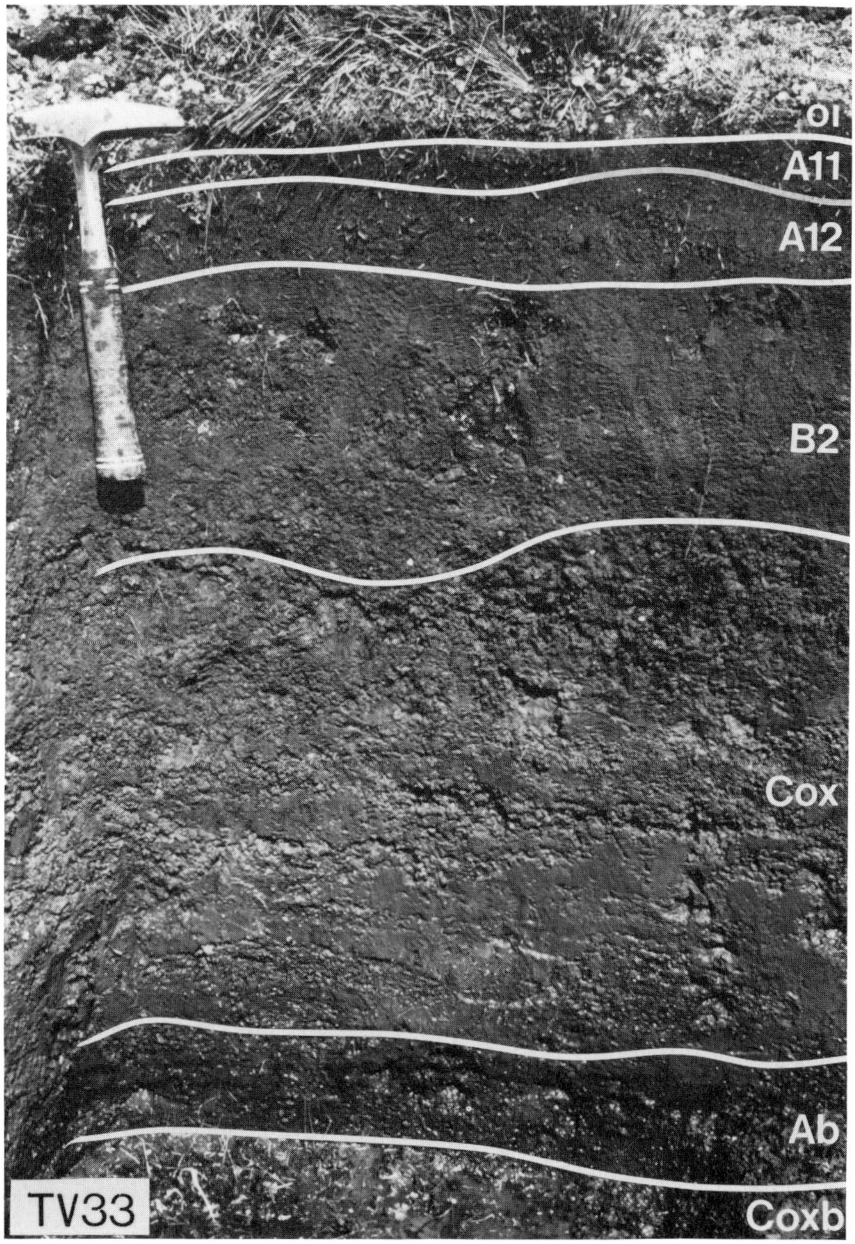

Figure 7.17: A, profile TV33 of middle-Holocene age in upper Teleki Valley.

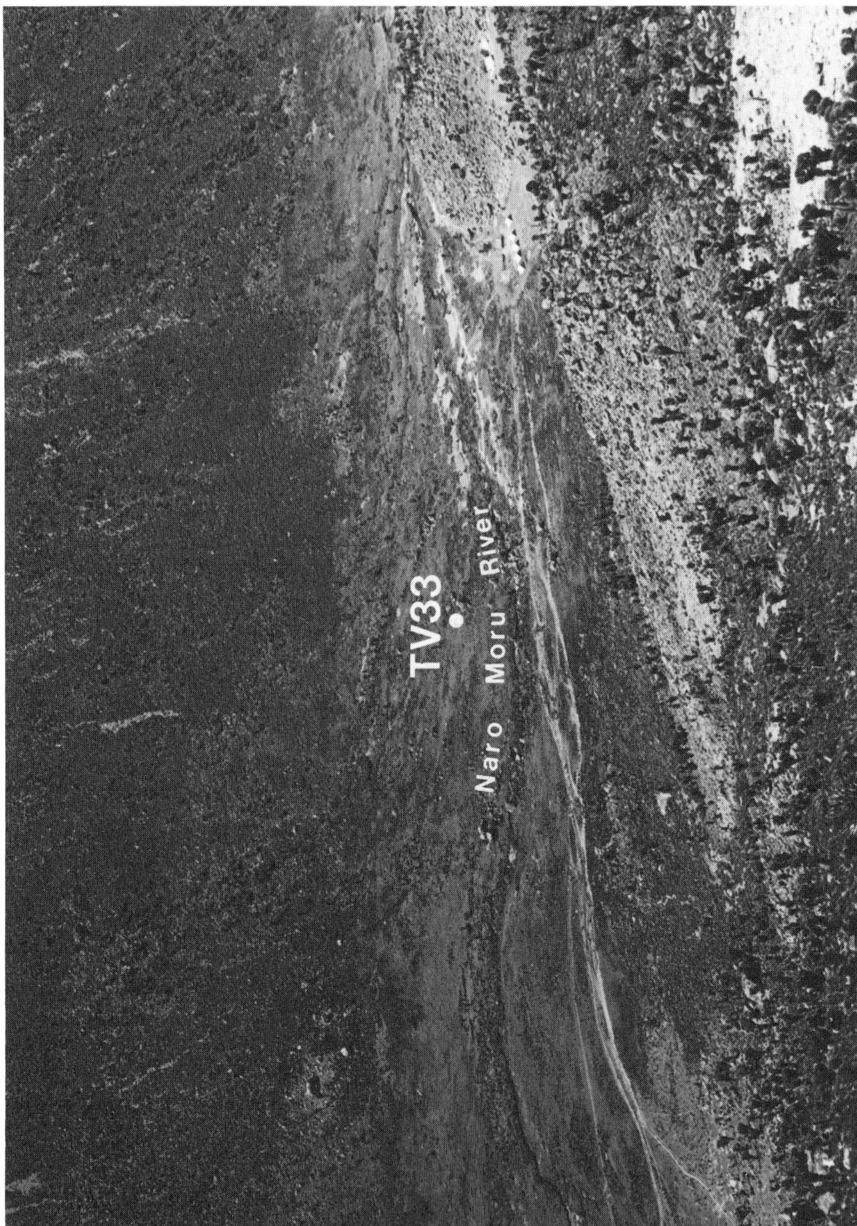

Figure 7.17: B, Profile TV33; location in alluvial fan.

Figure 7.18: Profile TV4 in fine-grained alluvium (unit II) and pebbly sandy outwash (unit I).

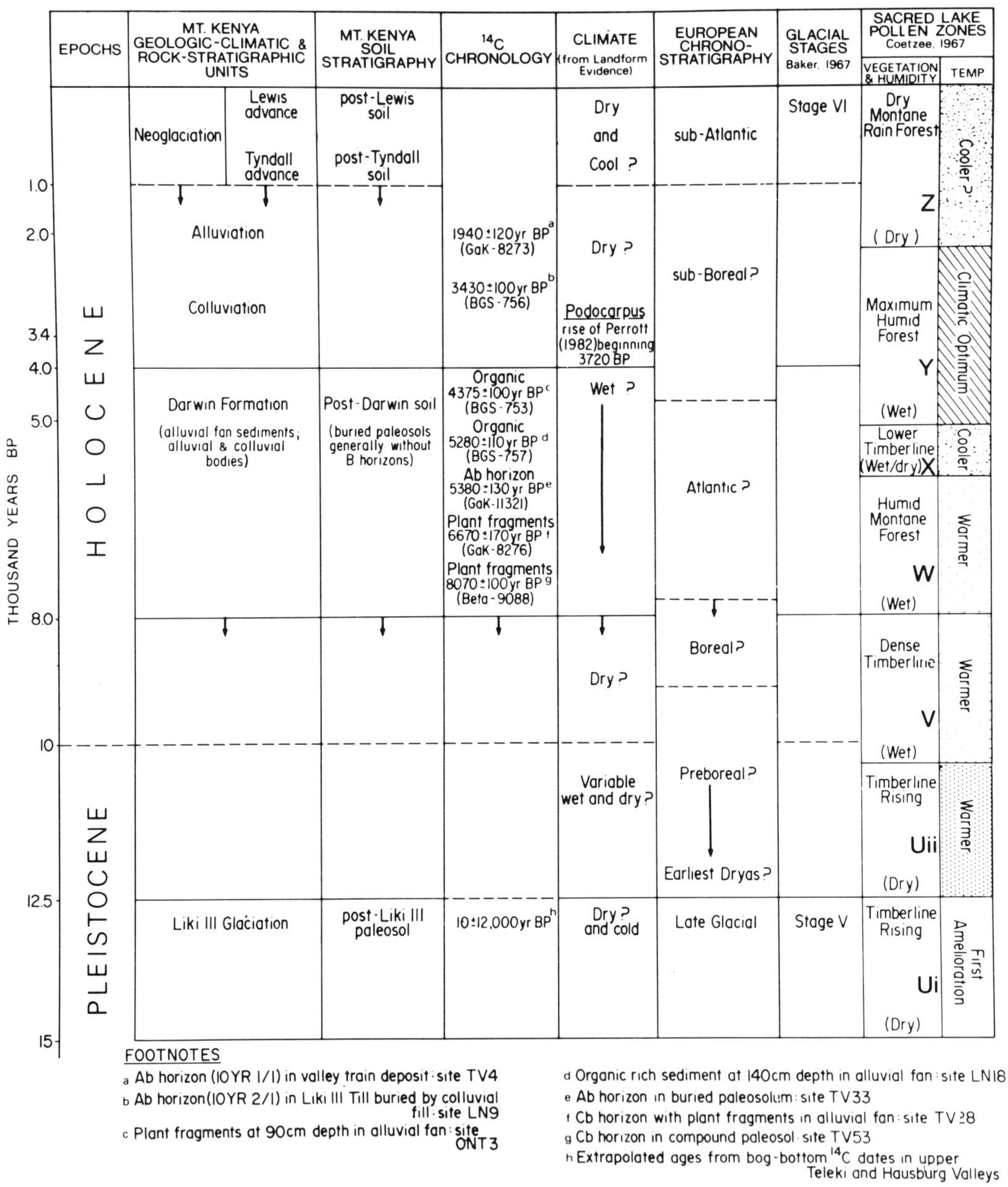

Figure 7.19: Stratigraphy showing late glacial and Holocene deposits, ^{14}C dates, pollen record, paleoclimate inferred from landform evidence and relationship to European chronostratigraphy.

in the European Chronology (Coetzee, 1967) being preceded by the Boreal and followed by the sub-Boreal.

During the ensuing 4000 yr paleoclimate tended to become dryer and cooler. Geomorphic evidence of decreased alluvial/colluvial fan construction, increased stream alluviation, followed by dune formation (northeast side) and ultimately by Neoglaciation (Chapter 8) tends to support the pollen sequence. The first 2500 yr of this paleoclimate is probably equivalent with the sub-Boreal in Europe; the last few hundred to a thousand yr is probably broadly coeval with the Atlantic Period (Coetzee, 1967). As discussed in the following section, the *Podocarpus* increase in bog core records starting ~3700 yr BP is one of the main paleoclimatic events of the later Holocene. In effect it marks the end of the middle Holocene and the start of drier conditions.

7.10 Pollen Sequence

The pollen stratigraphy shown on Figure 7.19, taken from Coetzee (1967) is based on pollen analyzed from Sacred Lake at 2200 m. on the northeast side of the mountain. The sequence starts with the first warming and approaching timberline beginnning ~14,000 yr BP or earlier. Paleoclimate was warming but dry during this period which includes the late glacial recessional stillstand. After 12,500 yr the timberline started to develop at Sacred Lake, and conditions became warmer and remained dry. Just before 10,000 yr BP the pollen record shows a dense timberline at Sacred Lake along with a wetter and warmer paleoclimate (pollen zone V). With the onset of humid montane forest after 8000 yr BP, paleoclimate became even wetter and warmer (pollen zone W). A somewhat cooler (although more humid) episode from ~4500 to ~5000 yr BP (pollen zone X) preceded the climatic optimum (pollen zone Y), which lasted to about 2000 yr BP. During the climatic optimum the maximum extent of humid montane rain forest developed at Sacred Lake and paleoclimate was wet. A cooler and drier paleoclimate after ~2000 yr BP gave rise to a dry montane rain forest at Sacred Lake.

At site HOB13 (4090 m) in a mire below Thompson Tarn on the southeast side of the mountain, a recovered radiocabon-dated core (Figure 7.20), gives a sequence of sandy, silty and clayey layers dated from 8125±600 yr BP (BGS-796) to the present. Sedimentation rates extrapolated from these data gave values of .06 mm/yr for the 78-105 cm level. When compared with sedimentation rates in Mahaney (1987a) the lower and upper units are out-of-phase with Teleki Valley and Hausberg Valley. In general, in other drainages on the mountain sedimentation rates are between the two extremes calculated for HOB13.

Analysis of the pollen spectrum shown in Figure 7.21 (personal communication, Emile Roche, 1986) showed some interesting trends throughout the postglacial. The relative decline in *Alchemilla* and *Artemisia* in early postglacial time, diminished count or disappearance in the middle Holocene and reappearance at the end follows the trend from colder to warmer to cooler as well as peaked moisture in the climatic optimum. *Podocarpus*, one of the main pollen indicators on the mountain slowly decreases from ~8000 yr BP to a peak following the climatic optimum. It is important to note the *Podocarpus* rise begins at 3540±150 yr BP (Figure 7.22). *Hypericum* and *Ranunculus* which are responsive to increasing moisture also increase up-core towards the climatic optimum while *Gramineae* decreases. While the pollen data indicate changes in species composition at 4090 m on the southeastern side of the mountain, there is absolutely no basis for inferring ice advances from the data.

7.11 Conclusions

This chapter summarizes available data for later Pleistocene and Holocene deposits on Mount Kenya. From 1890 to the present a number of workers have attempted to reconstruct Holocene glacial and paleoclimatic records with varying degrees of success. Most of the mountain is

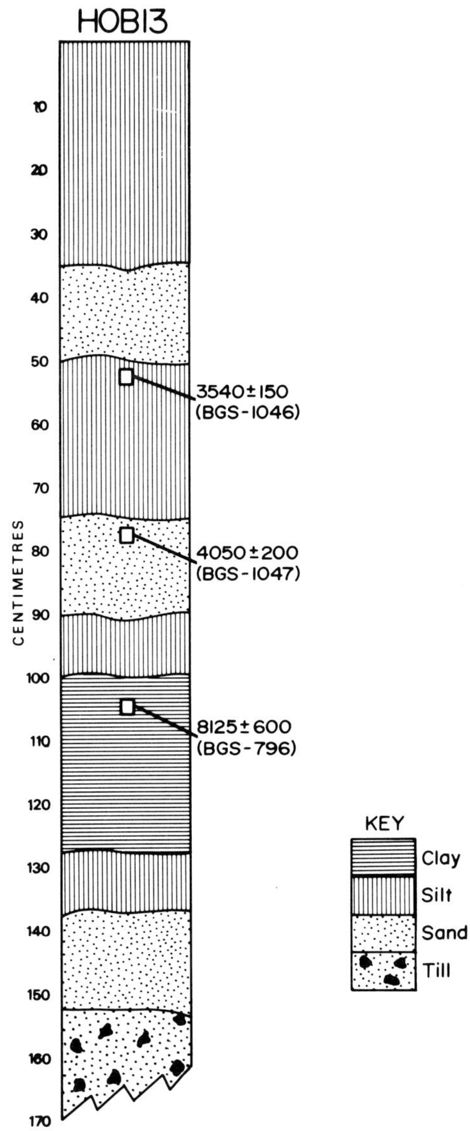

Figure 7.20: Bog core HOB13 from Hobley Valley

surfaced with later Pleistocene and late Holocene glacial sediments which are poorly dated; in some cases different stratigraphic units are recognized only by their relative topographic positions. Middle Holocene deposits that collectively form the Darwin Formation are soundly dated by radiocarbon.

Both late Holocene and later Pleistocene glacier fluctuations appear to have been broadly synchronous among the different drainages on Mount Kenya. Recessional moraines on Mount

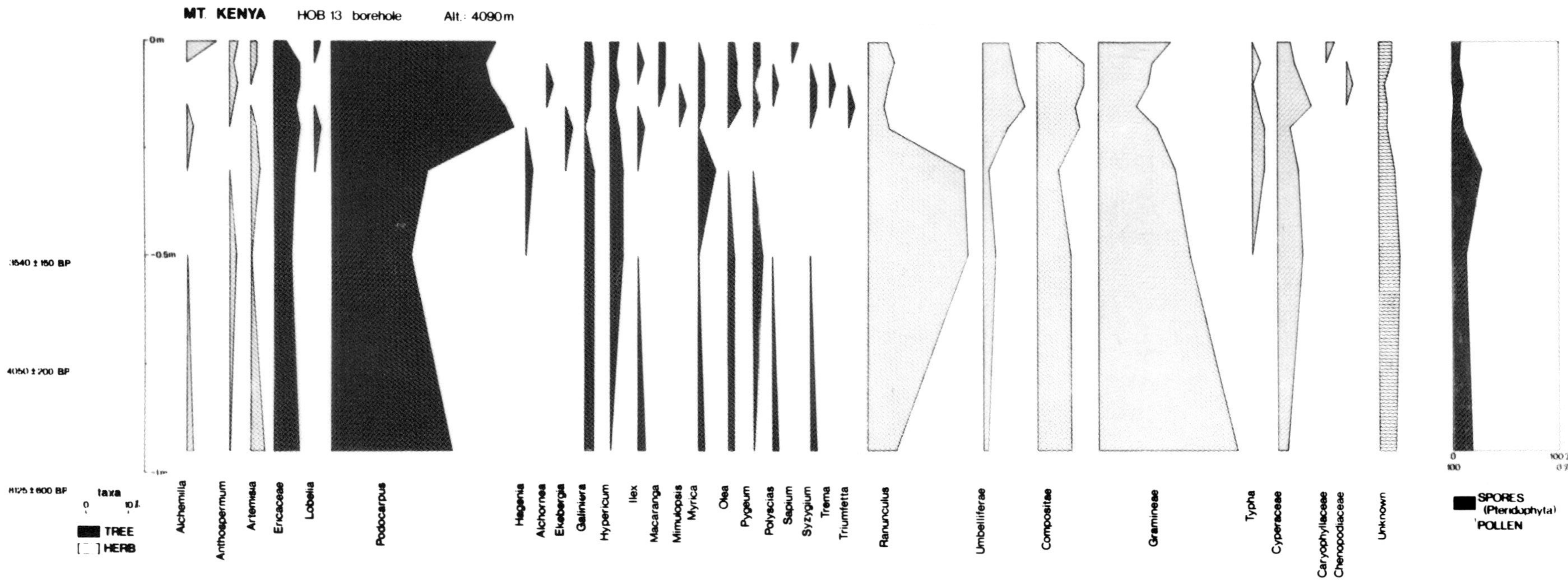

Figure 7.21: Pollen diagram from HOB13 core.

Kenya at ~4000 m in most drainages yield minimum ^{14}C ages of ~10,000 to 12,000 yr BP. Late Holocene moraines are either "Little Ice Age" equivalents, or somewhat older deposits emplaced during the Tyndall advance. The Neoglacial sequence on Mount Kenya is younger than 1000 ^{14}C yr BP, while on other East African mountains it may have begun *circa* 2000 yr BP. The major part of Holocene time appears to have been warmer and wetter than today with glaciers disappearing altogether, or distributed much as they are today.

CHAPTER 8

NEOGLACIATION

8.1 Introduction

Eleven cirque glaciers in the Central Peaks area (Figure 8.1) exist in an environment that is marginally within the present-day snowline (*c.* 4650 m). These cirque glaciers (Figures 8.2, 8.3 and 8.4) are remnants of much larger Pleistocene valley glaciers that flowed from a small ice cap completely covering Lenana (4985 m) (Figure 8.5), and reaching to ~3000 m in valleys on the western and eastern flanks of the mountain. Neoglacial climatic history lies recorded in cirques on all sides of the mountain. Because it is difficult to directly radiocarbon date these deposits, very little is known about the times at which Neoglacial fluctuations occurred. Using ^{14}C dates for two buried soils in outwash and debris-flow deposits, however, it proved possible to construct a lichen growth curve (Mahaney and Spence, 1989; Spence and Mahaney, 1988a) which allows dating of Neoglacial deposits on the southwestern flank of the mountain. Moreover, weathering rind data proved useful in differentiating deposits in the Neoglacial succession.

8.2 Previous Lichenometric Work

Roland Beschel pioneered the use of lichenometry in dating late Holocene deposits. He summarized the principles, methods, and applications of lichenometry in several papers (1950, 1955, 1957). Despite many objections to the method (Stork, 1963; Jochimsen, 1973) the assumption is that the maximum diameter of the largest lichen thallus on a surface is proportional to post-depositional time (or time since colonization) (Beschel, 1961; Porter, 1981; Mahaney and Spence, 1984, 1985, 1989; Innes 1985a; Mahaney, 1987a). Because lichen growth is thought to be individualistic even under nonvarying climatic conditions, some workers (Stork, 1963) argue that lichen growth is too erratic to provide accurate dates. Growth rates are known to respond to microenvironmental changes (Benedict, 1967) and long term (*circa* 1000 yr) fluctuations in climate (Currey, 1969; Benedict, 1985). The microenvironmental factors and others, including lithology (Mahaney, 1973; Porter, 1981) and substrate stability and moisture (Mahaney and Spence, 1984; Innes, 1985b, 1985c) may influence the maximum diameter of the largest (and presumably the oldest) lichen thallus growing on a deposit surface. The overall effect of these microenvironmental/ climate factors upon age determinations can be reduced by: choice of lichen species, employment of precision in measuring lichen diameters, and by close inspection of ecological factors in the field area (Spence and Mahaney, 1988a).

Traditionally lichenometry involved the use of growth curves in areas far removed (100 km) from their original place of employment (*e.g.* Kiver, 1974). In many alpine areas it proves difficult to find radiocarbon-datable materials, so that lichen size may only be used as a relative dating method to differentiate deposits (Kiver, 1974; Birkeland, 1973; Miller, 1973; Mahaney, 1974, 1978, 1984b; Mahaney and Spence, 1984; Mahaney *et al.*, 1984a, 1984b). On Mount Kenya, lichen size and percent cover have been used as a *relative* dating method for the last decade (Mahaney, 1979a, 1982, 1984a, 1985, 1987b). With the development of a growth curve it proved possible for the first time (Mahaney and Spence, 1989) to provide absolute ages for substrates in the Tyndall, Darwin and Lewis forelands. This greatly assisted in deposit correlation with other cirques.

8.3 Selection of Species

One fruticose, three crustose and four foliose lichen taxa were considered for lichen dating when this study began in 1976. The two foliose species *Umbilicaria africana* (Jatta) and

Figure 8.1: The Central Peaks area showing the Diamond, Darwin and Lewis glaciers from Teleki Valley.

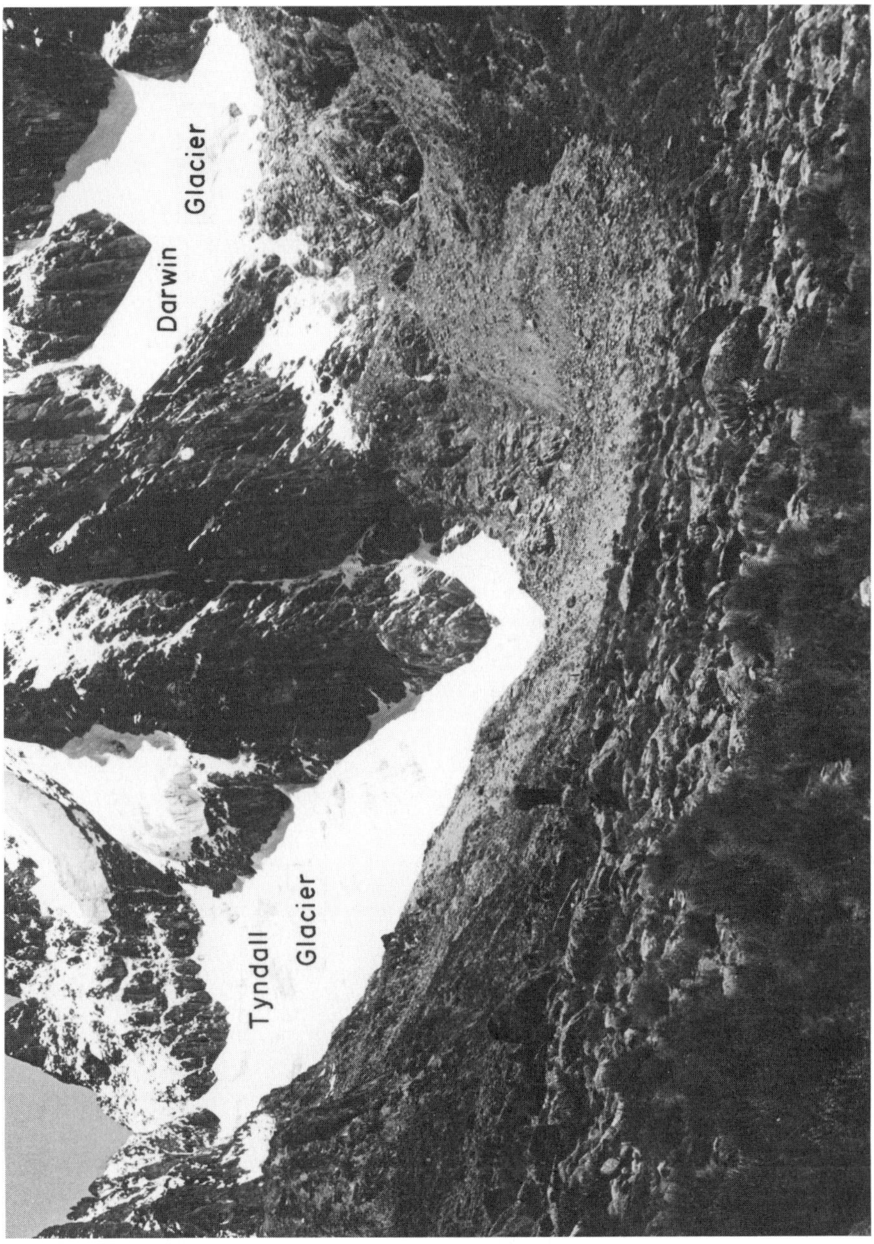

Figure 8.2: Forelands in front of the Tyndall and Darwin glaciers.

Figure 8.3: Foreland in front of the Lewis Glacier in 1983. Lewis Tarn and the proximal slopes of the Lewis Moraine (bottom right) give some indication of the degree of ablation over the last few decades.

Figure 8.4: Foreland in front of the Krapf Glacier on the northern flank of Mount Kenya.

Figure 8.5: Point Lenana and the left lateral moraines of the Lewis Glacier (photography in 1983).

U. umbilicariodes (B. Stein) (Figure 8.6) (Krog & Swinscow, 1986) were eliminated because they are slow colonizers and do not tolerate substate instability. Foliose species including *Parmelia* and *Xanthoria* are also slow colonizers, and the latter taxa may be fertilized by birds (Spence and Mahaney, 1988a). The *Parmelia* species may be useful for differentiating older from younger Neoglacial deposits, because they are nearly absent on Little Ice Age surfaces. Fruticose species of *Usnea* (Figure 8.7) were eliminated because they are not found on the younger deposits dating from the Lewis advance which occurred during the classical Little Ice Age (Mahaney, 1982b, 1985).

The most suitable lichens are members of the crustose group and include *Rhizocarpon* section *Rhizocarpon,* as well as species of *Lecidea* and *Buellia.* The lichen keys available today (Dodge, 1953, 1957) make it impossible to identify species of *Buellia* (Figures 8.8 and 8.9) and *Lecidea* in the tropical mountains. Because of these problems it was soon apparent that *Rhizocarpon* section *Rhizocarpon* (Figures 8.10 and 8.11) was the only reliable taxon useful in constructing a lichen growth curve.

The taxonomy of the individual species of *Rhizocarpon* is still incomplete (Mahaney and Spence, 1989; Spence and Mahaney, 1988a) despite considerable taxonomic research carried out since the pioneering work of Runemark (1956) thirty years ago (summarized in Innes, 1985c). The identification of *Rhizocarpon* to individual species involves analysis of spores and use of staining techniques, that are not practical in lichenometrical studies of this type. Taxa of *Rhizocarpon* (formerly *R. geographicum s.l.*) are common on Mount Kenya. Because of species identification problems (Innes, 1985c), *R. geographicum s.l.* becomes *Rhizocarpon* section *Rhizocarpon*. To maintain consistent methods in the field only *maximum* diameters of each thalli were measured.

Rhizocarpon section *Rhizocarpon* is abundantly and widely distributed in the high Afroalpine zone of Mount Kenya. Both a wide spatial distribution, a long-life span of 1000 yr or more, and consistent size-time relationships, make it the most suitable taxon for absolute dating of surface deposits.

The selected study sites in the Tyndall and Lewis cirques (Figures 8.1 and 8.2) are situated on the southwestern slope of Mount Kenya. The cirque floors lie at the upper end of Teleki Valley at about 4400 m (Tyndall) and 4325 m (Lewis). The climate, geology, geomorphology and glacial history have been summarized by Baker (1967), and Mahaney (1979a, 1982a, 1985, 1989a). The cirques lie within the upper Afroalpine and lower nival vegetation zones. The vegetation of the area and the biotic succession in front of the Lewis and Tyndall glaciers have been summarized by Coe (1967), Hedberg (1964), and Spence (1989); historically-dated moraines are shown in Figure 8.12.

A walking transect, established on June 11, 1984, from the ice front of the Tyndall Glacier towards Tyndall Tarn was established and designed to measure the largest diameters of lichen thalli on each aspect of individual boulders. Distance from the ice front, aspect, and substrate inclination were measured for each thallus. The objective was to find the largest thalli along this transect. Because boulders of different lithology exist on the moraines, only porphyritic phonolite clasts were sampled. The transect was terminated near Tyndall Tarn (Figure 8.12) at 160 m from the ice front. Four additional sampling points were established below this point. These were located at the positions of the ice front of 1926 (Figure 8.12); the crests of the first and second moraines of Lewis age below the tarn (Figure 8.12), and the crest of the Tyndall-age moraine 100 m to the south. The details of this sampling scheme, additional lichen data, and the age determination of distances from the ice front, can be found in Spence (1989), Mahaney and Spence (1989) and Spence and Mahaney (1988a), in Table 8.1, and in Figure 8.13.

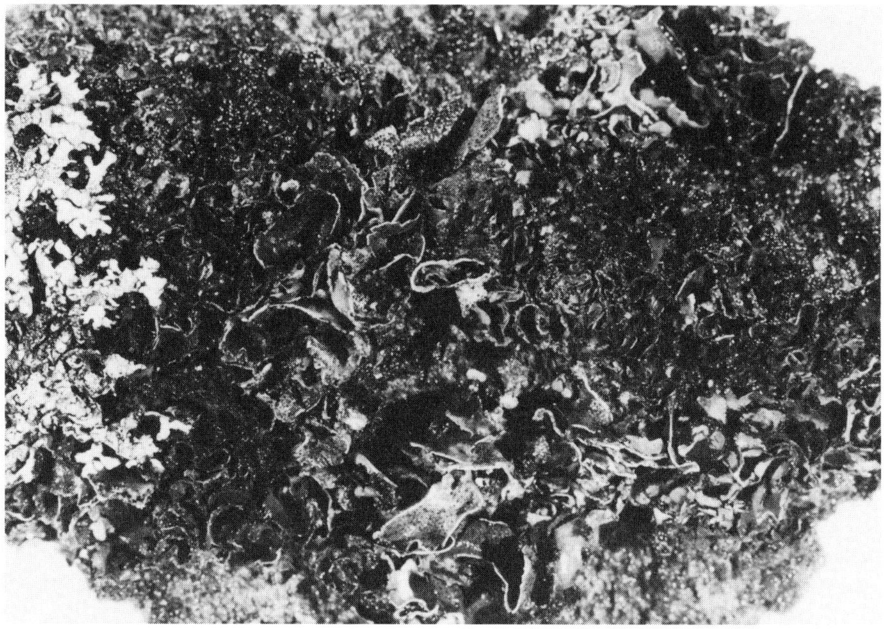

Figure 8.6: The foliose lichen *Umbilicaria haumania* (Beschel, 1961) in Tyndall cirque on deposits of the Tyndall advance.

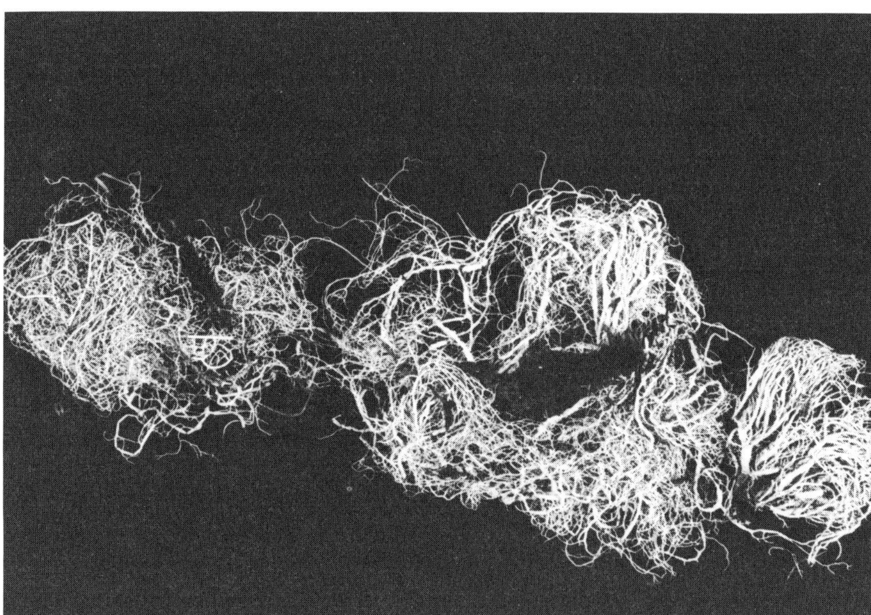

Figure 8.7: The Fruticose lichen *Usnea* in Tyndall Cirque. This lichen, common on early Neoglacial and Liki III substrates, is not present on younger Little Ice Age surfaces.

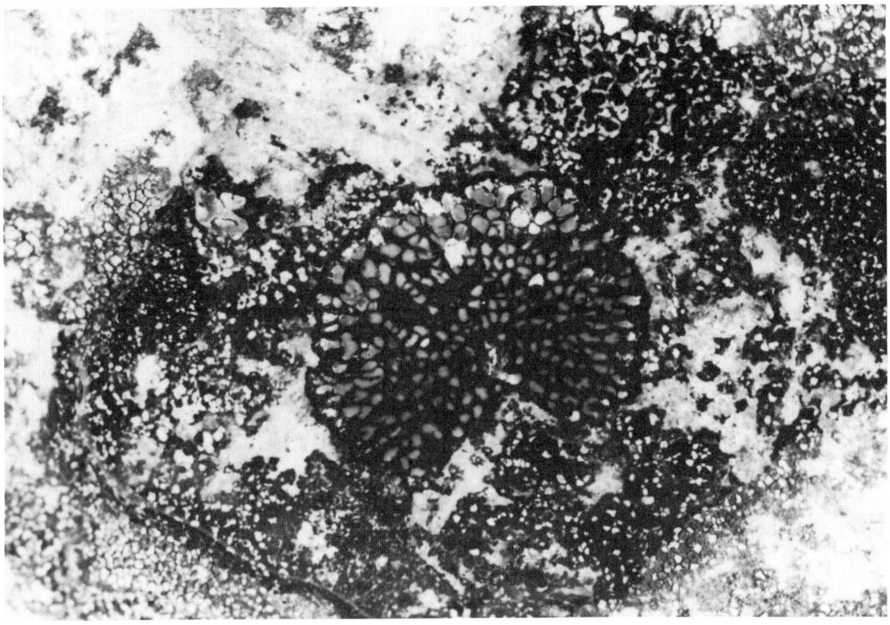

Figure 8.8: *Buellia* (center) on Lewis I moraine surface in Lewis Cirque.

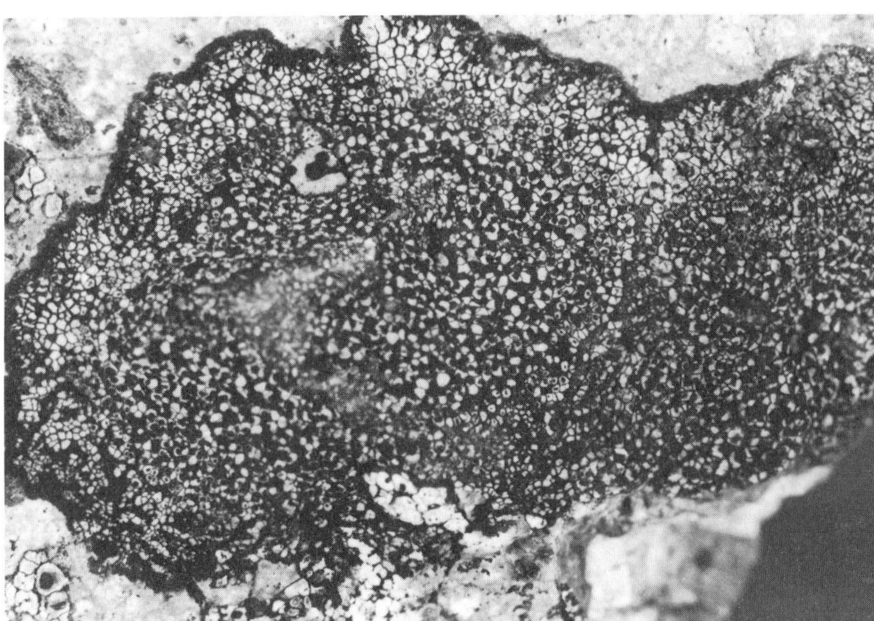

Figure 8.9: *Buellia* in Tyndall Cirque.

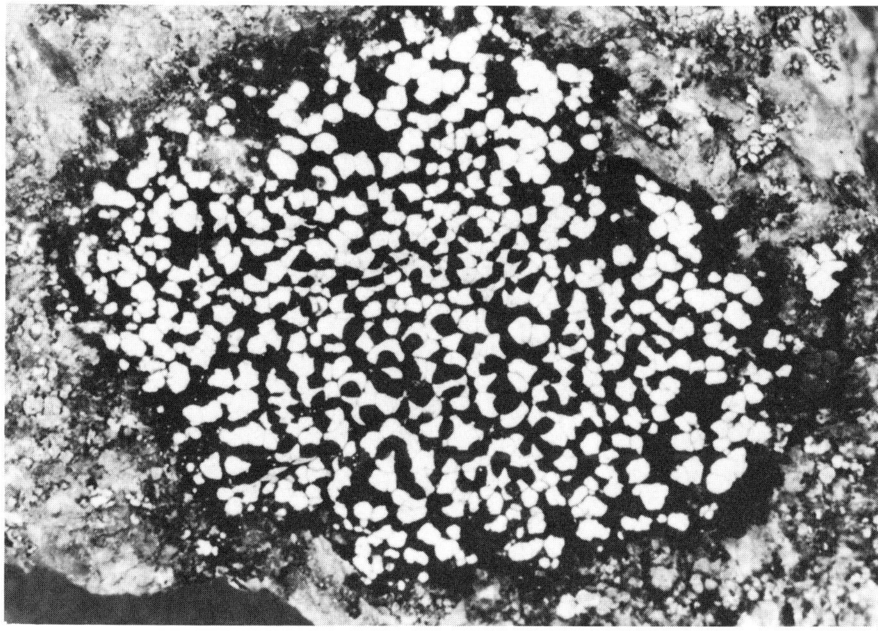

Figure 8.10: *Rhizocarpon* section *Rhizocarpon* on Neoglacial substrate in Lewis Cirque.

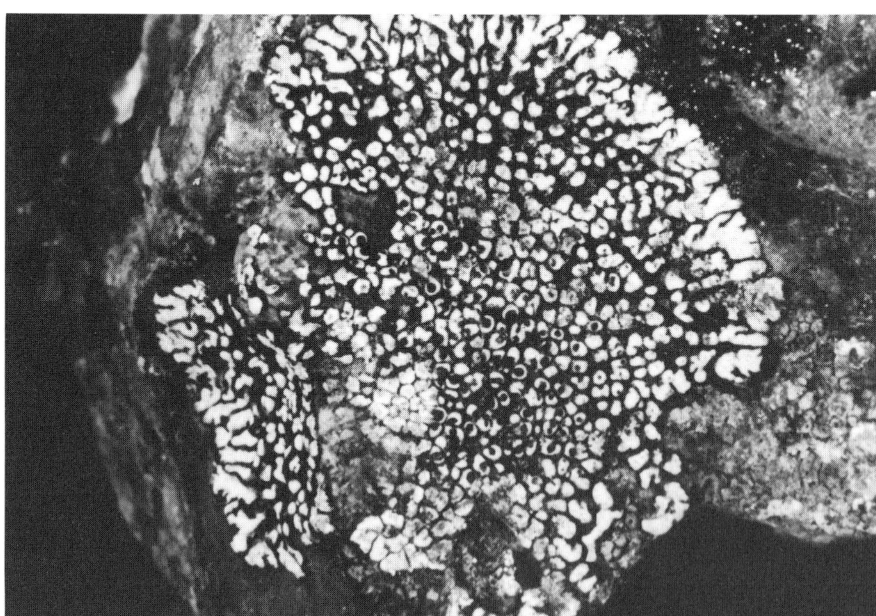

Figure 8.11: *R.* section *Rhizocarpon* on Neoglacial substrate in Tyndall Cirque.

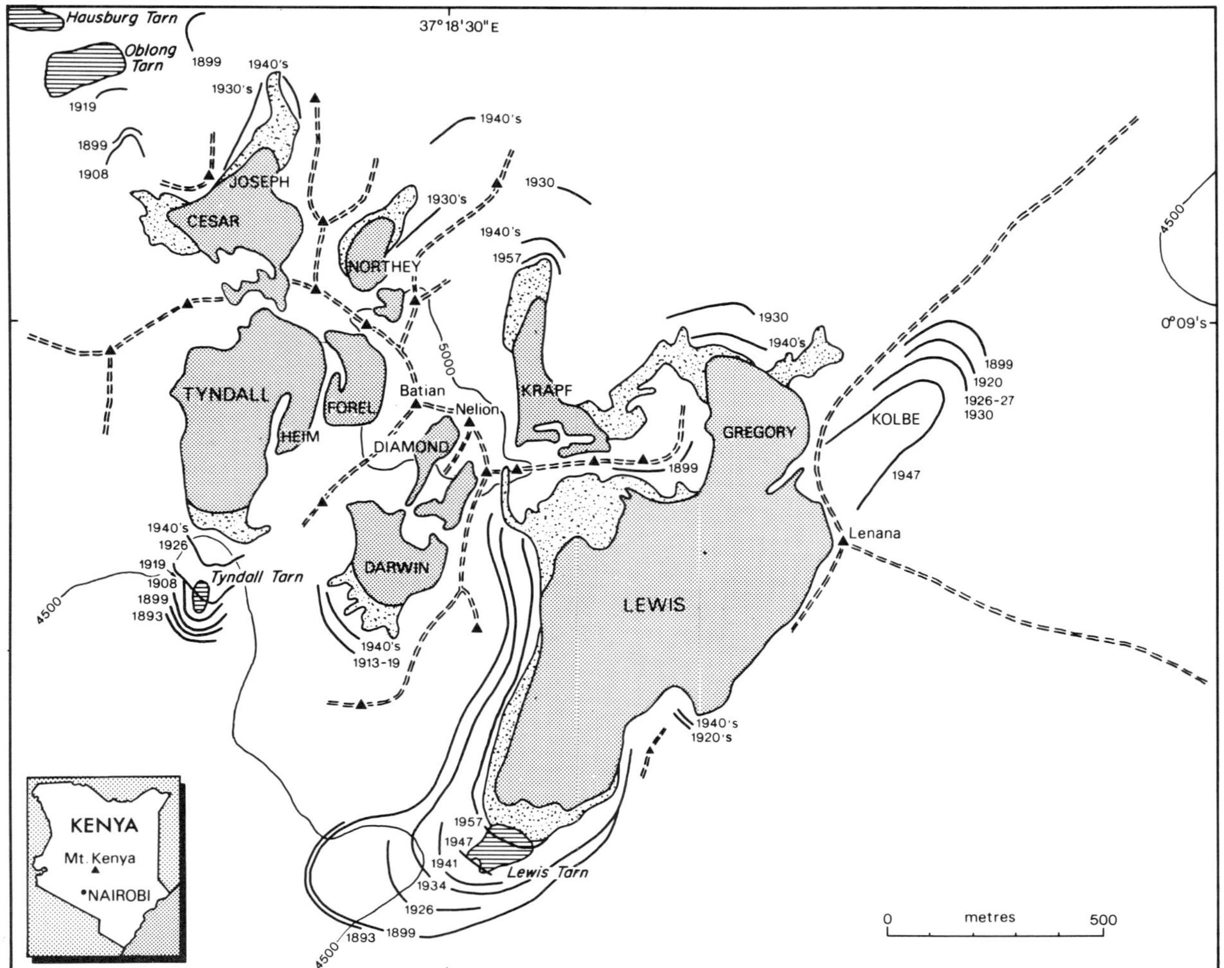

Figure 8.12: Map of historically dated ice positions from the last century (in part, from Hastenrath, 1984).

Table 8.1:

Lichen characteristics for late-Holocene deposits in the Central Peaks area, Mount Kenya Afroalpine area.

Site[a]	Age	*Rhizocarpon* section *Rhizocarpon*	*Lecidea*	*Buellia*	% Cover
LG13	Lewis IV	10	35	17	nil-5
LG14		12	15	nil	nil-2
LG12		25	42	20	1-<10
TT6		5	nil	nil	nil
LG10	Lewis III	24	51	24	1-20
TT12		20	54	50	<1-20
TT13		18	47	27	1-10
MK5		41	70	40	<5
HA8		40	61	24	<5-10
LG11	Lewis II	23	54	33	nil-20
TT1		23	82	35	10-35
TT5		12	34	28	5-20
TT7		23	82	35	<5-15
TT8		33	50	55	1-20
HA9		29	79	52	1-20
MK15		15	45	24	<1-3
MK17		12	37	20	<1-2
MK18		40	43	19	<1
HOB5		19	32	21	nil-10
GOR27		15	21	16	nil <5
GOR28		18	20	21	nil <2
LG1	Lewis I	30	60	47	3-50
LG6		34	94	56	5-10
LG7		32	112	38	3-15
LG8		18	64	39	<10
LG9		34	94	56	5-10
TT4		40	90	65	10-40
TT10		35	109	67	5-50
TT11		24	48	35	<5-30
HA10		45	112	75	5-15
HA12		21	66	49	3-25
HA13		20	54	40	1-10
MK4		82	51	75	25-40
MK16		37	63	24	<1-25
MK19		41	52	26	1-15
DAR1		50	71	33	3-25
GOR26		19	49	13	<1-20
GOR29		21	62	16	nil-10
HOB3		15	31	8	nil-10
HOB4		18	39	18	nil-15

a. Sites are located on Plate 5.
TT = Tyndall Tarn; LG = Lewis Glacier; HA = Hausberg cirque; MK = Mackinder Valley; HOB = Hobley Valley; GOR = Gorges Valley; DAR = Darwin cirque.

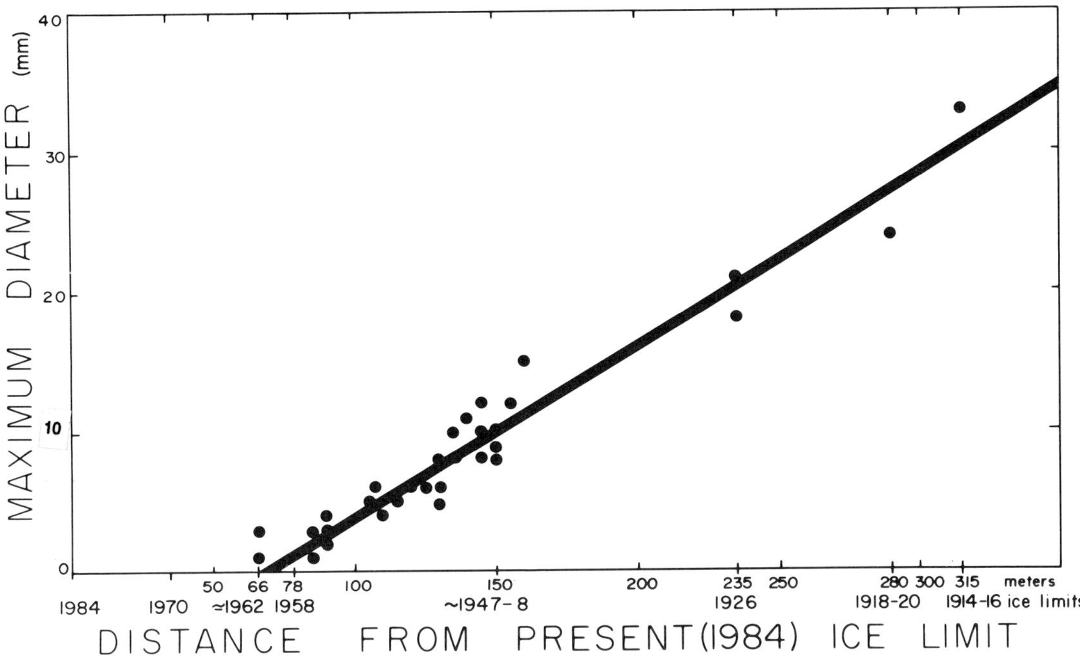

Figure 8.13: Maximum diameters of *R.* section *Rhizocarpon* at different distances from the 1984 front of the Tyndall Glacier (adapted from Mahaney and Spence, 1989).

In Lewis Cirque maximum diameters were measured on moraine, outwash, and debris-flow surfaces. Because some of these surfaces could be radiocarbon dated it proved possible to construct a growth-rate curve for the last *c*. 1000 yr BP.

8.4 Environmental Factors

Various environmental factors known to affect the growth of *Rhizocarpon* section *Rhizocarpon* in arctic and alpine areas were also examined on Mount Kenya. Rocks exposed at deposit surfaces range from syenite and porphyritic phonolite to trachytic types. While chemical compositions vary somewhat between these different rocks, *Rhizocarpon* section *Rhizocarpon* showed no preference for a particular rock chemistry, which is consistent with the findings of other workers (*cf.*, Benedict, 1967; Mahaney, 1973). The variation in rock texture produced a slight preference for faster colonization of porphyritic phonolite and syenite over trachytic rocks, which resulted in the elimination of the latter type from the sampling scheme. As pointed out by Benedict (1967), it may be easier for *Rhizocarpon* section *Rhizocarpon* to colonize coarser- textured stones, but unlike the central Rocky Mountains, the time frame here is short < 1000 yr BP. Most clasts (even porphyries) have very fresh and smooth surfaces (unless preweathered), and hence surface textural differences as they control moisture retention and crevice development are minimal.

Wind-blown ice, snow, firn, and fine particles may produce abrasion effects, which play a role in reducing lichen growth on particular aspects of individual boulders. Approximately 300 boulders were studied to determine if lichen growth varied at individual sites (Spence and Mahaney, 1988a). At over fifty percent of the sites studied, lichen growth dropped off by up to 25 percent on the downvalley side of boulders. This may indicate that upvalley winds, which sometimes reach 50 m/sec. are sufficient to blast lichen-encrusted rock surfaces with small particles thereby reducing growth to some degree. On the whole, however, aspect does not affect colonization and early growth (~ 70 yr) of thalli in front of the Tyndall Glacier (Spence and Mahaney, 1988a).

The elevation (4500 m) of the sites studied are close to the 500 mb level and hence receive considerable ultraviolet radiation. Increasing radiation on the northern side of the mountain might retard lichen growth to a considerable degree, while Lewis and Tyndall cirques (Figures 8.1 and 8.2), with higher cloud cover and somewhat more moist microenvironments, might show increased lichen growth. Although temperature is known to have profound effects on lichen growth (Beschel, 1961) in other areas (especially the arctic) it is not possible to evaluate its importance here. All lichen measurements were made across vertical elevation gradients of *c.* 250 m which should give a mean temperature difference of *c.* 0.6 degrees C (Mahaney, 1987a). There is no *apparent* relationship between temperature and lichen growth across this gradient (Spence and Mahaney, 1988a).

Moisture variability may have a much more profound effect on lichen growth (Benedict, 1967; Mahaney, 1973; Mahaney and Spence, 1984; Innes, 1985c). Lichens are capable of absorbing water from the air directly or from precipitation, and as a result increased growth rates often occur from irrigation near waterfalls, wind spray near lakes, temporary water storage in depressions, water seepage from springs, and fluctuating water levels in streams. On Mount Kenya, anomalous lichen growth occurs only near streams, seepage from alluvial and colluvial fans, and from local waterfalls (*e.g.* below the Darwin Glacier, Figure 8.1). Since all measurements were taken on moraine crests, little variation in growth could be attributed to moisture.

Substrate stability (Beschel, 1961) and its effect on lichen growth is not an insurmountable problem on Mount Kenya. All samples were measured to the nearest mm on crests that are presumably creeping but not under the influence of excessive frost movement. The depth of penetration of the freezing plane in these surfaces is unknown, but it must be limited to a few centimeters which would produce only minimal substrate movement, probably as a result of needle ice. Permafrost is unknown on Mount Kenya.

8.5 Neoglacial Deposits

Neoglacial deposits in the Central Peaks are shown in Figure 8.14. On the southwestern side of the mountain the Lewis Glacier was fully extended to the Lewis I position (Figure 8.15) when Gregory first visited the area in 1893 (Gregory, 1900). Recession of the Lewis Glacier over the last ~ 94 years exposed a sequence of new deposits and bedrock surfaces for lichen colonization. Because Gregory (1900) failed to mention the Tyndall Glacier and its terminus, it is possible to only approximate its position at the turn of the century. Presumably, it was at the crest of the outer moraine (Lewis I shown on Plate 5). Subsequent studies by Coe (1967) show the historically-documented episodic recession of the Tyndall Glacier paralleling that of the Lewis Glacier.

Deposits of till in front of the younger suite of Neoglacial moraines in Tyndall and Lewis cirques are considered to have been emplaced during an advance prior to the classical Little Ice Age advances of the last three hundred years (Mahaney, 1979a, 1982b, 1984b, 1985, 1989a, 1989b). These older deposits, dating from the Tyndall advance, have somwhat deeper and better-developed soils (Mahaney, 1982b, 1984b), thicker weathering rinds, better developed vegetation (Spence, 1989) and contain boulders nearly completely covered with lichens, especially *Usnea*, which are essentially absent on Little Ice Age moraine surfaces. Considered together all the evidence points towards a greater age for the Tyndall deposits.

A succession of outwash and debris-flow deposits below the classical Little Ice Age deposits were studied for radiocarbon-datable material, soil development, weathering features and lichen

Figure 8.14: Left lateral moraine of Lewis I age above the older Tyndall-age deposits.

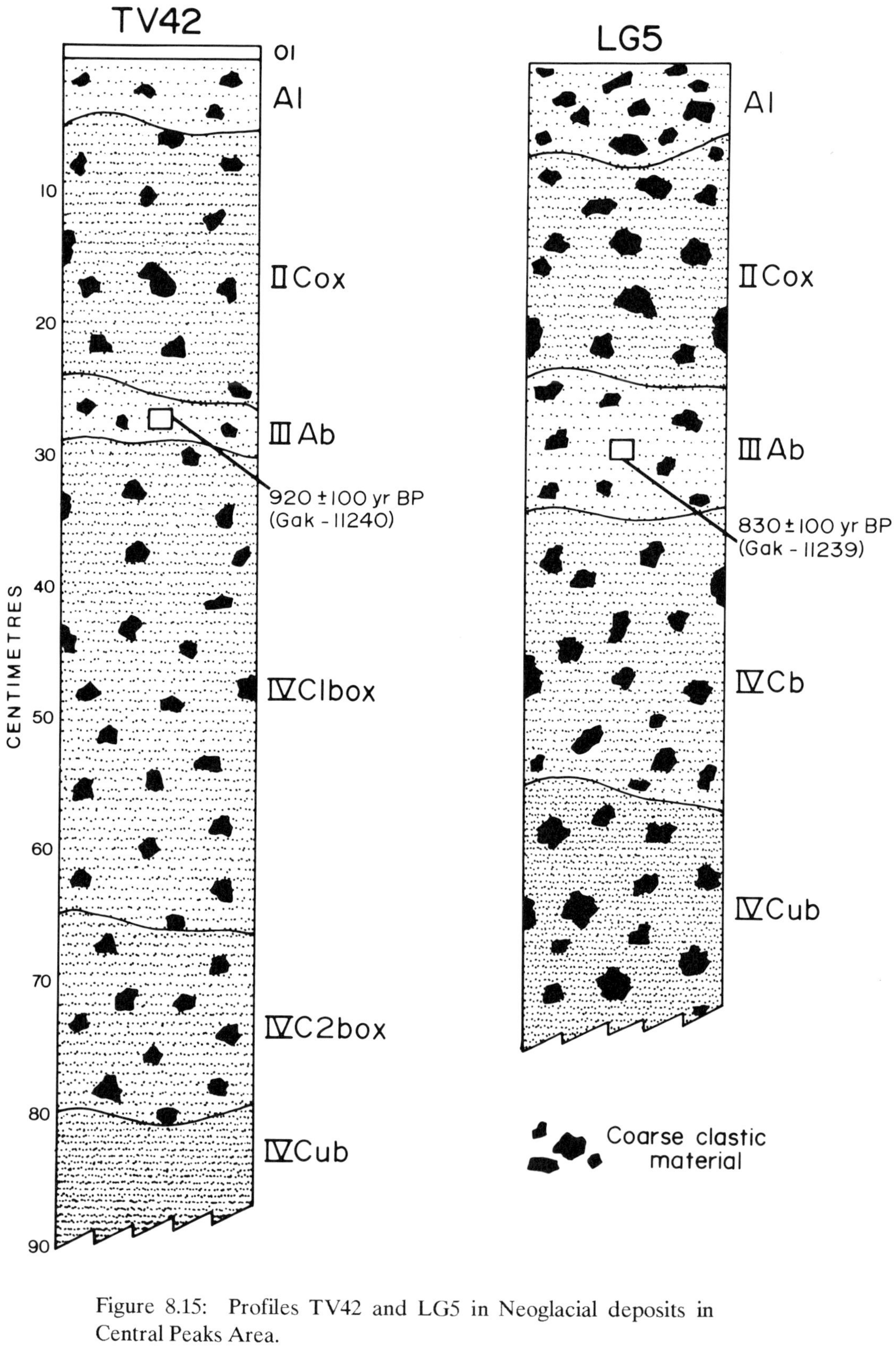

Figure 8.15: Profiles TV42 and LG5 in Neoglacial deposits in Central Peaks Area.

size and cover in an effort to correlate them with the moraine succession above. On the basis of topographic position, little weathering, and very thin soils, only the outer till in Tyndall Cirque, the younger outwash (LG5) (Figure 8.14), and debris-flow (TV42) (Figure 8.15) deposits correlate with the Tyndall advance that preceded the Little Ice Age.

8.6 Radiocarbon Dated Surfaces

Profile LG5 (Figure 8.16), in older outwash below the Tyndall-age moraines in Lewis Cirque, consists of an Entisol that developed in outwash (location see Figure 8.16) some time prior to 830 ± 100 yr BP (GaK-11239). Because the time for development of a cambic B horizon in these soils is limited to 2000 radiocarbon years (Mahaney and Boyer, 1986a), it is probable that this buried soil formed in outwash from an ancestor of the Lewis Glacier that was emplaced during pollen zone Z or some time after 2000 yr BP (Mahaney, 1985). Since the older Neoglacial moraines are located below the Lewis and Tyndall glaciers, and since they do not overlie datable materials, it is difficult to determine precisely the time at which this early Neoglacial advance occurred.

Below the outwash succession a sequence of debris-flow deposits (TV42, Figure 8.17) yielded a minimum ^{14}C age of 920 ± 100 yr BP (GaK-11240). This profile (TV42), while similar to LG5, consists of coarser debris carried by meltwater from an ancestor of the Lewis Glacier. Because the Ab horizon in this profile is thinner, it is possible that it formed over a shorter time than that in LG5, although the difference was probably less than a century (Mahaney and Boyer, 1986a). The time of burial is approximately the same in both Ab horizons, and because the surface soils are weathered to similar depths, with the same horizon complexities, in rather similar materials, it is likely the time of burial approximates the time of advance of glaciers prior to the Little Ice Age (*e.g.*, Tyndall advance). Presumably, the younger Lewis advance was sufficient to only produce a younger succession of outwash deposits that do not appear to have buried older surfaces (Mahaney and Spence, 1989).

8.7 Lichen Data

Measurements of maximum diameters of crustose lichens including *Rhizocarpon* section *Rhizocarpon* are given in Table 8.1. Sites are located on deposits shown on Plate 5 and are given relative ages (*e.g.* Lewis I, II) corresponding to the ice position over the last century. Percent lichen cover was approximated using percentage charts in Oyama and Takehara (1970).

In general the data show considerable variability in lichen cover on surfaces of the same age; however, the highest cover is found on the older deposits, which provides a useful index of age. Along with lichen cover, maximum diameters of *Lecidea, spp.* and *Buellia*, spp. (Table 8.1), show considerable variability on deposits of similar age, making them useful only for relative age determination. Some of these anomalously high diameters could result from transport of lichens of appreciable size across snow surfaces or even on ice surfaces during the time the moraines were emplaced. Such a situation has been reported in other areas (Mahaney, 1973) and if it happened here some lichens would have anomalously large diameters (*e.g.* HA10, TT10, and LG7 on Plate 5 and Table 8.1). An alternative hypothesis to explain these anomalously large diameters might be that the distal edges of some Neoglacial deposits are older (e.g. pre Little Ice Age), as is the case in some Rocky Mountain cirques (Mahaney, 1975).

The data for *Rhizocarpon* section *Rhizocarpon* shows a steady increase with age (Table 8.1) giving a ~100 yr great-growth of 34 mm. A steady decline in size to 10-12 mm on the Lewis IV moraine shows that it takes this species half a century (see Figure 8.18) to reach approximately one-third of its great-growth size. The full great-growth size of 34 mm is based on data from Lewis Cirque. The overall lower maximum diameters for other cirques probably reflect increased

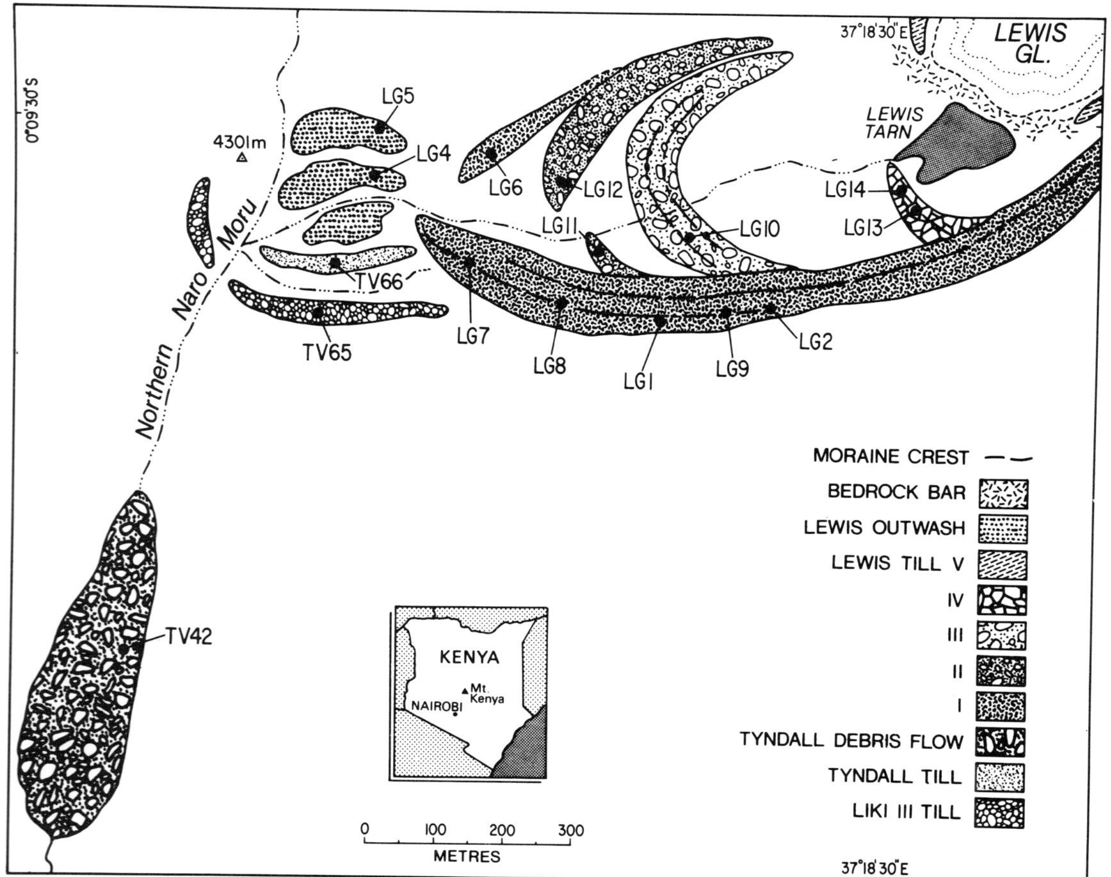

Figure 8.16 Location of dated Neoglacial deposits in Central Peaks Area.

Figure 8.17: Site TV42 on debris-flow deposit, upper Teleki Valley.

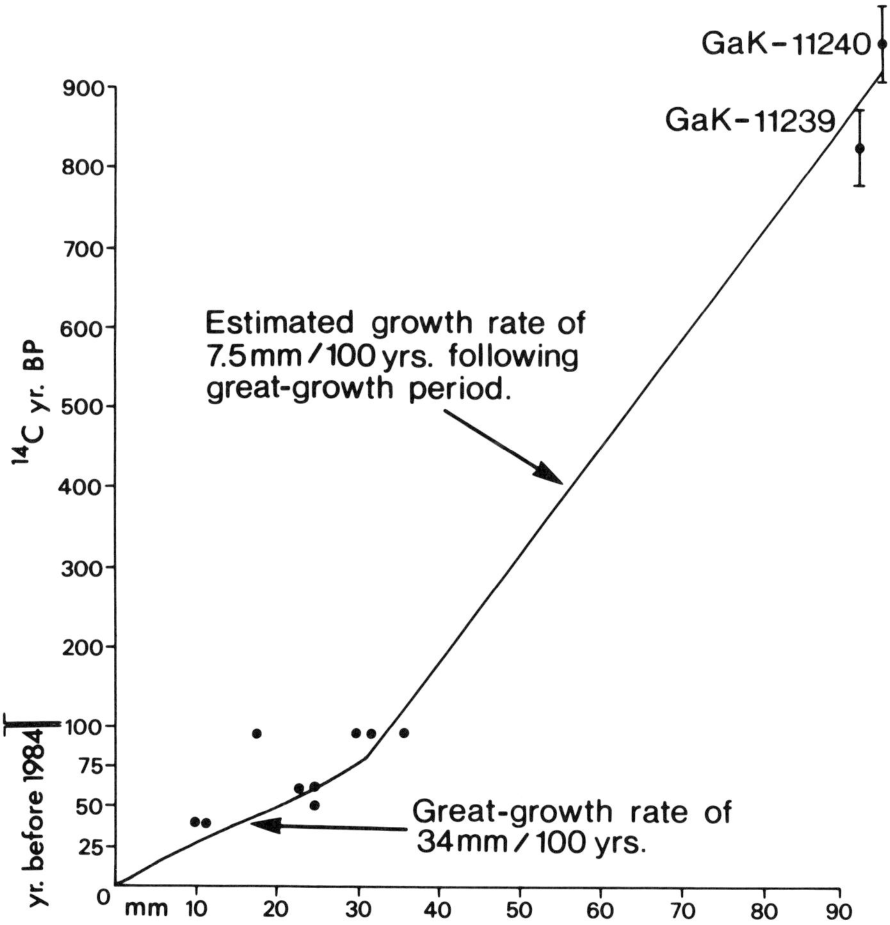

Figure 8.18: Lichen-growth curve for *Rhizocarpon* section *Rhizocarpon* in the Lewis foreland.

microenvironmental effects (decreased moisture). Higher values may indicate movement of somewhat greater lichen-encrusted rocks over snow or on ice surfaces.

8.8 Lichen Growth Curve

Because Tyndall and Lewis glaciers have receded steadily since 1892 (Gregory, 1900) we have accurate historical data on the termini positions over nearly a century (Figure 8.12 and 8.14, details in Spence and Mahaney, 1988a; and Spence, 1989). In Tyndall Cirque, Mahaney and Spence (1989) measured *Rhizocarpon* section *Rhizocarpon* thalli at twenty historically dated positions and produced the growth curve shown in Figure 8.18 depicting a great-growth value of *c.* 34 mm. The closest boulder to the glacier on which thalli were found was at 66 m, which corresponds to an approximate age since ice release of *c.* 23 yr (from 1984). A 3 mm diameter thallus was found on this boulder, suggesting that colonization takes place somewhat earlier, perhaps on substrates less than 20 yr old. Least squares regression of 180 maximum thalli measured along the transect (Figure 8.13) in front of Tyndall Glacier gave the equation:

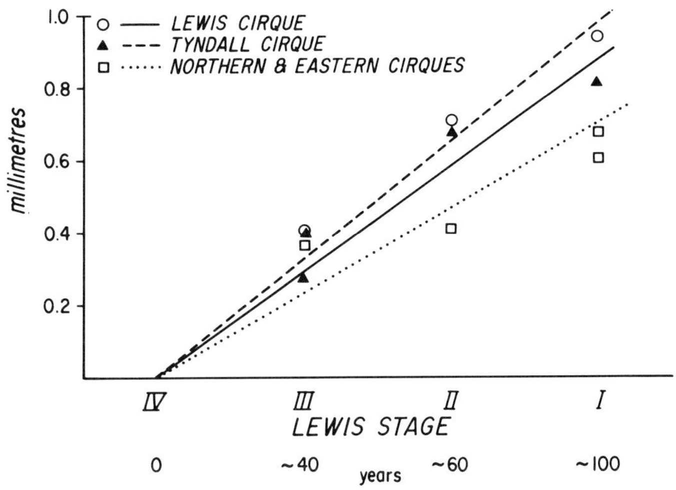

Figure 8.19: Weathering rind curves on late Neoglacial deposits on Mount Kenya. Rind development is slightly faster in the southwestern cirques where moist microclimate prevails, and slower on the northern and eastern flanks.

$$Y = 2.28X + 31.2 \qquad (2)$$

$$r^2 = 0.75$$

where Y is age since ice release, and X is thallus diameter in mm.

Below the Lewis Glacier measurements of *Rhizocarpon* section *Rhizocarpon* on historically and radiocarbon dated surfaces gives a curve shown in Figure 8.18. The great-growth period of this curve compares very closely with the Tyndall Glacier curve (Figure 8.13) indicating little variation in growth rates between the two forelands. Extension of the curve to *circa* 900 yr BP is based on maximum diameters measured at these two radiocarbon-dated sites. A least squares regression for the period 100 - *c.* 900 yr BP yielded the equation:

$$Y = 13.9X - 397.3 \qquad (3)$$

$$r^2 = 0.99$$

where Y is radiocarbon yr BP and X is thallus diameter in mm. This equation is only based on two data points at the older end and 4-6 data points at the younger end, so it should be used with some caution. Additionally, the change from the great-growth rate to a slower rate of growth, although probably around 100 years after ice release, may occur later (Mahaney and Spence, 1984). It is possible that growth rate slows down gradually so that the age of relatively young deposits (*e.g.* 100-500 yr) may be overestimated by the curve, while the age of older deposits may actually be underestimated. Future research on dating forelands in the area and correlations with thallus diameters will hopefully improve the accuracy of this curve.

Table 8.2:

Weathering rind data from cirques in the Central Peaks region, Mount Kenya, Afroalpine area.

Site[a]	Age[b]	Mean Maximum (mm)	Mean Minimum (mm)	n
TT6	Lewis IV	nil	nil	nil
LG10	Lewis III	0.28	nil	50
TT5		0.39	nil	50
TT12		0.47	nil	50
TT13		0.52	nil	50
HA8		0.46	nil	50
MK5		0.40	nil	50
LG11	Lewis II	0.68	nil	50
TT1		0.80	nil	50
TT7		0.78	nil	50
TT8		0.65	nil	50
TT9		0.61	nil	50
HA9		0.52	nil	50
MK15		0.36	nil	50
MK17		0.30	nil	50
MK18		0.41	nil	50
HOB5		0.62	nil	50
GOR27		0.36	nil	50
GOR28		0.38	nil	50
LG1	Lewis I	0.81	nil	50
LG6		0.78	nil	50
LG7		0.60	nil	50
LG8		0.46	nil	50
LG9		0.52	nil	50
TT4		0.68	nil	50
TT10		0.98	nil	50
TT11		0.92	nil	50
HA10		0.62	nil	50
HA12		0.78	nil	50
HA13		0.50	nil	50
MK4		0.66	nil	50
MK16		0.64	nil	50
MK19		0.48	nil	50
DAR1		0.60	nil	50
GOR26		0.54	nil	50
GOR29		0.56	nil	50
HOB1		0.68	nil	50
HOB2		0.68	nil	50
HOB3		0.54	nil	50
HOB4		0.55	nil	50
TV66	Tyndall	1.09	nil	50

a. Sites are located on Plate 5.
b. Age is determined from historically documented positions. See Figure 8.12.

8.9 Weathering Characteristics

Weathering rind measurements on Neoglacial surfaces are shown in Table 8.2. Both the mean maximum and mean minimum rinds were measured to the nearest mm with the mean value determined on counts of 50 stones at each site. In general, the rinds increase to ~0.9 mm in just under a century. The mean minimum rinds are all nil on Neoglacial substrates, which is similar to measurements of rinds on Neoglacial deposits in western Wyoming (Mahaney, 1978; Mahaney and Spence, 1984).

The mean maximum rinds on Neoglacial deposits (Table 8.2) show a nice clear increase in size with time (Figure 8.20) increasing from nil at time zero to 0.7-0.9 mm in a century, and ~1.09 mm in ~1000 yr. Rind growth is faster in Lewis and Tyndall cirques on the southwestern side of the mountain where microclimate is cooler and moisture more prevalent. On the northern and eastern flanks microclimate is warmer and drier giving rise to a difference in rind development of 0.2 mm over *circa* one century. Before the development of a lichen growth curve in 1984 one of the easiest methods of differentiating deposits in the Neoglacial succession was to use weathering rind measurements.

The importance of weathering rinds in distinguishing Liki III (late Pleistocene) from Neoglacial deposits is discussed in Chapter 7. All Liki III deposits have rinds above 1.5 mm with most clasts giving mean maximum diameters of ~2.0 and mean minimum rinds of .12 to .50. Even in the Afroalpine climate where weathering is a diurnal event, the rind record is not so different from the middle-latitude alpine zone, where at ~42°N in the Rocky Mountains the differences between later Pleistocene and Neoglacial deposits are similar. However, in the Afroalpine weathering rinds seem to develop faster in the first 100 yr. After that weathering slows down and gives values comparable with other middle latitude areas.

8.10 Soil Stratigraphy

Several Neoglacial soils in the Central Peaks area were described and analyzed with the objectives of: a) determining their significance in differentiating late-Holocene deposits; and b) assessing weathering of surficial sediments over the last thousand years. The most important soil sites are located in Lewis, Tyndall, and Harris cirques (Plate 5).

Deposits emplaced during the Lewis advance consist of an open network of boulders lacking in fine-grained sediments and soil cover. In places, however, thin soils appear where fine-grain material accumulated or where plants have colonized the substrates. These soils are generally thin weathering profiles lacking A horizons and giving rise to C/Cu profiles with faint oxidation features (10YR 5,6 hues) in the surface C horizon. Typical profiles are at sites TT9 (Lewis Cirque) and LG2 (below Lewis Tarn). The profile descriptions are given in Table 8.3.

In general particle size distributions are similar between soil horizons and parent materials. Mean phi calculations for both soil and parent material samples give values <2.0 (Mahaney, 1982b). No clay minerals are present in these young soils although small amounts of feldspar are common. Quartz is generally absent in most tills (see Vortisch, Mahaney and Fecher, 1987).

The older Neoglacial deposits have variable soil covers generally with A/Cox/Cu profiles as discussed by Mahaney (1982b). A typical post-Tyndall soil in till is found in the "apron" of till located due south of the Lewis I moraine ridge in Tyndall Cirque (Plate 5). The soil in this till gives a profile (Table 8.4) remarkably similar to profile TV66 (Figure 10.9). This profile is coarse textured and thin (41 cm depth). There is some evidence for a minor amount of clay translocation from the Al to the Cox horizon. Mean phi values increase somewhat giving values of 3.4 in the Al horizon and 3.1 in the Cox. As with the younger Neoglacial profiles insufficient time has elapsed for the genesis of clay minerals; however, primary minerals such as quartz and feldspar are present

Table 8.3:

Weathering profiles at sites TT9 and LG2, Central Peaks area.

Horizon	Depth (cm)	Description[a]
TT9[b]		
Cl	0-20	Dull yellow-orange (10YR 7/3m,d) pebbly coarse sandy loam, massive, loose consistence, nonsticky, and nonplastic.
Cu	20+	Light grey (2.5Y 8/1m,d) and pale yellow (2.5Y 8/3m,d) pebbly loamy sand, massive, loose consistence, nonsticky and nonplastic.
LG2[b]		
Cl	0-10	Dull yellow-orange (10YR 7/3m,7/2d) pebbly coarse loamy sand, massive structure, loose consistence, nonplastic and nonsticky.
Cu	10+	Pale yellow (2.5Y 8/3m, 8/2d) pebbly coarse sandy loam, massive, loose consistence, nonsticky and nonplastic.

a. Soil nomenclature follows Birkeland (1984) and Hodgson (1976). Cu is considered the best designation for unweathered undifferentiated parent material.
b. Profile locations are on Figure 8.14.

Table 8.4:

Post-Tyndall soil in Tyndall Cirque,

Central Peaks area.

Horizon	Depth (cm)	Description[a]
01	2.5-0	Black (10YR 2/1m).
Al	0-20	Brownish black (10YR 3/2m) and greyish yellow brown (10YR 4/2d) pebbly sandy loam, friable, nonplastic, and nonsticky.
Cox	20-41	Dull yellowish brown (10YR 4/3m, 10YR 5/3d) pebbly sandy loam, friable, nonplastic, and slightly sticky.
Cu	41+	Light grey (2.5Y 7/2m, 7/1d) pebbly loamy sand, loose, nonplastic and nonsticky.

a. Colors are from Oyama and Takehara (1970) and are given as moist (m) and dry (d). Nomenclature follows Birkeland (1984) and Hodgson (1976).

Table 8.5:

Compound soil profile (LG4) in lower Lewis Cirque

Horizon	Depth (cm)	Description
Al	0-7	Brownish black (10YR 3/2m, 3/2d) pebbly fine sandy loam, weak granular structure, friable consistence, nonsticky and nonplastic.
C	7-24	Greyish yellow brown (10YR 6/2m, 6/2d) pebbly loamy sand, massive, loose consistence, nonsticky and nonplastic.
Ab	24-34	Dull yellowish brown 10YR 5/4m, 5/3d) loamy sand, massive, very friable consistence, nonsticky and nonplastic.
Cb	34+	Greyish yellow brown (10YR 6/2m, 6/2d) pebbly fine sandy loam, massive, friable consistence, slightly plastic and slightly sticky.

in minor amounts. The pH in the solum of this profile is a moderately acidic 5.2 compared with neutral pH's in the younger Neoglacial profiles (Mahaney, 1982b).

Compound Neoglacial soils including ground soils and buried paleosols are common in lower Lewis Cirque (see section 8.6, profile LG5; Plate 5 for location). Profile LG4 (Table 8.5) is representative of soils formed in outwash older than early Neoglacial age and covered with younger outwash dating from the Tyndall advance. This profile, while similar in appearance to profile LG5 with respect to horizon sequence and thickness, shows somewhat more clay in the lower profile. In all, lower profile LG4 probably formed prior to the Tyndall advance after emplacement of outwash prior to ~1000 yr BP. Charcoal recovered from the Ab horizon could not be dated by radiocarbon (insufficient carbon content), but may be related either to prehistoric occupation or to a dry phase following the climatic optimum (pollen zone Z) and grass fires.

8.11 Conclusions

A growth-curve for *Rhizocarpon* section *Rhizocarpon* on Mount Kenya is used to date deposits emplaced over the last millenium on the southwestern flank. Following the initial great-growth period of *c.* 90 yr (34 mm/100 yr), growth slows to an average diameter increase of 7.5 mm/100 yr for at least 1000 yr. These values are high, compared to middle-latitude alpine and arctic localities where lichen longevity is 3000 yrs (Benedict, 1967, 1968; Beschel, 1961, 1965), but compare favorably with curves from oceanic areas, such as northwestern Europe and North America. The effects of climatic change on growth rate cannot be determined from the growth curve, and it is unlikely that it will be possible to date additional deposits and refine the curve, so as to determine the effects of short-term moisture trends on lichen-growth rates.

The growth curve may be used to date surfaces emplaced over ~900 years and *only* on the southern and western flanks of Mount Kenya. Because the northern and eastern sides are drier one can only assume that growth rates are slower, perhaps by a considerable amount.

Lichen sizes and cover prove useful in differentiating deposits and together with weathering rinds provide important information on relative ages. Overall the data for weathering rind development show that rinds thicken rapidly at first (*i.e.* within the first century) and then slow down. In all, weathering rind development reaches ~1.1 mm in 1000 yr.

Radiocarbon-dated Neoglacial soils are rare but those that are available were used to construct a lichen growth curve and to calibrate the timing of the early Neoglacial advance. From the data available it appears that the Tyndall advance occurred about ~ 1000 yr BP.

Chapter 9

SOIL MORPHOGENESIS

9.1 Introduction

The time factor in soil genesis can be studied using the chronosequence concept of Jenny (1941, 1980). This is defined as a sequence of different soils formed in similar parent materials and topographic settings, under nonvarying climate and biota, whose change in state is related to time since deposition. Jenny's general equation,

$$s = f\,(cl, o, r, p, t, \ldots) \tag{4}$$

where s = soil system
cl = climate r = relief
o = biota p = parent material
t = time

treats all factors of soil formation as independent variables and can be differentiated to analyze partial derivatives for each factor.

$$\text{Hence, } \left(\frac{ds}{dt}\right)_{cl,\,o,\,r,\,p} = dt \tag{5}$$

where the partial derivative can be used to assess rates of change in soil systems over time. Because "s" in equation 4 consists of a number of discrete factors, it is nearly impossible to solve (Mahaney, 1974; Birkeland, 1984); however, it is possible to solve equation 5 provided the soils are relict units[1] forming under similar environmental conditions in uniform parent materials.

On Mount Kenya, as in other mountain areas of the world, it is possible to hold parent material and topographic factors nearly constant, but quite impossible to hold climate and biota constant for periods longer than 2.0×10^3 yr. Because of climatic and biotic variations even within one altitudinal zone (*e.g.* Upper Afroalpine belt) it is possible to look only at *trends* in the development of soil properties over time to the attainment of near steady state conditions. Mature soils are in steady state or dynamic equilibrium with their environments as defined by a condition in which a particular soil factor (*e.g.* "s") no longer fluctuates or changes with time (t). The equation

$$\frac{\Delta s}{\Delta t} = 0 \tag{6}$$

defines a situation where a change in soil property (s) over changing time (t) is equal to zero. Because all soil properties approach steady state at different rates, we must arbitrarily decide on which properties are most diagnostic of steady-state conditions. Once steady state is achieved, the macrorate scale of change is equal to zero, the time scale being so large that "s" and "t" go unnoticed on the microscale.

---

1. Relict soils are ground soils that have formed over various lengths of time exposed to subaerial weathering at the earth's surface (see Morrison, 1978, for discussion).

On Mount Kenya it is possible to study changes in soil properties in the Upper Afroalpine (~4000 m) from the present day to the late glacial (~12,000 yr BP). At lower elevations around the margins of glaciation (in the ericaceous belt and Hagenia forest) it is possible to analyze relict surface soils that have developed since the early and middle Quaternary as well as buried paleosols that were exposed to subaerial weathering over periods of ~100,000 yr before burial due to renewed geological activity.

9.2 Upper Afroalpine Soil/Paleosol Chronosequence

Soil morphology, particle size, and selected soil chemical properties appear closely related to the age of Holocene and later Pleistocene substrates in the Upper Afroalpine area (~4000 m.a.s.l.). The parent materials for these deposits range from outwash and alluvial-colluvial fan sediments to tills; however, all consist mainly of coarse clastic materials that have sandy loam textures. The greatest soil differentiation occurs in younger Neoglacial and middle to late Holocene deposits. Depth of solum, degree of horizonation, development of surface and subsurface color and grade of soil structure are closely associated with the age of deposits. Horizons develop slowly becoming well differentiated within ~2000 yr, which is the time required for a B horizon to form (Mahaney, 1986). Particle size distributions and the porosity of the original parent materials tend to determine soil depth, which is somewhat greater in outwash and colluvial-alluvial fan sediments and lower in tills. Soil colors range from 2.5Y 5 and 10YR 5 and 6 values in younger Neoglacial weathered horizons and parent materials, and 10YR 4-5 values in B horizons, and 10YR 2,1 values in surface A horizons. Subsurface colors in the older profiles of middle Holocene (TV28) and later Pleistocene age (TV1, TV2, TV3, TV6 and TV14; see Chapter 7 for sections) show some evidence of mottling and oxidation-reduction effects in sediments with fine sandy and silty loam textures.

The particle size data available for soils in this chronosequence show that loamy sand textures, common in younger Neoglacial soils, develop sandy loam and silty loam textures by middle Holocene and later Pleistocene time, as a result of weathering and aeolian influx of fine-grained sediment. The B2 horizons, developed in older later Pleistocene pedons, may have slight increases of clay size material, but not sufficiently high to qualify as argillic subsurface horizons. Hence, all mature paleosols at ~4000 m.a.s.l. are Inceptisols belonging to the suborder of Humitropepts (Soil Survey Staff, 1975).

9.3 Chemical Trends

The soil chemistry for the Upper Afroalpine chronosequence is given in Tables 9.1-9.4. Locations for profiles LG2 and TT1 are shown on Plate 5; TV28 is in Figure 7.16B, and TV4 is in Figure 7.18. The morphological descriptions of the older later Pleistocene paleosols are in Figures 7.9 and 7.7c.

The younger soils and paleosols in the sequence [from LG2 to TV28 (Tables 9.1 and 9.2] show overall increases in extractable cations and cation exchange capacity with increasing time. Calcium and sodium, inherited from the weathering of alkaline volcanic feldspars and nepheline syenite, tend to show different degrees of removal depending on ion mobility in a humid climate. Sodium shows the highest degree of cation translocation over time (Table 9.1), followed by calcium. Magnesium and potassium remain relatively uniform or increase slightly over time. The cation exchange capacity increases dramatically over the first few thousand years from values of 5 to 55-60 meq/100g (Table 9.1).

Following changes in surface soil color over time (section 9.2), organic carbon and total Kjeldahl nitrogen (TKN) (Table 9.2) increases from nil to relatively high over the first few thousand

Table 9.1
Selected chemical parameters for Holocene soils and paleosols in upper Teleki Valley

Site	Horizon	Depth (cm)	Extractable Cations (meq/100g)				Cation Exchange Capacity (CEC) (meq/100g)	Base Saturation (%)	E.C. (mmhos/cm 25°C)
			Ca	Mg	Na	K			
LG2	C	0-8	1.20	0.35	2.68	0.39	5.0	100	0.20
TT1	C	0-15	0.65	0.12	1.00	0.31	5.9	35	0.25
TV4	A1	0-6	10.54	4.69	0.84	1.42	57.2	30.5	3.00
	B2	6-27	2.05	0.69	0.35	0.51	11.1	32.4	0.30
	Cox	27-55	1.30	0.33	2.95	0.18	22.5	21.1	0.10
	Cu	55-63	-[a]	-	-	-	-	-	-
	IIAb	63-68	2.35	0.30	2.44	0.12	38.2	13.6	0.15
	IICb	68-72	1.30	0.16	2.95	0.13	24.6	18.4	0.15
	IICub	72 +	0.65	0.10	2.35	0.28	13.4	25.2	0.15
TV28	A11	0-5	3.99	0.77	0.00	0.12	17.7	27.5	1.15
	A12	5-10	10.82	2.16	1.23	0.71	60.6	24.6	1.15
	B21	10-26	4.50	1.30	0.50	0.20	34.0	19.1	0.50
	B22b	26-34	3.75	1.05	0.40	0.24	37.8	14.4	0.50
	C1box	34-44	3.75	1.00	0.40	0.23	24.6	21.9	0.25
	C2box	44-54	10.00	1.50	0.50	0.33	38.0	32.4	0.20
	C3box	54-61	6.75	1.85	0.35	0.26	25.0	36.8	0.15
	C4box	61-81	7.75	1.25	0.45	0.29	28.4	34.2	0.15
	Cub	81 +	8.90	2.20	0.66	0.24	31.1	38.6	0.20

a. not available

Table 9.2
Carbon and nitrogen data for Holocene soils and paleosols, upper Teleki Valley, Mount Kenya

Site	Horizon	Depth (cm)	Organic Carbon (%)	Total Kjeldahl Nitrogen (%)	C/N
LG2	C	0-8	0.42	nil	nil
TT1	C	0-16	0.32	0.021	15.2
TV66	A1	0-6	9.25	0.610	15.2
	Cox	6-36	1.84	0.130	-
TV4	A1	0-6	16.56	1.390	11.9
	B2	6-27	1.85	0.128	14.5
	Cox	27-55	0.62	0.341	1.8
	Cu	-a	-	-	-
	IIAb	63-68	7.59	0.577	13.2
	IICb	68-72	3.70	0.059	-
	IICub	72+	1.69	0.126	-
TV28	A11	0-5	24.60	2.240	11.0
	A12	5-10	11.96	1.150	10.4
	B21	10-26	4.88	0.443	11.0
	B22b	26-34	5.91	0.572	10.3
	C1box	34-44	1.51	0.124	12.2
	C2box	44-54	3.21	0.271	11.9
	C3box	54-61	0.98	0.091	10.7
	C4box	61-81	1.30	0.090	-
	Cub	81+	3.50	0.260	-

a. not available

years. There appears to be successive increases of organic carbon, TKN and C/N ratios over Neoglacial and middle Holocene time (Table 9.2).

The chemical parameters shown in Tables 9.3 and 9.4 are representative of soils (paleosols) forming in later Pleistocene rock glacier (TV1) and till (TV2, 3, 6, and 14) deposits. Overall, the distribution of extractable cations reflects more time for leaching (hence, the lower Ca^{+2}); variations in Na^{+1} may reflect on the prevalence of syenite clasts in different deposits. Cation exchange capacity ranges from ~14 to ~59 meq/100 g, with the highest values in ground moraine (TV6) probably reflecting available soil moisture, and hence, a higher ground cover of vegetation. Base saturation (Table 9.3) data suggest that leaching over ~12,000 yr tends towards movement of soluble cations through the paleosol sola to the B/C horizon contacts.

Analysis of the data in Table 9.4 shows that later Pleistocene profiles have variable amounts of organic carbon and nitrogen (Figure 9.1), which tend toward lower values after the middle Holocene. In part this reflects on the drier nature of the moraine and rock glacier crests and their relatively impoverished and unstructured floral composition (personal communication, John Spence, 1984).

The pH in these soils/paleosols declines from ~7.0 to ~5.0 over relative lengths of postglacial time. Considering that the abrasion pH of most of these minerals (alkali feldspars) is ap-

Table 9.3
Selected chemical parameters for late glacial (Liki III) paleosols in upper Teleki Valley

Site	Horizon	Depth (cm)	Extractable Cations (meq/100g)				Cation Exchange Capacity (CEC) (meq/100g)	Base Saturation (%)	E.C. . (mmhos/cm 25°C)
			Ca	Mg	Na	K			
TV1	A11	0-10	0.80	0.25	0.26	0.23	29.5	5.2	0.45
	A12	10-18	0.25	0.11	0.12	0.12	27.5	2.1	0.30
	B2	18-58	0.88	0.13	0.24	0.26	22.7	6.6	0.10
	IICox	58-78	1.23	0.15	0.30	0.26	20.0	9.7	0.10
TV2	A1	0-10	1.48	0.25	1.00	0.41	26.1	12.0	0.15
	B2	10-40	1.60	0.12	2.58	1.15	13.0	41.9	0.15
	IICox	40-78	1.80	0.25	2.62	1.70	10.5	60.6	0.15
	IICu	78 +	1.80	0.40	2.62	1.65	9.6	67.4	0.15
TV3	A1	0-6	0.50	0.07	1.88	0.32	13.9	19.9	0.25
	B2	6-14	1.10	0.14	2.37	0.17	15.0	25.2	0.15
	IIC1ox	14-39	0.95	0.10	2.70	0.17	7.1	55.2	0.15
	IIC2ox	39-59	0.95	0.11	2.90	0.14	10.0	41.0	0.15
	IICu	59 +	4.00	0.35	2.53	0.40	19.5	37.3	0.15
TV6	A1	0-13	0.65	0.69	1.00	1.25	59.3	6.1	0.35
	B2	13-30	0.25	0.03	0.10	0.15	18.0	2.9	0.10
	IICox	30-65	0.25	0.03	0.50	0.18	19.3	5.0	0.20
	IICu	65 +	0.30	0.03	0.12	0.15	12.3	4.9	0.10
TV14	A11	0-5	1.20	0.36	0.30	0.34	30.3	7.3	1.40
	A12	5-24	0.65	0.12	0.10	0.10	19.8	4.6	0.35
	IIB21	24-35	0.35	0.04	0.10	0.05	20.9	2.6	0.15
	IIB22	35-45	0.25	0.04	0.12	0.05	20.0	2.3	0.10
	IICox	45-75	0.35	0.05	0.12	0.06	25.7	2.3	0.10

Table 9.4
Carbon and nitrogen distributions in late glacial (Liki III) paleosols, in upper Teleki Valley

Site	Horizon	Depth (cm)	Organic Carbon (%)	Total Kjeldahl Nitrogen (%)	C/N
TV1	A11	0-10	10.99	0.616	17.8
	A12	10-18	5.78	0.381	15.2
	B2	18-58	1.23	0.108	11.4
	IICox	58-78	0.75	0.055	-
TV2	A1	0-10	3.62	0.343	10.6
	B2	10-40	0.45	0.038	11.8
	IICox	40-78	0.39	0.020	-
	IICu	78+	0.16	0.006	-
TV3	A1	0-6	2.04	0.163	12.5
	B2	6-14	1.01	0.078	13.0
	IIC1ox	14-39	0.33	0.011	30.0
	IIC2ox	39-59	2.62	0.021	-
	IICu	59+	0.65	0.006	-
TV6	A1	0-13	7.66	1.510	5.1
	B2	13-30	1.53	0.101	15.2
	IICox	30-65	1.85	0.141	-
	IICu	65+	0.49	0.030	-
TV14	A11	0-5	18.72	0.713	15.2
	A12	5-24	5.25	0.439	7.0
	IIB21	24-35	8.25	0.316	15.2
	IIB22	35-45	6.36	0.298	12.4
	IICox	45-75	7.70	0.324	-

proximately 10 to 11, the greatest drop in pH occurs from time zero over the first one hundred years (Figure 9.1). As it is, the pH drops nearly 1.5 units over the first ~2000 years, and is in or close to steady state within ~6000 yrs. As found in other chronosequence studies (Mahaney, 1974) the increase in organic carbon is greater than nitrogen with some fluctuation (Figure. 9.1) after the first few thousand years. This fluctuation amounts to ~6 % for organic carbon and 0.8% for N. Similar fluctuations with respect to organic carbon and nitrogen distributions in other cold regions have been reported (see Crocker and Major, 1955). In other studies in northern Sweden (Stork, 1963) and in the Colorado Front Range (Mahaney, 1974), organic carbon and nitrogen increase steadily over the first few hundred or thousand years and remain relatively constant after achieving steady state. The fluctuation in values of organic carbon and nitrogen on Mount Kenya may be due to variations in microorganism populations that are known to be relatively impoverished in numbers per gram compared to other middle-latitude alpine and arctic ecosystems (Mahaney & Boyer, 1986a).

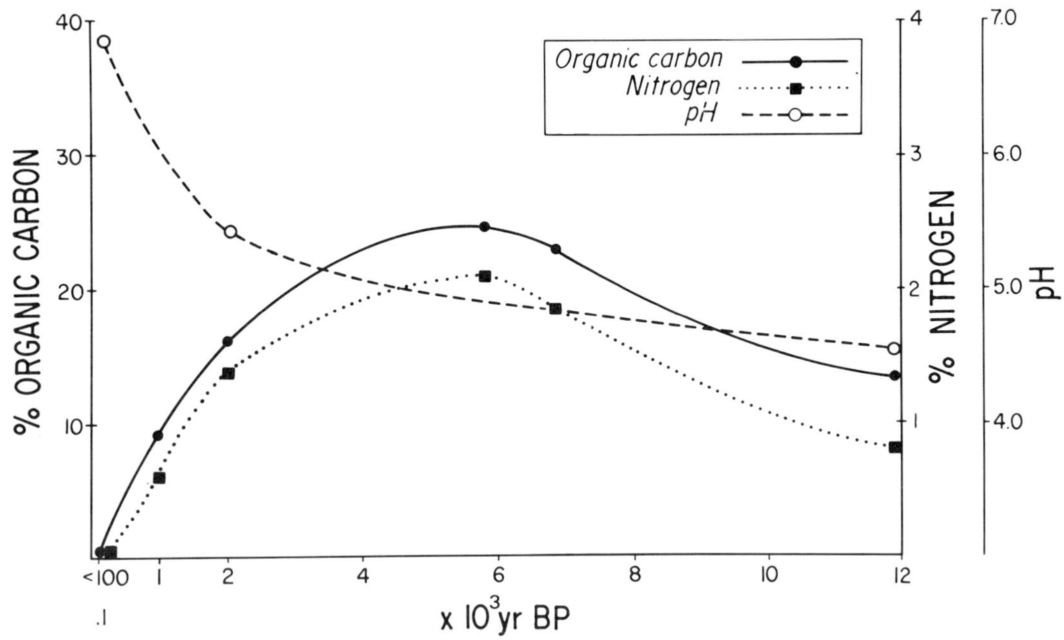

Figure 9.1: Trends for pH, organic carbon and nitrogen in the upper Afroalpine chronosequence.

9.4 Liki (Würm) Paleosol Chronosequence

Paleosols in deposits of Liki II age (~15,000 yr BP) provide important information about weathering since the end of the last glaciation. These paleosols formed in deposits at elevations of ~3200-3500 m in most major drainages, first during the colder and drier late glacial period (Mahaney, 1985) and later during the variable paleoclimates (including the wetter and warmer climatic optimum) of the Holocene. Close comparison of paleosols forming in these deposits (Figure 9.2) with older paleosols (Figure 9.3) of Liki I age (early Liki, isotope stage 4) show that time and paleoclimate play important roles in determining paleosol development.

Profiles LN14 and TV22 (Figure 9.4) are typical of paleosols with histic epipedons in the moorland areas on the western and northwestern flank of the mountain. Both profiles are formed in till with depth of weathering increasing from the northwest (LN14) to west (TV22) giving an overall profile depth of ~55 to ~80 cm. In general, these profiles fine upwards as a result of the airfall influx of fine-grained sediments. Accordingly, textures vary from sandy loam till in the lower sola and subsoil to silty loam in the upper sola (Mahaney, 1979a). The coarse clastic material consists mainly of fresh stones of trachytic, phonolitic and syenitic composition; most weathered clasts are found in the B2 horizons. Root penetration is greatest in paleosols on the western flanks where climate is wetter giving rise to a higher cover of tussock grass and other lower-story biotic elements of the lower Afroalpine belt. Profiles similar to TV22 stretch over a large area known as the "vertical bog", from ~3200-3700 m, in upper Teleki Valley.

In Gorges Valley, on the eastern flank of Mt. Kenya a sequence (for location see Chapter 6) of paleosols, in early and late Würm moraines (Figure 9.5), provides an interesting contrast with profiles in western and northwestern drainages. The oldest profiles, GOR1 and 2, formed in early Liki moraines, immediately upvalley from "the Gates", are weathered to greater depths when compared with the younger paleosols in Teleki Valley and Liki North Valleys. Many clasts in B

Figure 9.2: Buried paleosol consisting of till and loess in profile LE3, Lake Ellis area.

Figure 9.3: Buried paleosol in loess (IIIAb below shovel handle) over till of Lake Ellis age. The surface material is outwash and loess.

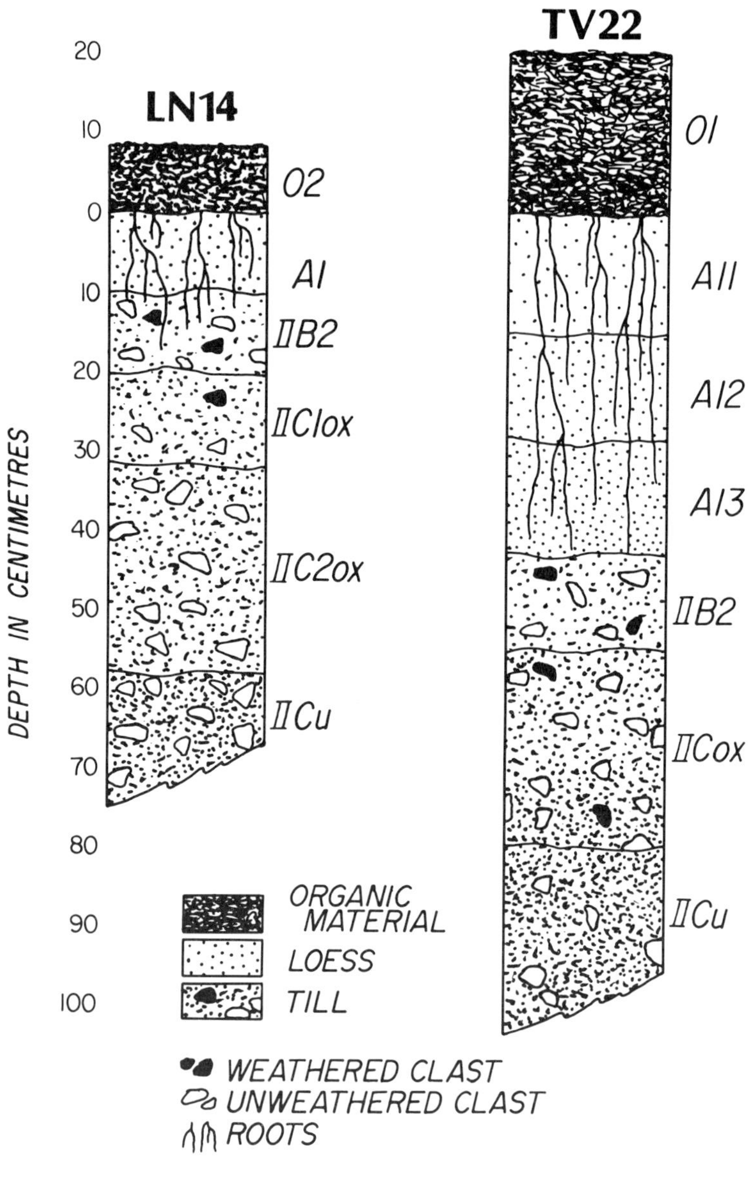

Figure 9.4: Paleosol profiles in Liki II moraines on the western flank of Mount Kenya. Both profiles are Inceptisols forming over the last ~15,000 yr.

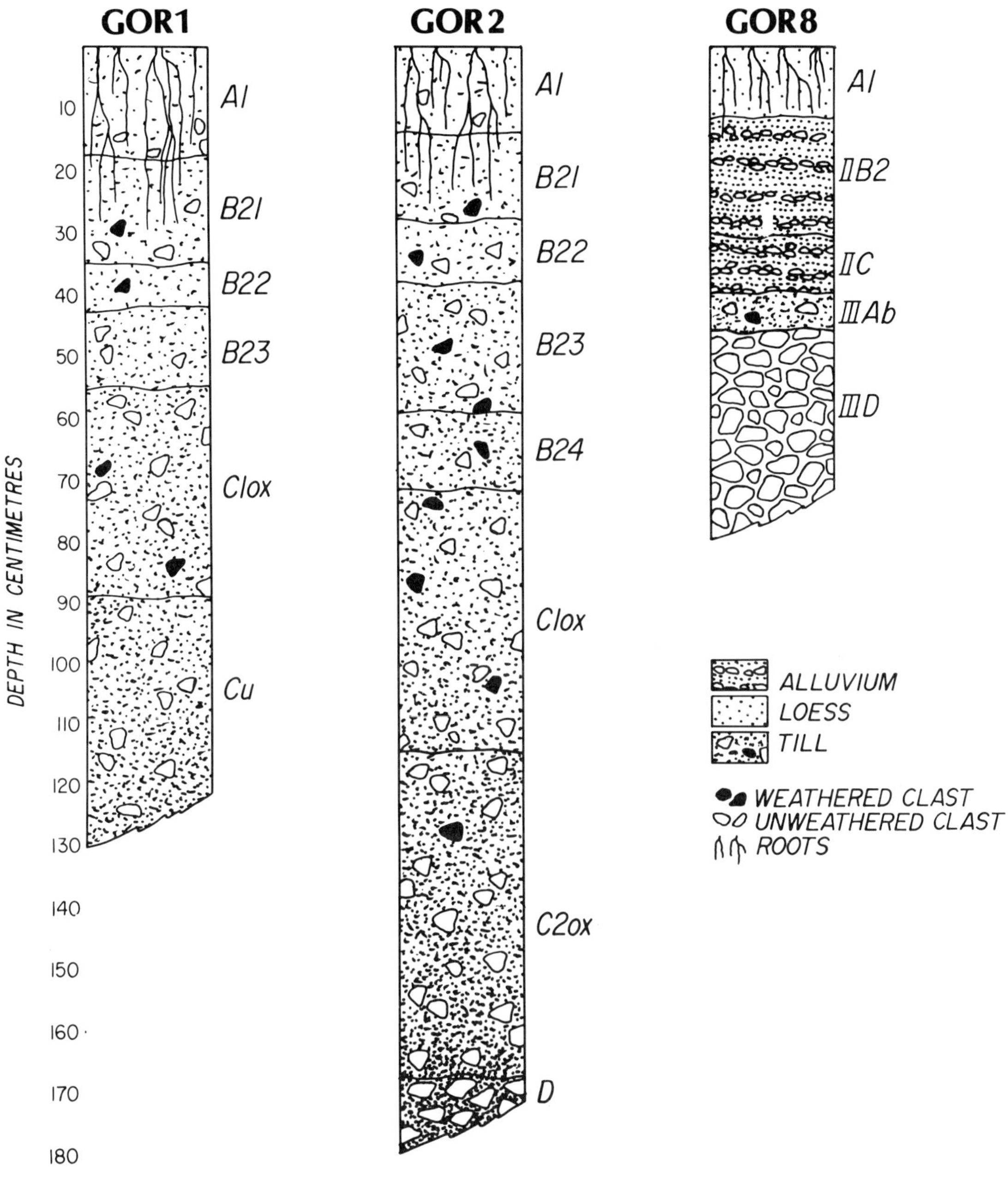

Figure 9.5: Paleosol profiles GOR1 and 2 in Liki I drift and paleosol profile GOR8 in Liki II drift, Gorges Valley.

horizons of the GOR1 and 2 profiles were exfoliated and a few could be cut with a knife. Roots tended to penetrate into the B horizon complexes, but were neither well developed nor as frequent as roots in paleosols in the western drainages. As with the younger late Liki paleosols (e.g. GOR8), this older group fined upwards, from coarser loamy sand and sandy loam textured tills, to silty loam textured loess covers. In some cases the surface A horizons contained a mix of till and loess.

A comparison of the B horizon thicknesses between the two sets of profiles shows a considerable increase in thickness from Liki II to I, which must be attributable to a combination of time and Liki Interstadial paleoclimate.

The younger GOR8 profile, a compound paleosol in a Liki II moraine in upper Gorges Valley, contains a lower portion which was buried by emplacement of alluvium at 5110 ±120 yr BP (Table 9.3 for details). The lower zone appears to show negligible development over a ~10,000 yr weathering period. Because this profile lacks a B/C horizon complex, one might conclude that paleoclimate was too dry and cold during the late glacial and early Holocene to allow pedogenesis to proceed. The overlying ground soil formed in loess and alluvium contains a reasonably well developed B horizon, about 20 cm thick, that may have been weathered primarily during the climatic optimum.

9.5 Paleosol Chemistry in Liki (Würm) Chronosequences

An analysis of selected soil chemical parameters in profiles TV22 and LN14 (Tables 9.5 and 9.6; Figure 9.4) and profiles GOR1, 2 and 8 (Tables 9.7 and 9.8; Figure 9.5) show some interesting relationships. Both profiles LN14 and TV22 show a certain amount of basic cation recycling, presumably by plant roots in the A horizons. The basic cation distributions in profile TV22 decrease downward in the profile to the subsoil/parent material contact, indicating a reasonably well leached profile. Paleosol LN14, on the other hand, shows the same downward pattern, with increases occurring below the B/C horizon contact. Presumably this increase is a response to decreased leaching in the subsoil, which is controlled by reduced soil moisture on the northwestern flank of the mountain.

Cation exchange capacities in profiles LN14 and TV22 (Tables 9.5 and 9.6) are higher (A12 horizon) in the Teleki Valley profile, a response presumably to higher organic matter input there. Overall the CEC's are higher in the subsoil of the TV22 profile in response to increased leaching and movement of organic carbon. The base saturation distributions tend to support this interpretation (Table 9.5) showing higher values in the subsoil of profile LN14. The distributions of total salts in both profiles are fairly low and uniform, with the exception of the epipedon in profile TV22. Increased total salt content in the A11 horizon of profile TV22 probably represents liberation of NO_3^- and SO_4^{-2} from decaying organic matter in the overlying 01 horizon.

The organic profile at sites LN14 and TV22 can be reconstructed from the data in Table 9.6. The amounts of organic matter and nitrogen are higher in TV22 compared with the LN14 profile. The downward distributions of C and N in the two profiles reflect increased vegetation density in the western drainages and more available soil moisture for translocation. Accordingly, carbon/nitrogen ratios vary from ~6 to 20 in the two profiles.

In Gorges Valley two Liki I profiles (GOR1 and 2; Table 9.7) show similar effects of base recycling in the surface epipedons. In addition, ratios of exchangeable Ca to Mg tend to increase downward by nearly one order of magnitude (profile GOR1). These ratios suggest that leaching is most effective within the paleosol sola, decreasing with greater depth. Cation exchange capacity is highest in the surface horizons of both profiles, first decreasing with depth and then increasing in response to fluctuations in either organic matter or clay. Base saturation percentage distributions show low values ($<10\%$) in the paleosol sola, increasing in the subsoil at depths greater

Table 9.5
Selected chemical parameters for Liki III paleosols in Teleki and Liki North Valleys

Site	Horizon	Depth (cm)	Extractable Cations (meq/100g)				Cation Exchange Capacity (CEC) (meq/100g)	Base Saturation (%)	E.C. (mmhos/cm 25°C)
			Ca	Mg	Na	K			
TV22	A11	0-16	0.98	0.88	0.21	1.16	34.8	9.3	1.20
	A12	16-28	0.48	0.52	0.20	0.79	88.2	2.3	0.40
	A13	28-43	0.15	0.10	0.13	0.25	55.4	1.1	0.15
	IIB2	43-57	0.05	0.02	0.05	0.09	39.8	0.5	<.10
	IICox	57-83	0.08	0.02	0.05	0.12	26.4	1.0	<.10
	IICu	83+	0.15	0.02	0.10	0.25	25.0	2.1	<.10
LN14	02	8-0	0.76	0.90	0.23	1.24	65.3	4.8	0.10
	A1	0-10	0.10	0.17	0.16	0.20	57.1	1.1	0.15
	IIB2	10-20	0.09	0.05	0.15	0.19	37.7	1.3	<.10
	IIC1ox	20-33	0.80	0.08	0.16	0.45	18.7	8.0	0.15
	IIC2ox	33-58	1.30	0.12	0.20	0.70	17.0	13.6	0.10
	IICu	58+	2.40	0.18	0.32	1.25	18.7	22.0	<.10

Table 9.6
Carbon and nitrogen data for Liki II paleosols, in Teleki and Liki North Valleys

Site	Horizon	Depth (cm)	Organic Carbon (%)	Total Kjeldahl Nitrogen (%)	C/N
TV22	A11	0-16	15.56	2.330	6.7
	A12	16-28	24.60	1.920	12.8
	A13	28-43	13.60	1.110	12.3
	IIB2	43-57	4.33	0.256	16.9
	IICox	57-83	1.15	0.089	-
	IICu	83+	0.33	0.025	-
LN14	02	8-0	22.61	1.620	14.0
	A1	0-10	14.21	0.718	20.0
	IIB2	10-20	3.39	0.279	12.2
	IIC1ox	20-33	0.42	0.019	-
	IIC2ox	33-58	0.32	0.009	-
	IICu	58+	0.26	0.003	-

than 55 cm. With the exception of the A horizons, total salts are low throughout, presumably in response to effective leaching over time.

In contrast with these two older profiles, GOR8 (Table 9.7) shows a near uniform distribution of extractable basic cations through the profile down to an interlocking network of boulders, cobbles and pebbles lacking any matrix material. Cation exchange capacities vary in response to organic matter contents in the A1 and Ab horizons. Base saturation and total salt distributions reflect the leaching record.

The organic matter distributions in these Gorges Valley profiles (Table 9.8) show overall lower values of organic carbon and nitrogen in the surface A horizons when compared with the western and northwestern flanks of the mountain. The distributions of organic carbon and total nitrogen with depth in profiles GOR1 and 2 show a secondary maxima in the lower paleosol sola. Because precipitation is lower on the eastern side of the mountain (~ 1000 mm), this trend is viewed as a relict feature, inherited either from an interstadial or climatic optimum paleoclimate with sufficient soil moisture to translocate organic matter. The fluctuations in organic matter parallel the variations of CEC in horizons that show little change in percent clay, supporting the hypothesis that CEC variations are controlled by humus (which is similar to the upper Afroalpine soils and paleosols; Mahaney and Rutter, 1989). Fluctuations in values of carbon/nitrogen with depth, while somewhat lower than on the western side of the mountain, tend to follow the organic carbon and nitrogen distributions with depth. The youngest profile (GOR8) yielded values in the A1 and Ab horizons that, perhaps, reflect on production of carbon and nitrogen in the colder and drier late glacial-early Holocene *vs.* the later Holocene (including the climatic optimum).

Clay minerals in these Liki I, II and Holocene paleosols are dominantly allophane with traces of halloysite-metahalloysite that are probably the result of preweathering. Most paleosols younger than isotope stage 4 have not had sufficient time to produce appreciable amounts of clay minerals or other weathering products (e.g. gibbsite) (Vortisch, Mahaney and Fecher, 1987).

Table 9.7
Selected chemical parameters for Liki I and II paleosols in Gorges Valley

Site	Horizon	Depth (cm)	Extractable Cations (meq/100g)				Cation Exchange Capacity (CEC) (meq/100g)	Base Saturation (%)	E.C. (mmhos/cm 25°C)
			Ca	Mg	Na	K			
GOR1	A1	0-17	0.67	0.22	0.12	0.23	19.4	6.4	.06
	B21	17-35	0.03	0.01	0.04	0.03	6.0	2.2	< .01
	B22	35-42	0.12	0.02	0.08	0.16	30.2	1.3	< .01
	B23	42-55	0.76	0.01	0.18	0.45	27.0	5.9	< .01
	C1ox	55-89	0.83	0.01	0.62	1.14	19.8	13.1	< .01
	Cu	~ 100	0.64	0.02	0.19	1.25	20.0	49.3	< .01
	Cu	~ 120	1.11	0.02	0.36	1.05	20.2	12.6	< .01
GOR2	A1	0-14	1.00	0.28	0.17	0.34	22.5	7.8	.03
	B21	14-27	0.04	0.01	0.06	0.04	6.2	6.8	.01
	B22	27-38	0.04	0.01	0.05	0.05	7.2	2.1	< .01
	B23	38-59	0.19	0.02	0.06	0.22	24.4	2.0	< .01
	B24	59-72	0.27	0.05	0.06	0.45	27.8	3.0	< .01
	C1ox	72-115	2.00	0.60	0.56	0.61	28.0	13.5	< .01
	C2ox	115-170	3.08	0.81	0.54	1.01	26.0	20.9	< .01
GOR8	A1	0-11	0.98	0.19	0.06	0.61	25.2	7.3	.27
	IIB2	11-31	0.26	0.03	0.03	0.05	7.4	5.0	.10
	IICox	31-40	0.20	0.04	0.06	0.07	7.2	5.1	.11
	IIAb	40-45	0.56	0.12	0.05	0.17	50.8	1.8	.10

Table 9.8
Carbon and nitrogen data for Liki I and II paleosols, in Gorges Valley

Site	Horizon	Depth (cm)	Total Carbon (%)	Organic Carbon (%)	Total Kjeldahl Nitrogen (%)	C/N
GOR1	A1	0-17	6.84	6.83	0.400	17.0
	B21	17-35	0.47	0.47	0.120	3.9
	B22	35-42	2.70	2.56	0.230	11.0
	B23	42-55	1.16	1.06	0.100	10.6
	C1ox	55-89	0.21	0.21	0.070	-
	Cu	100	0.13	0.13	0.060	-
	Cu	120	0.13	0.13	0.050	-
GOR2	A1	0-14	8.01	7.24	0.550	13.2
	B21	14-27	0.95	0.95	0.130	7.3
	B22	27-38	0.40	0.40	0.120	3.3
	B23	38-59	1.41	1.30	0.140	9.3
	B24	59-72	0.99	0.89	0.120	7.4
	C1ox	72-115	0.45	0.41	0.100	-
	C2ox	115+	0.23	0.23	0.100	-
GOR8	A1	0-11	6.72	6.72	0.390	17.2
	IIB2	11-31	0.91	0.61	0.150	4.1
	IICox	31-40	0.89	0.81	0.170	-
	IIAb	40-45	11.29	8.40	0.850	9.9

9.6 Early to Late Pleistocene Buried Paleosols

A number of early to late Quaternary buried paleosols are known on Mt. Kenya (Mahaney, 1979a, 1980, 1981, 1984a). Collectively they provide important information on the weathering and air-fall deposition that occurred during the time they were exposed at the land surface (see Chapter 11 for paleoclimatic significance). Profiles were selected to provide representative data on interglacial vs. interstadial paleosols.

Profile LE3 is a compound paleosol near Lake Ellis that consists of early Liki lacustrine sediment (163-210 cm; Figure 9.2) overlain with till and loess (63-163 cm; Figure 9.2). The ground paleosol (0-63 cm; Figure 9.2) is an Inceptisol formed in outwash of late Liki age. Both the ground and buried paleosols have reddish 7.5YR colors that are more a function of lithology (Fe-rich trachytic rocks) than age. Soil colors usually approach 7.5YR 4/1-4/4 in the surface and buried B2 horizons. Pebbles and cobbles dominate in the till (IVB21bm-IVC3boxm). Because of the location of profile LE3 on the drier eastern side of the mountain and relatively young age of the profile, clay production has been minimal giving rise to friable and loose consistence throughout along with nonplastic and nonsticky properties.

The carbon and nitrogen distributions in Table 9.9 provide important information on translocation of organic components during early Liki interstadial time and in late Liki stadial and postglacial time. The buried paleosol shows smaller concentrations of C and N in the IVB21bm and IVB22bm horizons indicating considerably less movement during Liki interstadial time

Table 9.9
Carbon and nitrogen distributions in a compound paleosol (LE3) profile of Würm age

Site	Horizon	Depth (cm)	Organic Carbon (%)	Total Kjeldahl Nitrogen (%)	C/N
LE3	A1	0-12	7.13	0.355	20.0
	IIB21	12-25	1.41	0.148	9.5
	IIB22	25-36	0.34	0.026	13.1
	IIC1ox	36-55	0.34	0.019	17.9
	IIC2ox	55-63	0.25	0.024	10.4
	IIIAb	63-69	6.94	0.448	15.5
	IVB21bm	69-83	0.06	0.014	4.3
	IVB22bm	83-97	0.10	0.005	20.0
	IVC1boxm	97-127	0.04	0.004	-
	IVC2boxm	127-142	0.08	0.004	-
	IVC3boxm	142-163	0.13	0.016	-
	VCub	163-169	0.08	0.006	-
	VCubm	169-180	0.06	0.004	-
	VCuboxm	180-210	0.22	0.009	-

(~40,000-~50,000 yr BP). Because the IIIAb horizon was buried by emplacement of outwash at >43,210 yr BP, the early Liki paleosol probably formed over several thousand years followed by burial of outwash prior to the advance of late Liki ice. The data suggest a cooler drier paleoclimate during the time the buried paleosol was exposed at the land surface.

The distribution of extractable cations, CEC, total salts, and base saturation in profile LE3 (Table 9.10) tend also to support the paleoclimatic interpretation. Cation exchange capacities increase in the buried paleosol in response to increases in percent clay rather than to variations in organic carbon. Removal of soluble bases by leaching is reflected in the base saturation trend which is lowest in the surface paleosol and highest in the buried paleosol. This serves to illustrate that the buried paleosol formed in a paleoclimate capable of producing minor translocations of clay and organic matter within the upper paleosolum, but drier and cooler than the present day. Base saturation in the ground soil is somewhat lower than in the paleosol solum indicating that somewhat wetter conditions prevailed during Liki interstadial paleoclimates.

Profile LE4 (Figure 9.3) of Lake Ellis age ($>7.9 \times 10^5$ yr) at Lake Ellis is a compound paleosol with the buried component formed in till and loess (below the shovel handle). Surface stones in this profile can be cut with a knife. The till unit in profile LE4 has pebbly clay loam and pebbly sandy clay textures suggesting great antiquity; the IIIAb horizon consists mainly of clay loam material. Remarkably well preserved granular structure in the IIIAb horizon and blocky structure in the B2b horizon, indicate that clay genesis, and combinations of wetting and drying and possibly freeze and thaw processes, led to the development of blocky peds. Stickiness and plasticity increase downward in this profile in response to increasing clay content.

The data for C and N in Table 9.11 show that considerably greater movement occurred in the buried paleosolum relative to the surface paleosol. Because the till has normal remanent magnetism with a reversed overprint it was likely emplaced in either the Jaramillo or Olduvai subchrons (R. W. Barendregt, personal communication, 1988). The distribution of extractable

Table 9.10
Selected chemical parameters for a compound paleosol of early Würm age, Lake Ellis area

Site	Horizon	Depth (cm)	Extractable Cations (meq/100g)				Cation Exchange Capacity (CEC) (meq/100g)	Base Saturation (%)	E.C. (mmhos/cm 25°C)
			Ca	Mg	Na	K			
LE3	A1	0-12	1.0	0.1	0.1	0.3	23.0	6.5	.04
	IIB21	12-25	0.3	0.1	0.1	0.1	10.6	5.7	.02
	IIB22	25-36	0.3	0.0	0.2	0.1	4.2	14.3	.02
	IIC1ox	36-55	0.5	0.1	0.2	0.2	6.9	14.5	.02
	IIC2ox	55-63	1.3	0.1	0.2	0.4	8.0	25.0	.02
	IIIAb	63-69	2.8	0.3	0.2	0.4	40.0	9.3	.02
	IVB21bm	69-83	5.5	1.2	0.3	1.2	22.7	36.1	.02
	IVB22bm	83-97	5.9	1.7	0.1	1.2	22.3	39.9	.02
	IVC1boxm	97-127	0.6	2.1	0.1	1.0	24.4	15.6	.02
	IVC2boxm	127-142	0.6	1.7	0.1	1.1	21.0	16.7	.02
	IVC3boxm	142-163	1.7	0.1	0.2	0.7	19.9	13.6	.01
	VCub	163-169	12.8	3.4	0.2	1.8	37.2	48.9	.03
	VCubm	169-180	6.9	1.9	0.6	1.1	23.9	43.9	.03
	VCuboxm	180-210	13.4	3.9	0.2	1.6	33.9	56.3	.02

Table 9.11
Carbon and nitrogen in a compound paleosol of Lake Ellis age

Site	Horizon	Depth (cm)	Organic Carbon (%)	Total Kjeldahl Nitrogen (%)	C/N
LE4	A11	0-13	8.28	0.490	16.9
	A12	13-21	2.78	0.156	17.8
	B2	21-33	0.50	0.050	10.0
	IIC1ox	33-42	0.56	0.044	-
	IIC2ox	42-53	0.46	0.044	-
	IIIAb	53-71	8.52	0.725	11.8
	IVB2b	71-85	2.16	0.173	12.4
	IVC1b	85 +	0.16	0.011	-

cations (Table 9.12) shows slightly higher levels in the buried paleosolum, but otherwise there is little difference between the two units. Variations in CEC with depth reflect increased C in the A11 horizon and increased clay in the buried paleosol. The distribution of base saturation downward in the profile shows some leaching in the A12 and B2 horizons of the ground paleosol and the IIIAb/IVB2b complex in the buried paleosol. Total salts vary slightly in response to the input of organic matter.

A multi-story compound paleosol (GOR53; Figure 9.6) in Gorges Valley contains a complex sequence of sediments beginning with till of Lake Ellis age (unit IIIB2b). This till has a 5YR 3/6 reddish brown color, clay loam texture; stones are all weathered and remanent magnetism is normal with reversed overprint. Because it is younger than the oldest glaciation it may have been deposited during the Jaramillo subchron. The overlying beds of colluvium (unit II) and slope wash (I) show only minor movement of C and N (Table 9.13). The upper B2b horizon in unit II has slightly higher concentrations of organic carbon, which is probably due to loss of C during diagenesis following weathering. It is very possible that lower unit II was emplaced during the Naro Moru Glaciation and weathered during the Naro Moru-Teleki Interglaciation. Upper unit II may have been emplaced during the Teleki Glaciation and weathered during the Teleki-Liki Interglaciation. Unit I is probably of Liki age; the absence of a B horizon would suggest it must be of late Liki age.

Analysis of the distribution of extractable cations (Table 9.14) shows some removal of Ca in the ground paleosol to approximately 50 cm depth followed by slight variations with depth. Variations in EC with depth reflect the particle size distributions in most cases, except in the A1 and A3 horizons. Distributions of bases support the exractable cation data where leaching is strongest in the A1-A3-C1 complex of horizons in unit I. Total salts are low throughout, increasing slightly in the A1 horizon of the ground paleosol.

Profile GOR41 (Figure 9.7) is a compound paleosol formed in basalt (lower unit) and outwash of Lake Ellis age. The buried paleosol formed just outside the advance of Lake Ellis ice, sometime after the retreat of the Gorges Glacier. Because the weathered basalt has reversed remanent magnetism with normal overprinting it is possible that weathering occurred during the middle Matuyama Chron, followed by remobilization of Fe, and normal overprinting during the Jaramillo subchron. Some time in the later Matuyama Chron the buried paleosol was covered with younger outwash of Lake Ellis age. The paleosol in the outwash unit has normal remanent magnetism that is probably Jaramillo in age.

Table 9.12
Selected chemical parameters of a compound paleosol of early Lake Ellis age

Site	Horizon	Depth (cm)	Extractable Cations (meq/100g)				Cation Exchange Capacity (CEC) (meq/100g)	Base Saturation (%)	E.C. (mmhos/cm 25°C)
			Ca	Mg	Na	K			
	A11	0-13	5.0	0.5	0.0	0.6	25.8	23.6	.08
LE4	A12	13-21	0.5	0.1	0.1	0.2	10.4	8.7	.03
	B2	21-33	0.2	0.0	0.2	0.1	5.8	8.6	.03
	IIC1ox	33-42	0.5	0.1	0.1	0.2	8.6	10.5	.02
	IIC2ox	42-53	0.5	0.1	0.1	0.2	7.2	12.5	.01
	IIIAb	53-71	2.4	0.4	0.0	0.3	49.2	6.3	.02
	IVB2b	71-85	2.3	0.3	0.3	0.8	34.1	10.9	.01
	IVC1b	85+	4.2	0.8	0.1	1.2	28.2	22.3	.01

Figure 9.6: A multi-story compound paleosol at site GOR53, Gorges Valley. The dark lower zone at the bottom of the section is till of Lake Ellis age overlain with younger colluvium. The more highly weathered upper colluvial unit was probably weathered during the Teleki-Liki Interglaciation which is considered to have a high intensity paleoclimate (see Vortisch, Mahaney and Fecher, 1987).

Figure 9.7: Profile GOR41 in weathered basalt and loess. Note rounding of weathered basalt near the rock hammer as a result of exfoliation. The Ab horizon is developed in loess of Lake Ellis age (> $7.3\text{-}7.9 \times 10^5$ yr) that was deposited and weathered prior to burial by overlying outwash.

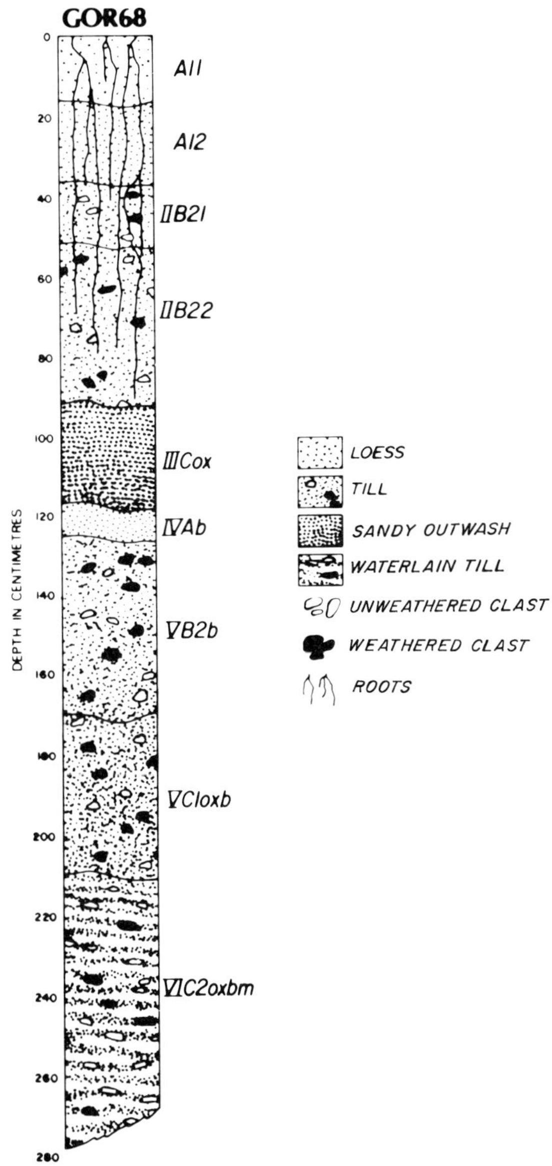

Figure 9.8 Section GOR68 with Gorges-age tills deposited during the Olduvai subchron on Mount Kenya. While this is the oldest known Quaternary glacial deposit on Mount Kenya, there is considerable scanning electron microscopy evidence for an even older pre-Olduvai glaciation on the mountain.

Table 9.13
Carbon and nitrogen distributions in a multi-story compound paleosol of middle- to late-Quaternary age in Gorges Valley

Site	Horizon	Depth (cm)	Total Carbon (%)	Organic Carbon (%)	Total Kjeldahl Nitrogen (%)	C/N
	A1	0-20	10.57	9.92	0.620	16.0
GOR53	A3	20-26	4.50	4.22	0.330	12.7
	C1	26-52	1.24	1.17	0.100	11.7
	C2	52-69	0.95	0.83	0.080	12.5
	Cu	69-82	0.12	0.18	0.050	-
	IIB2b	82-92	0.44	0.40	0.090	4.4
	IIC1b	92-106	0.27	0.23	0.060	-
	IIC2b	106-110	0.37	0.30	0.050	-
	IIB2b	110-125	0.21	0.20	0.060	3.3
	IIC1b	125-136	0.08	0.08	0.040	-
	IIC2b	136-156	0.04	0.07	0.060	-
	IIIB2b	156+	0.51	0.45	0.100	4.5

The buried paleosol has clay loam textures and 5YR and 7.5YR colors indicating it probably formed under fairly high intensity paleoclimate. Consistence is firm throughout due to high percentages of clay; most horizons are sticky and plastic. The one exception is the Ab horizon that has a silty loam texture derived mainly from reworking of local loess. The overlying ground paleosol, an Alfisol, has 7.5YR and 10YR colors, variable consistence and considerably lower stickiness and plasticity.

The data in Table 9.15 for profile GOR41 show some movement of C and N into the lower paleosolum presumably during the middle Matuyama Chron. The loess in the IIAb horizon is presumably younger than the weathered bedrock in unit III. The overlying outwash (unit I) was probably deposited during the end of the Lake Ellis Glaciation ($>7.9 \times 10^5$ yr). There is considerable evidence for movement of C and N in the ground paleosol of this pedocomplex which presumably is the result of increased leaching stages that occurred during two major interglaciations and the postglacial period (or the entire Brunhes Chron).

Profile GOR68 (Figure 9.8) is representative of multistory compound paleosols in the oldest moraines of Gorges age. The lower waterlain and lodgement tills (units V and VI) show either a reversed remanent magnetism or a reversal with normal overprinting suggesting a pre-Olduvai subchron age. The overlying till (unit II) carries a normal remanent magnetism that was probably formed during the Olduvai subchron. The very fact that in this two-till complex, the upper till gives a normal with reversed overprint suggests it was probably emplaced during the ~200,000 yr long Olduvai event.

The data in Table 9.16 show that total C is close to the values for organic C in most horizons excepting the buried B horizons of unit V. The precise quantitative, distinction between organic C and inorganic C, and increase of the latter, is probably related to the higher frequency of alkali volcanic rocks in unit V, in relation to other constituents, which gives a greater abundance of soda amphiboles (calc-alkali volcanic series, see Baker, 1967).

Overall there appears to have been greater movement in the buried paleosolum relative to the ground paleosol, which presumably results from a more intensive paleoclimate. The upper

Table 9.14
Selected chemical parameters of a multi-story compound paleosol in Gorges Valley

Site	Horizon	Depth (cm)	Extractable Cations (meq/100g)				Cation Exchange Capacity (CEC) (meq/100g)	Base Saturation (%)	E.C. (mmhos/cm 25°C)
			Ca	Mg	Na	K			
GOR53	A1	0-20	0.04	0.12	0.07	0.24	32.4	1.5	.15
	A3	20-26	0.06	0.08	0.08	0.18	27.7	1.4	.12
	C1	26-52	0.37	0.07	0.13	0.28	22.3	3.8	.08
	C2	52-69	2.67	0.59	0.13	0.51	20.7	18.8	.08
	Cu	69-82	4.43	1.38	0.15	0.66	16.4	40.4	.10
	IIB2b	82-92	2.39	0.84	0.15	0.64	17.4	23.1	.08
	IIC1b	92-106	2.72	0.83	0.14	0.61	11.9	36.1	.09
	IIC2b	106-110	3.43	0.94	0.15	0.79	17.6	30.1	.11
	IIB2b	110-125	2.66	0.79	0.18	0.69	17.9	24.1	.12
	IIC1b	125-136	2.36	0.71	0.15	0.69	17.6	22.2	.12
	IIC2b	136-156	4.13	1.05	0.15	1.02	23.7	26.8	.14
	IIIB2b	156 +	3.60	0.88	0.24	1.56	22.8	27.5	.11

Table 9.15
Carbon and nitrogen distributions of a multi-story compound paleosol in Gorges Valley

Site	Horizon	Depth (cm)	Organic Carbon (%)	Total Kjeldahl Nitrogen (%)	C/N
	A1	0-14	7.89	0.500	15.8
	B21	14-28	-	-	-
GOR41	B22	28-41	2.41	0.196	12.3
	C1ox	41-63	0.58	0.076	7.6
	C2ox	63-79	0.39	0.045	8.7
	IIAb	79-92	7.57	0.600	12.6
	IIIB2b	92-103	0.63	0.045	14.0
	IIIC1oxb	103-128	0.11	0.013	8.5
	IIIC2oxb	128-155	0.08	0.006	-
	IIIC3oxb	155-175	0.08	0.006	-

Table 9.16
Distributions of total and organic carbon and nitrogen in the oldest paleosol complex (GOR68) in Gorges Valley

Site	Horizon	Depth (cm)	Total Carbon (%)	Organic Carbon (%)	Total Kjeldahl Nitrogen (%)	C/N
	A11	0-16	7.81	7.36	0.780	9.4
GOR68	A12	16-36	4.60	5.12	0.450	11.4
	IIB21	36-51	1.31	1.09	0.120	9.1
	IIB22	51-91	0.63	0.58	0.100	5.8
	IIICox	91-117	0.09	0.06	0.080	-
	IVAb	117-125	6.74	6.56	0.580	11.3
	VB2b	125-170	4.33	3.46	0.340	10.2
	VC1oxb	170-210	5.78	4.56	0.430	10.6
	VIC2oxbm	210+	0.14	0.18	0.090	-

unit (I), with its rather thick (0.3 m) dark colored (10YR 1/1) moorland-like surface epipedon, is representative of those paleosols within the present Hagenia forest limits that obtained their present morphology during the last glacial maximum when the timberline was depressed at least 700 m. These relict features are difficult to remove because of the impoverishment of numbers of bacteria and fungi in paleosols above ~3000 m (Mahaney and Boyer, 1986a).

9.7 Conclusions

Soils and paleosols on Mount Kenya provide important information on trends of soil profile development within the Afroalpine environment at ~4000 m.a.s.l. The morphogenesis of these soils follows pathways that are probably close to the processes that occurred lower down on the mountain at elevations of ~3000 m.a.s.l., near or at the end of each glacial episode when fresh moraine material was left behind by retreating glaciers. Presumably within a few thousand years, at the end of each glacial period, Afroalpine pedocomplexes came into or approached near equilibrium (steady state) conditions with the ameliorating paleoclimate.

In older pre-Pleniglacial paleosol complexes it is possible to study brief or long lasting "time windows" that give considerable insight into interglacial and interstadial paleoenvironments. Many of the compound paleosol complexes discussed in this chapter were exposed to subaerial weathering processes for variable lengths of time, but usually for not less than ~100,000 yr. With the increasing array of dating methods at our disposal it is becoming increasingly possible to establish the time of exposure with some precision. Comparing buried paleosols with their relict counterparts (ground paleosols) still exposed to the subaerial environment for different lengths of time, under different paleoclimates, it is possible to determine what leads to the development of "welded" (Birkeland, 1984) paleosol properties.

The very fact that many buried paleosols on Mount Kenya have cemented subsurface horizons prompted a study of the relationships between weathering, age, and paleoclimate that is more fully addressed in Chapter 11. Briefly, it appears that any tendency to produce warmer and wetter conditions (even low intensity interstadial or postglacial paleoclimates), leads to increased soil water movement, ground water fluctuations, and hence liberation of Si and Mn + Fe that eventually reprecipitate as a coating cementing matrix material together (see Derbyshire and Mahaney, 1989 for a discussion).

CHAPTER 10

AMINOSTRATIGRAPHY

10.1 Introduction

The fate of amino acids in buried paleosol horizons is of considerable interest to Quaternary geologists and pedologists, particularly because the racemization of aspartic acid and glutamic acid, among others, provides a means of dating organic-rich buried paleosol horizons (Mahaney, Boyer, and Rutter, 1986). Dated paleosols ranging from ~960 to >42,000 yr BP have been described on Mount Kenya (Figures 10.1 and 10.2; for location of site TV23 see Figure 5.2) (0°10′S; 37°15′E) in several drainages between 2990 and 4250 m elevation (Mahaney, 1979a, 1981, 1982a, 1982b, 1984a, 1984b and 1985). Most paleosols (Figures 10.3) have developed in transported regoliths of variable lithology (e.g., till, debris-flow, outwash, loess, tephra, etc.). Paleosol sequences at different elevations (*e.g.* ~3000 and ~4000 m), calibrated with radiocarbon, provide a convenient field setting in which to test the reliability of amino acid racemization for age determination. Moreover, chronostratigraphic refinement of amino acid racemization assists in increasing our understanding of the timing of different geologic events before, during, and after the last (Liki) glaciation on Mount Kenya.

Many varieties of fossils (including paleosols) are routinely analyzed to determine D/L ratios of amino acids for relative age determination. The amino acid dating method is particularly useful in cases where fossils and/or buried paleosols have experienced similar thermal histories and diagenetic conditions. As pointed out in Chapter 3, marine molluscs have been successfully and widely utilized for age determination, whereas freshwater and terrestrial molluscs are difficult to analyze (Andrews *et al.*, 1979; Belknap and Wehmiller, 1970; Miller and Hare, 1975; Mitterer,

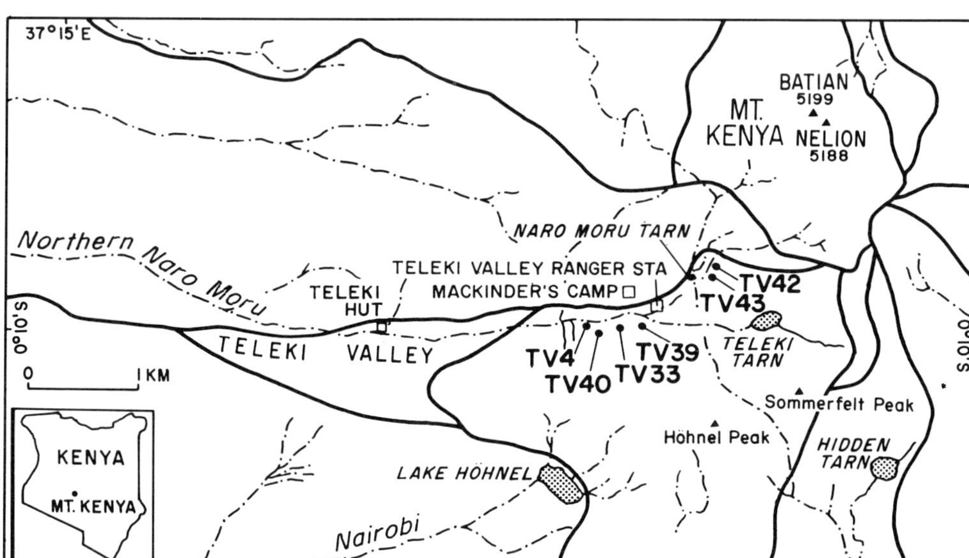

Figure 10.1: Location of sites on the western flank of Mount Kenya from which samples were collected for amino acid dating.

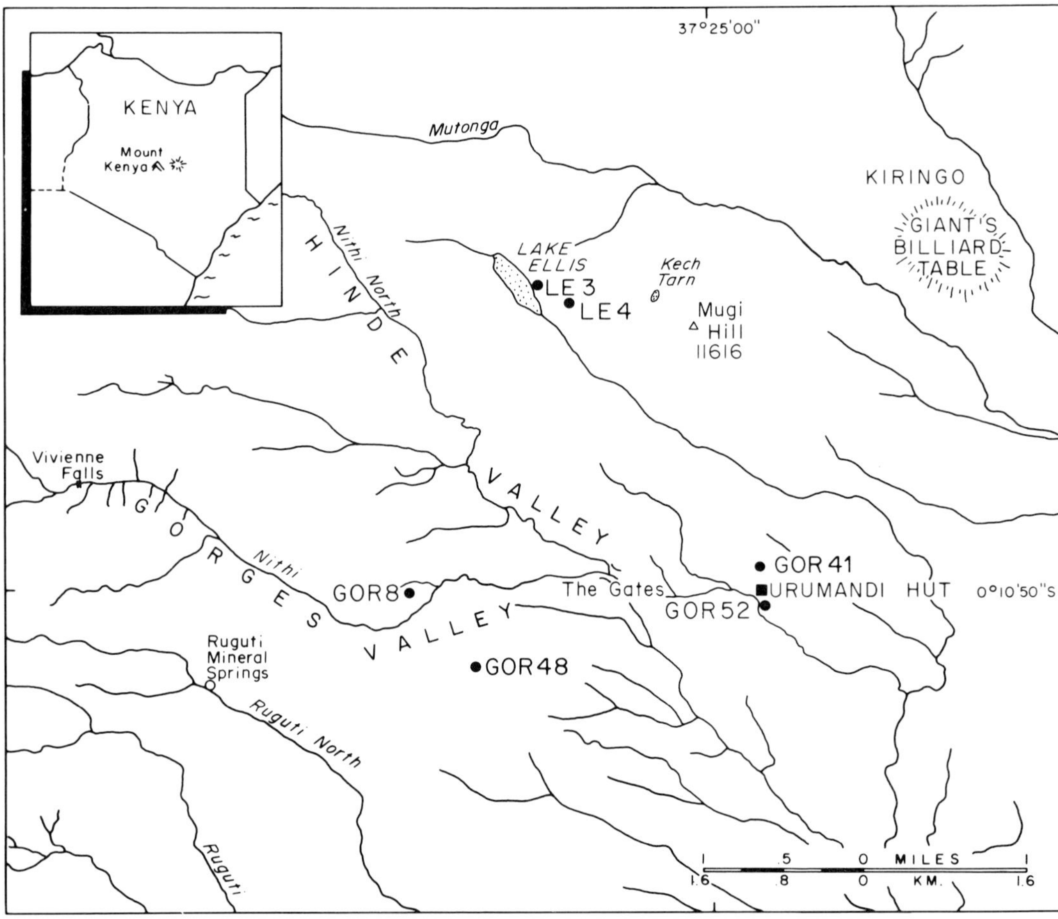

Figure 10.2: Location of sites on the eastern flank of Mount Kenya from which samples were collected for amino acid analysis.

1974, 1975; Rutter *et al.*, 1980; Wehmiller, 1982). Bone and wood have proved most useful as well, although numerous problems have been encountered (Bada, 1972; Bender, 1974; Hare, 1974; Rutter and Crawford, 1984). In addition, marine sediments (Bada *et al.*, 1970; Kvenvolden *et al.*, 1970), foraminifera (King, 1980; Wehmiller and Hare, 1971), coral (Wehmiller and Hare, 1970), and paleosols (Limmer and Wilson, 1980; Mahaney, Boyer and Rutter, 1986; and Mahaney, 1989a) have all been utilized for amino acid racemization dating.

Racemization (or epimerization) of amino acids in geological specimens is a complex process involving a number of factors. The rates of interconversion processes are dependent upon the type of amino acid, ambient temperatures, types of materials in which the amino acids are embedded (shell, bone, wood, *etc.*), as well as the "state" in which the amino acids occur, *i.e.*, whether they are terminally bound, internally bound, or free amino acids. In addition, other external diagenetic factors including moisture content, acidity, and oxidation-reduction conditions may play important roles affecting the rate of racemization. The accuracy of D/L determinations depends upon accurate field descriptions, laboratory preparations, and the reliability of the analytical equipment employed (see Figure 10.4). These factors also contribute to problems using amino acid D/L ratios to date paleosols. The complex nature of soil genesis

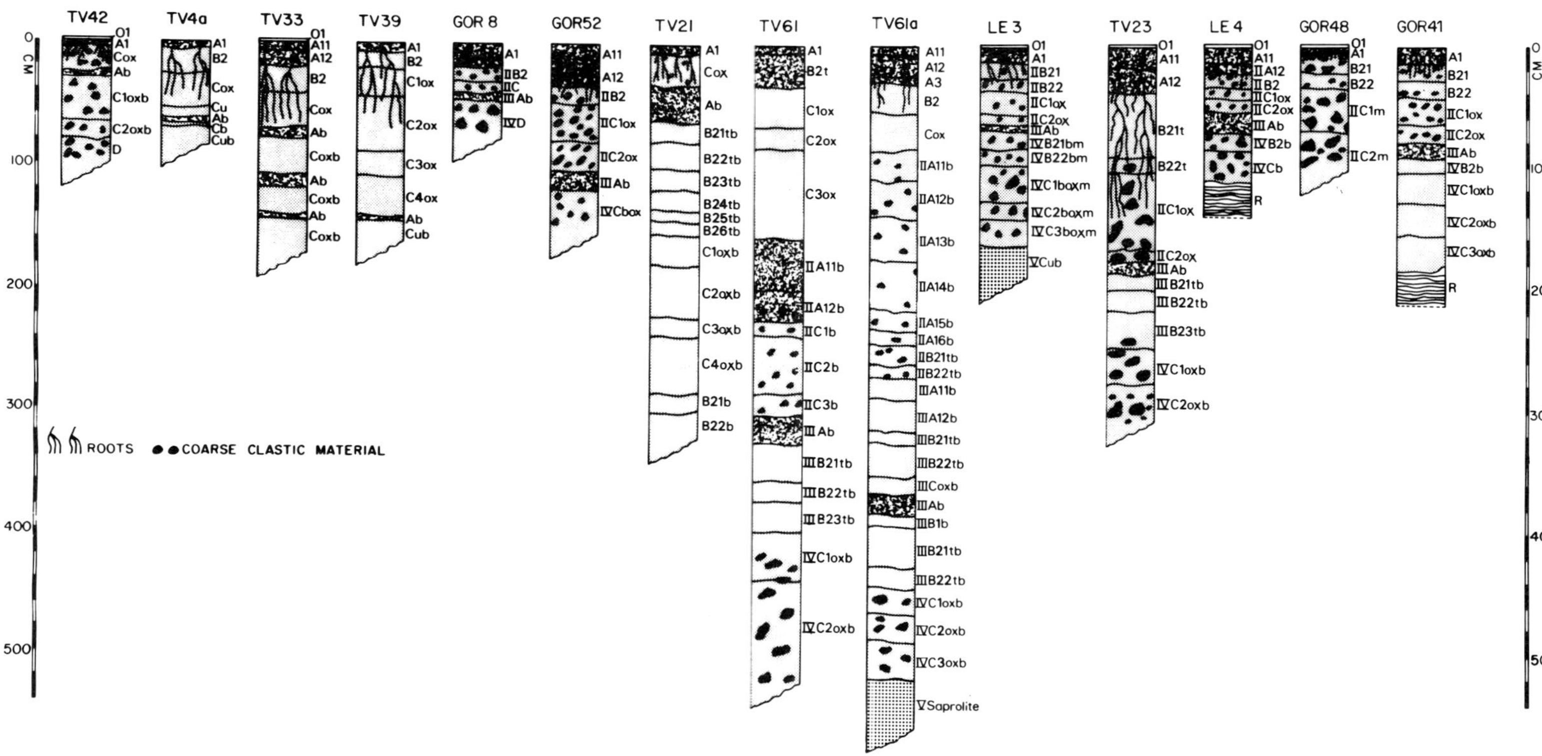

Figure 10.3: Sequence of dated paleosols on the Mount Kenya Massif.

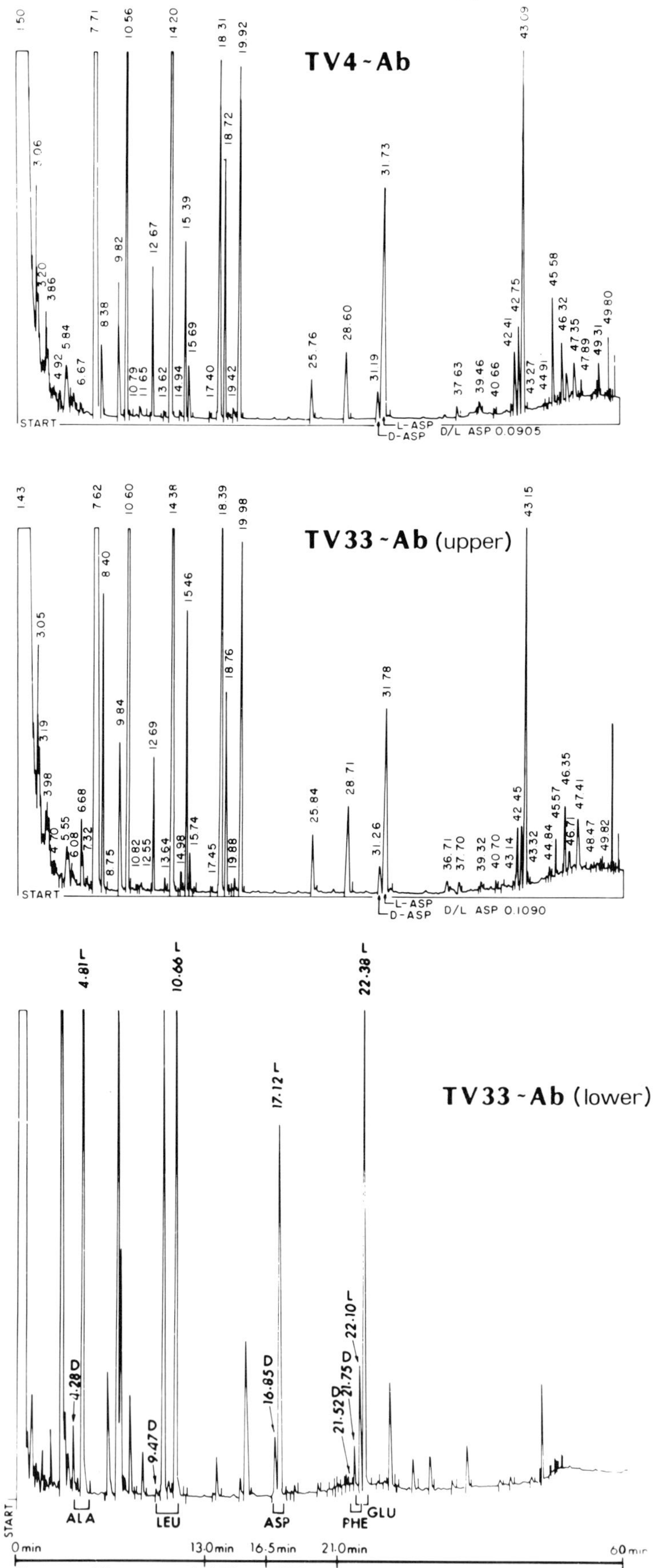

Figure 10.4: Amino acid spectra for a suite of middle to late Holocene paleosols dated by radiocarbon; Profile TV4-Ab (1940 ± 100 yr BP); Profile TV33-Ab (upper) (3610 ± 180 yr BP); Profile TV33-Ab (lower) (3400 ± 170 yr BP). The lower sample in this sequence is probably contaminated with amino acids brought in by ground water influx (Mahaney and Boyer, 1986a). See Table 10.1 for information on radiocarbon dates and D/L ratios of amino acids.

and diagenesis that occur at the surface and, after burial, compound the problems already inherent in the method. However, the data presented in this chapter show a trend for D/L ratios, of aspartic acid and others, to increase with age of the paleosol. Moreover, the depth distributions in two paleosols of late glacial and Neoglacial ages show some interesting differences that are related to age and geobiochemical contamination.

10.2 Sampling Sites

A number of glacial and periglacial deposits of Pleistocene and Holocene age are found in the Afroalpine and *Hagenia* woodland/bamboo forest belts of Mount Kenya (Figure 10.3). These glacial and nonglacial deposits have been assigned ages on the basis of topographic position (Baker, 1967), as well as weathering features, degree of soil expression, radiocarbon, and thermoluminescence dating (Mahaney, 1979a, 1980, 1981, 1982a, 1982b, 1983a, 1984a, 1984b, 1984c and 1986).

10.3 Climatic Limitations

The Afroalpine area rises from ~3200 m to the summit of Mount Kenya at 5199 m (Figs. 10.1 and 10.2). Precipitation derived from the Indian Ocean is delivered by the tropical easterly wind belt as well as by the classical and Atlantic monsoon systems. The southeastern slopes are wettest with precipitation reaching 2500 mm (at 2250 m); the northern slopes are dry with ~1000 mm annually. On the western flank of the mountain, precipitation increases from *circa* 500 mm at Naro Moru (2000 m), to *circa* 1500 mm on the high moorland (*circa* 3500 m), and then decreases in the central peaks (*circa* 4300 m) to 750-800 mm (Coetzee, 1967). Temperature extrapolations (Coe, 1967, Thompson, 1966) suggest that temperatures at 4200 m average 3.1 to 3.6°C (from observations taken during two periods in 1948; Coetzee, 1967; Hedberg, 1964). At present, temperatures at *circa* 3000 m approximate 10°C (mean annual); the temperature depression during the last glaciation is estimated at 7°C (Coetzee, 1967, p. 98) giving an estimated mean annual temperature of 3.0°C from 15,000 to 25,000 yr BP.

The depth of frost penetration in Afroalpine sites while limited to a few cm, is diurnal in nature, and controlled largely by variations in thermal conductivity of different cover sediments. Variations in subsurface (~1.0 m depth) temperatures, as they might affect racemization rates of amino acids are considered small even between Pleistocene cold periods and warmer Holocene climate. Frost penetration in surface deposits in the *Hagenia* woodland/bamboo forest (<3200 m), while nil at present, may have occurred to depths of a few cm during the last glaciation when glaciers reached ~3200 m. However, in the Afroalpine zone variations in surface temperature during the Quaternary probably did not affect buried paleosols insofar as racemization rates are concerned.

10.4 Vegetation Influences

Below timberline, the subalpine forest consists of *Hagenia* woodland (*Hagenia abyssinica*) and bamboo (*Arundinaria alpina*). Above the timberline and the heather zone (~3200 m), the lower Afroalpine zone, dominated by *D. keniensis*, extends upwards to ~3700 m where it merges with the upper Afroalpine zone characterized by *D. keniodendron*. Vegetation elements of the Upper Afroalpine zone merge with species of the nival zone at *circa* 4500 m (Coe, 1967; Hedberg, 1964, Figs. 10.1 and 10.2). The probability that different species of plants produce variable ratios of amino acids was considered insofar as they might affect the amount of each L-form amino acid that might be present in a paleosol at time zero, although not the D/L ratios which should be close to zero.

10.5 Buried Paleosols

Buried paleosols at a number of sites formed in different parent materials are shown in sequence from youngest to oldest (Figure 10.3). Some of these paleosols have been described in previous publications (Mahaney 1982a, 1984b, 1985, 1986, 1989a; Mahaney & Boyer, 1986a; Mahaney, Boyer and Rutter, 1986); however, older profiles such as TV61A were described by Mahaney (1989b). The ^{14}C ages of these buried paleosols were determined using whole-soil samples taken from the center of each buried Ah or Al horizon, air dried on aluminum foil in the field, and stored and shipped frozen to laboratories in Canada, Japan and the U.S.A. The clay mineral and aluminosilicate composition is dominantly allophane in the younger Holocene and late-pleistocene units (< 100,000 yr BP), with halloysite/metahalloysite and gibbsite increasing in the older pre-Pleniglacial systems (Vortisch, Mahaney, and Fecher, 1987).

The youngest buried paleosol (site TV42), collected from a prominent debris flow in upper Teleki Valley (Figure 10.1; Table 10.1), formed in pebbly and cobbly material which subsequently was buried by debris flow sediment from renewed mass wasting activity that probably coincided with the end of the Tyndall advance (see Chapter 8) (Mahaney, 1984a, 1985, Mahaney, Boyer and Rutter, 1986).

With increasing age the next oldest paleosol was described in a valley train deposit in upper Teleki Valley (site TV4a; Figure 10.1). This buried paleosol in clayey silty materials was covered by pebbly sandy alluvium approximately 1940 $\pm$120 yr BP; the Ab horizon in the buried profile has been dated by thermoluminescence at 1170 $\pm$270 yr BP (Mahaney, 1985; pers. comm., D. Huntley, 1985).

Alluvial fans, collectively called the Darwin Formation (Mahaney, 1984a, 1985) formed between 4000 and 8000 radiocarbon years ago. Profile TV33 (Figure 10.1) consists of an Inceptisol (ground soil) over three buried Entisols that radiocarbon date between 3600 $\pm$180 yr BP (GaK-11320) and 5380 $\pm$130 yr BP (GaK-11321). The lowermost Ab horizon in this sequence radiocarbon dates at 3400 $\pm$170 yr BP (GaK-11322) and is known to be contaminated by microorganisms (Mahaney and Boyer, 1986a). Profile TV39 (Figure 10.1 for location) contains an Inceptisol (ground soil) over a buried paleosol in slopewash. The Ab horizon was covered by slopewash 4540 $\pm$130 yr BP.

On the eastern slopes profile GOR8 (Figure 10.2) in Liki I Till was covered with alluvium at 5110 $\pm$120 yr BP. Profile GOR52 in Gorges Valley (Figure 10.2) contains alluvium deposited sometime in the late Pleistocene or early Holocene. An Entisol forming in these sediments was covered by younger alluvium 5100 $\pm$100 yr BP.

Profile TV21 on the western flank of the mountain at 2850 m (Figure 10.1) contains a sequence of buried paleosols in slope wash (Figure 10.3). The Ab horizon in this sequence dates at 12,560 $\pm$170 yr BP (BGS-760) (Table 10.1). At slightly higher elevation (2960 m) profile TV61 contains a sequence of buried A horizons in colluvium and loess yielding ^{14}C dates between 21,530 $\pm$430 (Beta-11371) and 25,410 $\pm$260 (Beta - 10823). Profile TV61A, located within 15 m of site TV61, contains a complex of paleosols in slopewash, colluvium, loess, till, and residual regolith; the upper part of this sequence has been dated by ^{14}C.

Profile LE3 (Figure 10.2) is a compound paleosol containing a buried component formed in loess over till. The Ab horizon radiocarbon dates at > 43,210 yr BP. The buried paleosol overlies a truncated unweathered deposit of lake sediment which probably predates the Liki Glaciation (Figure 10.2). The surface paleosol is formed in outwash and loess of late Liki age. Both the ground and buried paleosols are Inceptisols, which are common above timberline on Mount Kenya. Coarse textures prevail in the ground paleosol suggesting only slight weathering; however, in the buried soil a silty loam texture (loess) grades downward into till with sandy loam textures. The cemented lower paleosolum and subsoil in the buried paleosol, while common in older

Table 10.1 D/L ratios for amino acids in buried paleosols on Mount Kenya, East Africa

Site[a] and Horizon	Alanine	Aspartic Acid	Glutamic Acid	Leucine	Phenylalanine	Valine	^{14}C Date (yr/BP)
TV40-Al Afroalpine	.05	.07	.07	.01	.03	-	modern
TV4-Al Afroalpine	.10	.08	.12	.02	.03	-	modern
TV43-Al Afroalpine	.11	.10	.12	.01	.04	.02	modern
TV42-Al Afroalpine	.10	.08	.11	.01	.03	-	modern
TV61A-A11 Hagenia forest	-	.09	-	-	-	-	modern
TV42-Ab	.19	.18	.14	-	-	.03	920 ± 100(GaK-11240)
TV4-Ab	.09	.09	.09	.01	.01	.01	1940 ± 100(GaK-8273)
TV33-Ab (upper)	.11	.13	.09	.02	-	-	3610 ± 180(GaK-11320)
TV33-Ab (lower)	.08	.16	.07	.01	.02	.02	3400 ± 170(GaK-11322)
TV39-Ab	.20	.16	.13	.12	-	.08	4540 ± 130(GaK-11241)
GOR8-Ab	.21	.26	.22	.04	-	.15	5110 ± 120(GaK-11570)
GOR52-Ab	.27	.29	.21	-	-	-	5110 ± 100(Beta-10821)
TV2l-Ab	.28	.49	.31	-	-	-	12,560 ± 170(BGS-760)
TV6l-IIAIIB	.32	.51	.29	-	-	-	21,530 ± 430(Beta-11371)
TV6l-IIAI2b	.27	.61	.31	-	.11	-	25,170 ± 450(Beta-11372)
TV6l-IIIAb	.40	-	-	-	-	-	25,410 ± 260(Beta-10823)
TV61A-IIA12b	-	.45	-	-	-	-	b
TV61A-IIA13b	-	.35	-	-	-	-	b
TV61A-IIA15b	-	.40	-	-	-	-	26,050 ± 410 (Beta-18501)
TV61A-IIIAb	-	.56	-			-	27,480 ± 290 (Beta-18495) 34,200 ± 3800 (BGS-1162)[c]
LE3-Ab	.22	.45	.22	.29	-	-	> 43,210 (Beta-9085)
TV23-Ab	-	-	-	-	-	-	> 40,000 (GaK-11319)
LE4-Ab	.23	.33	.21	.05	-	.13	> 42,629 (Beta-9086)
GOR48-carbon in B2 horizon	.21	.27	.19	.08	-	.18	b
GOR4l-Ab	.21	.32	.17	.10	-	-	b

a. Sites are located on Figures 10.1 and 10.2
b. Not dated
c. Charcoal
Nil

pre-Pleniglacial paleosols on Mount Kenya, is not so common in younger Liki systems. The coarse-textured surface paleosol and finer-textured and cemented buried paleosol may result from wetter conditions following emplacement of Liki Till. Because the paleosol formed in buried till and loess is weathered to a depth of 100 cm, at least several thousand years would be required for its development. The surface paleosol must have formed under a dry climate throughout the late Pleistocene and Holocene.

Paleosol profile TV23 (Figure 10.3) is a compound buried paleosol overlain with a ground paleosol with somewhat similar characteristics (Mahaney and Boyer, 1986a). Both the buried and surface paleosols are Alfisols, which are more common below timberline on Mount Kenya. Heavy textures in the B21t and B21tb horizons are the result of downward movement of clay in the profiles, which directly influences the development of weak blocky structures and clay films. Reddish colors in this profile result partly from the chemical composition of the parent material which is ultrabasic in nature and high in iron (Mahaney 1982a, and 1984b). The surface paleosol, formed in loess and till of Teleki age, and the buried paleosol formed in older loess plus tephra and till of Naro Moru age, are representative of paleosols formed in older tills of pre-Liki age found elsewhere along the western flank of Mount Kenya (Vortisch, Mahaney and Fecher, 1987). Along the northern and eastern flanks of the mountain, tills of similar age occur in similar stratigraphic positions, but the prevailing vegetation is alpine moorland, not bamboo forest. Despite a dark color in the IIIAb horizon, attempts at radiocarbon dating provided only an infinite date of >40,000 yr BP. Dating by thermoluminesence places this horizon at 62 $\pm 20 \times 10^3$ yr (personal communication, G. Berger, January 1985).

Site LE4 (Figure 10.3), located in a moraine ridge at 3520 m, contains pre-Teleki Till and outwash over bedrock. The ground paleosol forming in loess and outwash to a depth of 54 cm is an Inceptisol. Between 54 and 126 cm depth the buried paleosol represents weathered pre-Teleki Till, the Ab horizon of which gives an infinite radiocarbon date of >42,629 yr BP (Beta-9086). The data suggest that following emplacement of Naro Moru Till a paleosol formed in cobbly material upon which 18 cm of loess was emplaced. Because this paleosol contains a finer texture throughout, it must have been exposed at the surface for a considerable time following till emplacement and prior to the deposition of the outwash body.

Profile GOR48 (Figure 10.2), located in Naro Moru Till in Gorges Valley does not contain any buried paleosols. However, cemented subsoil horizons together with its topographic position provide evidence for considerable age. Baker (1967) considered this surface to represent the earliest glaciation on the mountain; however, it is known to have normal magnetic polarity placing it within the lower Brunhes Chron (<730,000 yr) (Mahaney, Barendregt and Vortisch, 1989).

Site GOR41 (Figure 10.2), situated on the highest strath terrace in the Nithi River Basin along the eastern slopes of Mount Kenya, contains two paleosols. The lowermost paleosol in the sequence was formed from bedrock emplaced ~2.0 my. This lower unit was later buried by stream activity, presumably during the Lake Ellis Glaciation (>730,000 yr).

10.6 Amino Acid Racemization

Paleosols represent one of the most difficult natural materials to analyze for D/L ratios of amino acids because they are subject to so many biogeochemical influences (Mahaney and Boyer, 1986a; Mahaney, Boyer and Rutter, 1986). Many aspects of contamination are reported by Limmer and Wilson (1980) in allophanous soils of New Zealand. In addition to their complex genesis, and diagenetic changes that take place at the surface, and after burial, soil organic matter is difficult to isolate for analysis with gas chromatography (Mahaney, Boyer and Rutter, 1986). To determine the probability of biogeochemical contamination the ratio of extractable Ca/Mg, and pH, were analyzed with depth in each paleosol to elicit information on downward movement. While some downward translocation occurred in ground profiles on the western flanks of the mountain, very

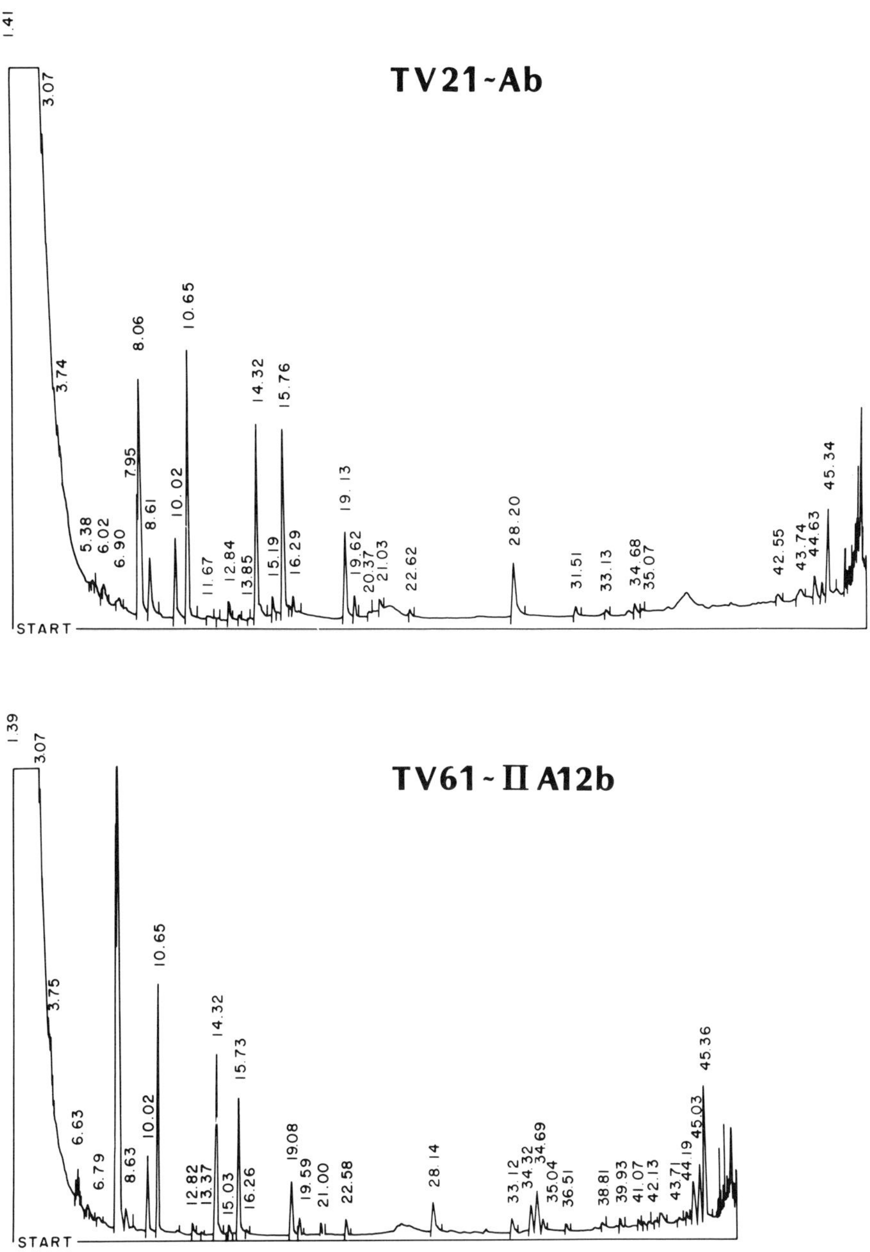

Figure 10.5: Amino acid spectra of two paleosols: Profile TV21-Ab (12,560 ± 170 yr BP) of late glacial age, and profile TV61-IIA12b (25,170 ± 450 yr BP) of Pleniglacial age.

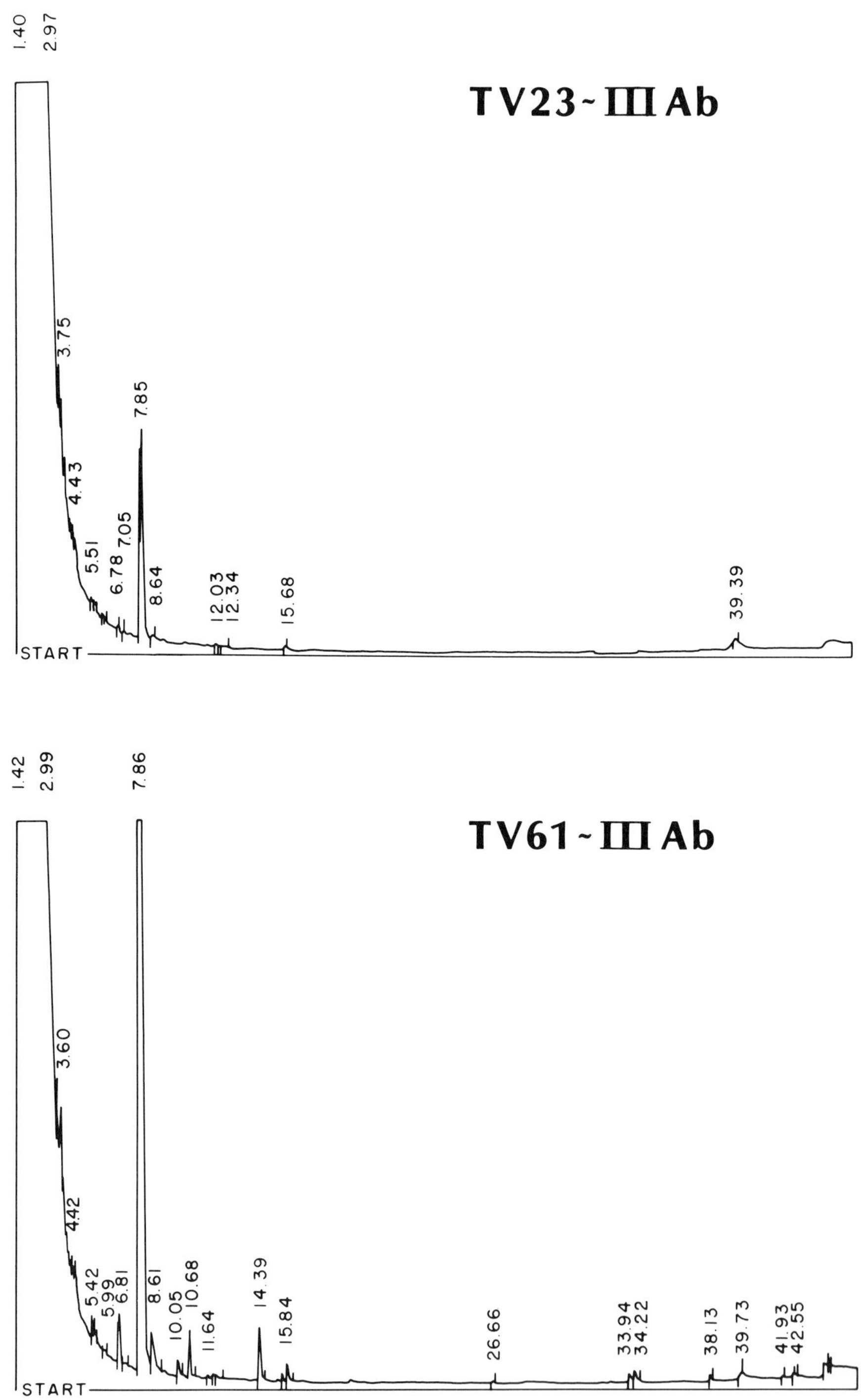

Figure 10.6: Amino acid spectra of two paleosols: Profile TV23-IIIAb (>40,000 yr BP) which is considered coeval with the Liki-Teleki Interglacial (>100,000 yr BP) (Mahaney 1982a, 1983a, 1986), and TV61-IIIAb (25,410± 260) contaminated with fungi and bacteria, suggesting the C-14 age is too young.

little occurred on the eastern flank, presumably in response to drier climate there. In all cases studied, downward translocations occurred mainly in surface paleosols, leaving the buried paleosols unaffected and with apparently valid radiocarbon dates as well as amino acid values.

The amino acid composition in these buried paleosols changes dramatically with age for the samples studied in different sediments. In the sequence shown in Figures 10.4, 10.5 and 10.6 and Table 10.1, the amino acid spectra for Holocene (at 4,000 m) buried paleosols yield similar values for all amino acids, with aspartic acid showing a sequential increase in D/L ratio with greater age (Table 10.1). In older paleosols (at ~3,000 m) in late glacial, Pleniglacial and pre-Pleniglacial age substrates, amino acid spectra show a marked decline in the number of acids. The oldest samples—TV23-IIIAb and TV61-IIIAb—give a plateau-like spectra with few acids present, although it is not always possible to obtain reliable D/L ratios for samples with such low concentrations.

The amino acids extracted from the samples are genuine fossil entities, as indicated by the data in Table 10.1. The evidence for this comes from the sequential decline in the number of amino acids present with increasing age (Figures 10.4, 10.5 and 10.6). Recent contamination of these paleosols would involve plant roots and downward movement, which would introduce new amino acids increasing the concentrations of the six acids shown in Table 10.1. While there does not appear to be any contamination in the samples shown in Figures 10.4, 10.5 and 10.6, it may prove possible to accept or reject some radiocarbon dates when amino acid compositions are known. For example, the TV23-Ab horizon is considered to have great antiquity of over one hundred thousand years which is supported by its amino acid composition (Table 10.1) and its spectra (Figure 10.6). Sample LE4-Ab is considered to have a similar age, but because the amino acid composition shows several amino acids present, and because the D/L ratio for aspartic acid is only .33 (Table 10.1), there is some possibility of contamination from above by root activity and/or airfall influx, or from below by ground water movement. That little contamination affected this horizon is supported by an infinite ^{14}C date and low microflora counts (Mahaney & Boyer, 1986a). Even though other amino acids such as leucine, phenylalanine and valine often behave erratically (Mahaney and Rutter, 1988), they can be used to identify horizons that may have been subjected to geobiochemical contamination. The data in Table 10.1 show a rapid disappearance for leucine, phenylalanine and valine followed by sporadic reappearances in older buried horizons that may signal contamination.

The D/L ratios for aspartic acid (Table 10.1) were plotted against time (Figures 10.7 and 10.8) at elevations of ~3000 and ~4000 m in an attempt to determine the effect of temperature on rates of recemization.

The curves for the Afroalpine zone at ~4000 m (Figure 10.7) and Hagenia forest at ~3000 m (Figure 10.8) show somewhat different racemization rates for the different elevations. Because buried paleosols in the Afroalpine are limited to the last few thousand years (Figure 10.7), it is only possible to compare extrapolated curves with the first ten thousand years of racemization at ~3000 m. The results show that aspartic acid gives a D/L value (at 10,000 yr) of .27 at ~3000 m and .21 at ~4000 m. Comparing alanine at the two elevations gives values (at 10,000 yr) of .17 at ~4000 m and .21 at ~3000 m. Glutamic acid, which appears useful for relative dating of younger Afroalpine buried horizons at 4000 m, is of less utility at 3,000 m (Table 10.1). Both in the Afroalpine and in the *Hagenia* woodland aspartic acid appears to give the most consistently useful D/L ratios (Mahaney, Boyer and Rutter, 1986; Mahaney, 1989b).

One factor that might contribute to producing consistent D/L ratios is that the samples are from the tropical Afroalpine zone where diurnal temperature changes are to the order of 20°C and do not penetrate ground soils to more than a few cm. Minimization of frost penetration by close proximity to the equator tends to eliminate the effect of frost penetration as a factor important in controlling the rate of recemization (Mahaney, Boyer and Rutter, 1986).

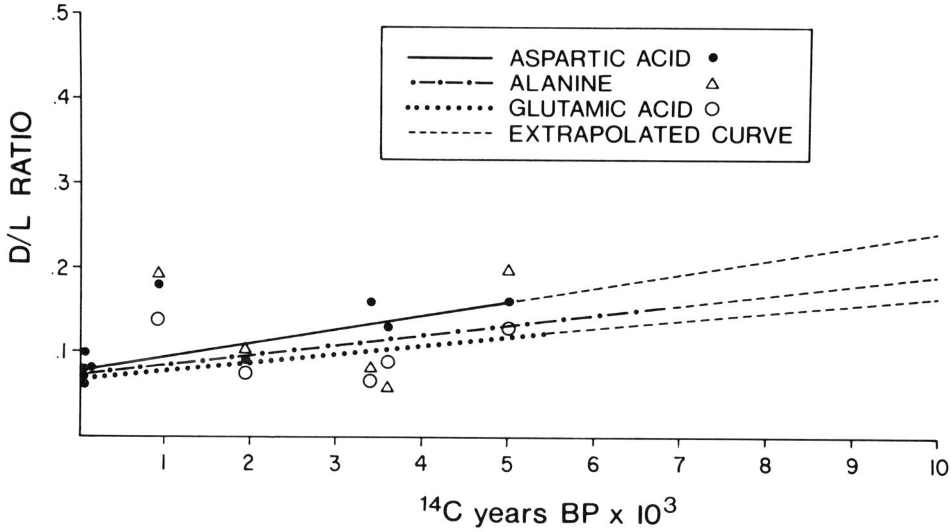

Figure 10.7: Curves for aspartic acid, glutamic acid, and alanine in the Mount Kenya paleosol sequence at ~4000 m.

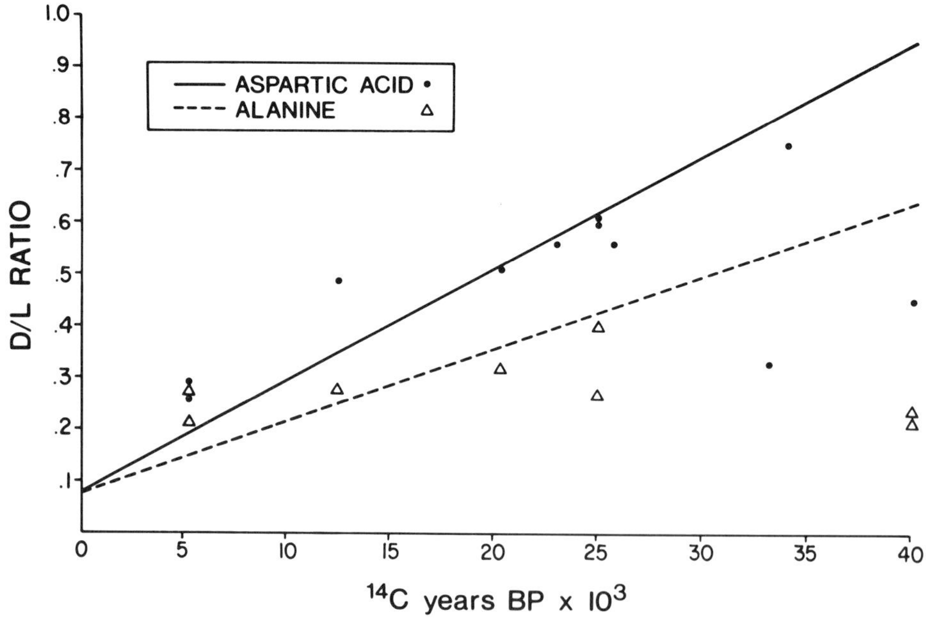

Figure 10.8: Curves for aspartic acid and alanine in the Mount Kenya paleosol sequence at ~3000 m.

Therefore, amino acid D/L ratios appear to provide relative ages for buried paleosols. In addition, D/L ratios often provide a useful criterion for acceptance or rejection of radiocarbon dates.

10.7 Particle Size

Particle size ratios were studied to determine if they might influence amino acid values either by clay adsorption or by moisture retention. The data in Table 10.2 reveal considerable variations in percentages of sand, silt and clay, reflecting differences in geological origin and age. Even percentages of clay vary widely among the buried A horizons in the paleosol sequence, and although some amounts are essentially nil (*e.g.* GOR48 and 41) there is no apparent relationship with the concentration of amino acids.

10.8 Distributions in Surface Soils

One paleosol of late glacial age and a soil of Neoglacial age in the Afroalpine belt (~ 4300 m) were analyzed to determine if D/L ratios of amino acids could be used for relative age determinations and as a means of assessing biogeochemical contamination. The two profiles (Figure 10.9) are prominent stratigraphic markers used to differentiate late glacial (*circa* 12,500 yr BP) from early Neoglacial (*circa* 1000 yr BP) glacial deposits which are commonly located in proximity to one another.

The sites (TV65 and 66) are located on the boundary between the nival and upper Afroalpine vegetation zones (Coe, 1967) below the Lewis Glacier (Figure 10.9). The Neoglacial sequence shown on Figure 8.16 is described in Mahaney (1979a, 1985 and 1987a), in Spence and Mahaney (1988a), and in Mahaney and Spence (1988). The vegetation at the field sites (Figure 2.20) belongs to the upper Afroalpine and consists mainly of tussocks of *Festuca pilgeri* and *Dendrosenecio. keniodendron* with scattered *Poa schimperiana (e.g.* similar to the upper Afroalpine vegetation 100 m below in Teleki Valley (Mahaney & Boyer 1986a) influenced by the tropical easterly wind belt (Pedgley, 1966; Thompson, 1966). Rainfall arrives with the monsoon giving dry conditions from December to March when northeasterly winds reach the mountain from the deserts to the north; wet conditions from July to October are accompanied by southeasterly winds from the Indian Ocean. Daily weather tends to bring katabatic (downhill) and anabatic (uphill) wind patterns producing precipitation estimated at 700-800mm; for a detailed discussion of the climate see Chapter 2.

The two profiles TV66 (Entisol) and TV65 (Inceptisol) belong to an important sequence of soils formed in Tyndall (Neoglacial) and Liki III (late glacial age) tills (Mahaney, 1985, 1987a). Root systems penetrate to shallow (~ 15 cm) depths in both profiles and coarse clastic materials at depth are typical of alpine tills found elsewhere (Mahaney, 1978; Mahaney & Spence, 1984). No attempt was made to differentiate the two deposits using counts of unweathered/weathered stones; however, TV66 had fresh stones throughout and TV65 contained only a few weathered stones (mainly phonolite) in the Cox horizon. The major differences between the two profiles are the relative depths of weathering (Figure 10.9) and the ~ 30 cm loess cap on the surface of the TV65 profile.

The distributions of sand, silt, and clay give loamy sand to sandy loam textures in the surface becoming fine sandy loam and loam with depth. Particle size distribution curves are linear which is typical for heterogeneous sediments that range from poorly sorted to nonsorted (Mahaney and Rutter, 1989).

The mineralogy of the $< 2\,\mu m$ fraction in the two profiles shows only trace amounts of illite in the Cox horizon of the oldest profile. No other crystalline clay minerals are present in the profiles. Quartz, an aeolian indicator on Mount Kenya (Vortisch, Mahaney & Fecher, 1987) is

Table 10.2 - Particle-size distributions for the buried paleosol horizons in the Mount Kenya Quaternary Sequence

Sample	Depth (cm)	< 2 mm Sand % (2mm-63μm)	Silt % (63-2μm)	Clay % (<2μm)
TV42-Ab	25-29	47.0	36.0	17.0
TV4-Ab	63-67	37.4	45.6	17.0
TV33-Ab (upper)	71-78	52.2	33.8	14.0
TV33-Ab (lower)	140-144	61.7	29.3	9.0
TV39-Ab	142-146	69.4	22.7	7.9
GOR8-Ab	40-45	-	-	-
GOR52-Ab	105-121	50.6	39.4	10.0
TV21-Ab	42-66	17.5	58.5	24.0
TV61-IIAllb	160-200	92.8	7.2	0.0
TV61-IIA12B	200-226	91.8	8.2	0.0
TV61-IIIAb	306-330	88.8	11.1	0.1
LE3-Ab	63-69	37.6	59.2	3.2
TV23-Ab	176-186	45.5	40.0	14.5
LE4-Ab	53-71	17.1	64.9	18.0
GOR48 (carbon in B horizon)	3-25	81.8	18.0	0.2
GOR41-Ab	79-92	72.8	27.0	0.2

\- not available

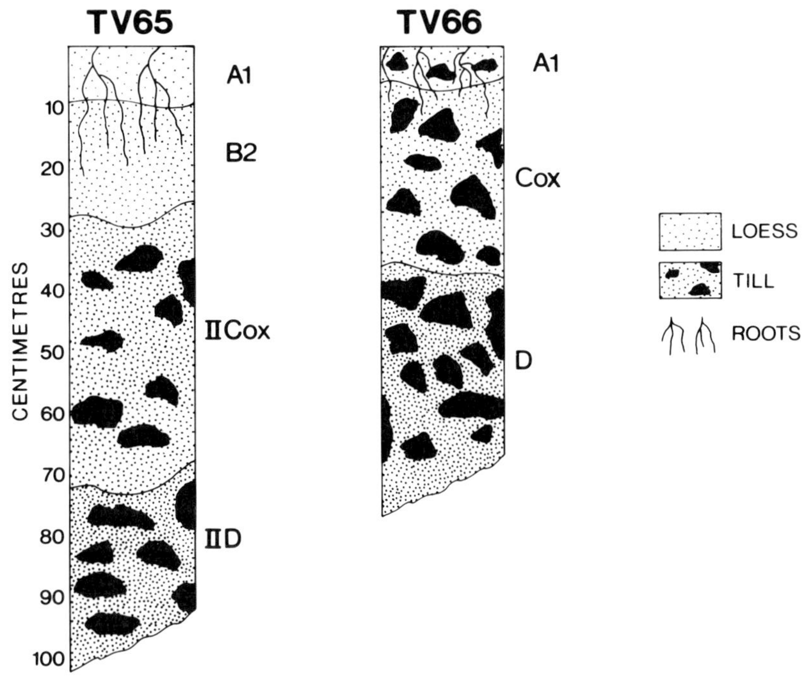

Figure 10.9: Relative development of profiles TV66 (Neoglacial) and TV65 (late glacial) in the foreland of the Lewis Glacier.

also higher in the A1 and B2 horizons of the TV65 profile relative to the TV66 profile. Both profiles have medium to small amounts of analbite, K-analbite, pyroxene, hornblende, and olivine, with the ferromagnetic minerals increasing downward into the till bodies.

The pH distributions in both profiles are similar as shown in Table 10.3. In the youngest profile pH values show medium acid conditions throughout, whereas in the older late glacial paleosol pH is strongly acid in the solum becoming medium acid at depth in the Cox horizon. On the basis of the pH depth distributions it appears that both profiles are not subject to much downward movement of H^+ ions; however, in profile TV65 there is a slight increase in acidity in the B2 horizon. If H^+ ion activity increases with depth, then amino acids might also have a tendency towards downward translocation in solution.

The total salt content of profiles TV66 and 65, as determined by electrical conductivity, shows somewhat lower and uniform values for the younger Neoglacial soil. The older late glacial paleosol contains a higher total salt content in the A1 horizon, which is probably derived from decaying organic matter. In general, the total salt distributions are similar to values determined by Mahaney (1982b, 1986) for young late Holocene and older late Pleistocene soils on the southwestern flank of Mount Kenya.

Cation exchange capacities (CEC) give some indication of the expected clay mineral composition in the soils and paleosol. In the younger profile (TV66) CEC is low but within the range to support 1:1 (Si : A1) clay minerals like kaolinite. In the older profile (TV65), CEC's in the paleosol solum are above 20, which is sufficient to support 2:1 clay minerals (*e.g.*, illite). The increase in the older TV65 profile probably results from the dissociation of H^+ ions from higher

Table 10.3 D/L ratios and standard deviations (σ) of amino acids in soil stratigraphic units of early Neoglacial (TV66; ~1000 yr BP) and late glacial (TV65; 12,000 yr BP) ages. The standard deviations are calculated on three GC analyses of a single sample.

Site	Horizon	Depth	Aspartic Acid		Alanine		Valine		Leucine		Proline		Phenylalanine		Glutamic Acid	
		(cm)	Ratio	σ	Ratio	σ	Ratio	σ	Ratio	σ	Ratio	σ	Ratio	σ	Ratio	σ
TV66	A1	0-6	.0581	.0069	.1185	.0051	.0633	.0396	.0324	.0110	----	----	.2804	----	.1553	.0597
	Cox	6-36	.1545	.0019	.1673	.0029	.0246	.0286	.0368	.0108	----	----	----	----	----	----
TV65	A1	0-9	.1456	.0269	.1321	.0011	.0441	.0198	.1754	.1047	----	----	----	----	----	----
	B2	9-28	.2695	.0027	.1783	.0109	.0374	.0307	.0283	.0199	.3688	.0589	----	----	----	----
	IICox	28-72	.2338	.0037	.1407	.0034	.1266	a	.0663	a	.2974	.0296	----	----	----	----

---- Not detected
a Not calculated

amounts of organic matter in the soil sola. Even with higher CEC's in the older profile the higher content of volcanic glass probably inhibits the development of crystalline minerals (see Mahaney, Fahey and Lloyd, 1981, for discussion of later Quaternary soils on Mount Adams Volcano in the Cascade Range of northwestern U.S.).

Distributions of organic carbon and nitrogen (Figure 10.10) were analyzed to determine differences between the profiles both with respect to the surface A horizons and to distribution with depth. Organic carbon is considerably higher in the older (TV65) profile, probably as a result of the low microbial numbers and the inability of microorganisms to utilize the available organic matter as an energy source (Mahaney and Boyer, 1986a). Lower organic carbon in the younger profile may result from younger ages and a lower frequency of plant cover. Both profiles have relatively low total nitrogen throughout, indicating little evidence for downward movement.

With respect to the amino acids, aspartic acid and alanine (Table 10.3) have depth distributions of increasing D/L ratios into the B2 horizon of TV65 and the Cox horizon of TV66. Both amino acid D/L ratios tend to decrease somewhat in the Cox horizon of the oldest profile (TV65) which may mean that amino acids in the Cox horizons are younger than in the B2 above. Taking the D/L ratios for both acids, and plotting them against known ^{14}C calibrated racemization rates in buried A horizons (Figure 10.7) gives a predicted age of ~7000 ± 1000 yr

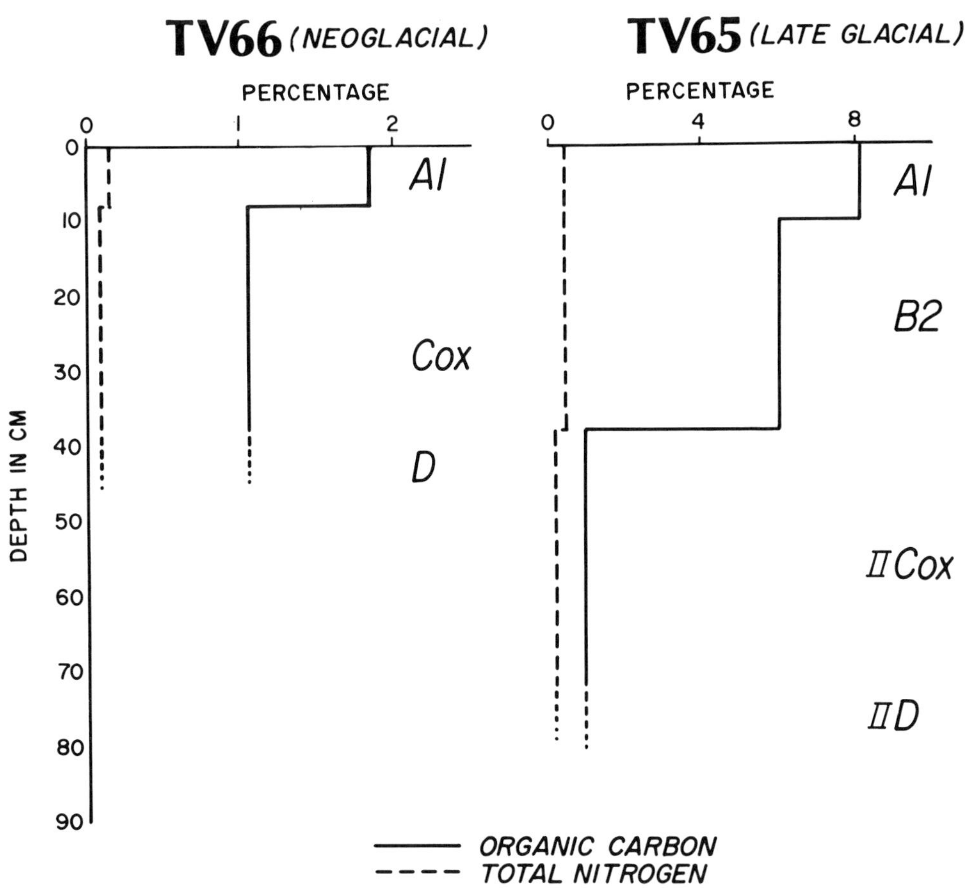

Figure 10.10: Organic carbon and nitrogen distributions in profiles TV66 and 65.

BP age for the B2 horizon in profile TV65. Aspartic acid D/L ratios in the Cox horizon of TV66 gives predicted ages of *circa* 5000 yr BP, which is considerably higher than radiocarbon dates on the deposits (Spence and Mahaney, 1988a).

The depth distributions for valine and leucine provide evidence for biogeochemical contamination. In the surface (A1 horizon) of profile TV65 higher D/L values for leucine and valine indicate the possibility of airfall influx. Because this profile has a silty epipedon (surface horizon) such airfall activity is not unexpected. The higher valine D/L ratio in the surface of the younger profile may also signal biogeochemical contamination.

Two additional amino acids – phenylalanine and glutamic acid – give concentrations only in the A1 horizon of the TV66 profile. Because these amino acids are generally found only in younger soils in the Mount Kenya Afroalpine zone, (Mahaney, Boyer and Rutter, 1986), their disappearance in the subsurface of TV66 and in the TV65 profile is not unexpected.

The D/L ratio depth distributions for aspartic acid and alanine tend to give reasonable differences between the older late glacial paleosol *vs.* the younger early Neoglacial soil (especially with respect to the subsurface horizons). Because these soils have sandy loam textures and reasonable porosity it is possible that some D-form amino acids might blow in from time to time and survive transport downward into the subsurface. Higher values of valine and leucine in the surface horizons may provide evidence for biogeochemical contamination, possibly as a result of local redistribution of amino acids from older sites downvalley. In general, the data analyzed here leads to the conclusion that amino acids (especially aspartic acid and alanine) can be expected to provide reliable relative ages for soil profiles in glacial deposits in the Afroalpine environment.

10.9 Conclusions

Buried and surface paleosols on Mount Kenya have been dated by radiocarbon and amino acid dating techniques. D/L ratios of amino acids, as determined for these samples, prove useful for relative age determinations and as an important crosscheck on the validity of radiocarbon dates. Aspartic acid, alanine and glutamic acid prove remarkably reliable as age indicators, giving progressively higher ratios for samples of increasing age. Several other amino acids including leucine showed distinct age-dependent trends, although not as reliable as the aspartic acid – alanine-glutamic acid group. The aspartic acid ratio/age relationships, verified by radiocarbon controls, allow construction of aspartic acid-glutamic acid-alanine time diagrams that can be used to date samples ranging from a few hundred to >42,000 yrs BP.

CHAPTER 11

PALEOCLIMATIC RECONSTRUCTIONS

11.1 Glacial *vs* Interglacial Paleoclimate

Quaternary glacial and interglacial deposits, in several drainages on Mount Kenya, have been discussed with respect to using relative (RD) and absolute dating (AD) methods to assist in reconstructing paleoclimate. Topographic position of deposits, weathering features, degree of paleosol development, amino acids, TL, and radiocarbon have proved useful in assigning ages to Quaternary deposits. AD and RD methods provide reasonably precise ages for late Pleistocene and Holocene deposits; however, early and middle Pleistocene deposits proved difficult to date with precision. Paleosols provide information on paleoclimate and allow paleoenvironmental reconstruction to and beyond the last interglacial ($>1 \times 10^5$ yr BP).

Six Pleistocene glaciations and two Neoglacial advances are known on Mount Kenya (Figure 2.27) in East Africa. The two oldest glaciations occurred prior to the Matuyama/Brunhes boundary (.73-.79 my) extending to ~2850 m on the eastern side of the mountain. The oldest group of deposits, emplaced during the Gorges Glaciation, is composed of tills with strong reversed remanent magnetism becoming normal (with reversed overprints) upwards. Loess covering these glacial deposits is sometimes reversed and other times normal indicating considerable differences in age. Tills emplaced during the Lake Ellis Glaciation have strong reversed remanent magnetism; however, overlying loess is sometimes reversed and other times normal. Tills documenting the Naro Moru Glaciation are found on both the western and eastern slopes of the mountain between ~2950 and 3150 m. They occur at lower elevations on the western flank suggesting that an asymmetrical icecap existed during the first glaciation of the middle Pleistocene. The strong normal remanent magnetism in these drifts suggests they date from the lower Brunhes Chron. Clast composition in these older deposits assists in unravelling the erosional history of the mountain from the removal, first, of the roof rocks of basaltic composition (Gorges Glaciation), and later of nepheline syenite and porphyritic phonolite, from the volcanic plug. Clasts of the plug rock are found in drifts emplaced during the Lake Ellis Glaciation. Higher percentages of intrusive rock in Naro Moru drifts indicate that erosion of the plug rock was well underway by the early Brunhes Chron and that glaciers were actively deepening their valleys. The paleosols in these older glacial deposits occur both as relict (surface) and buried entities. Collectively they provide important evidence of the totality of weathering (and hence paleoclimate) that occurred over varying time periods during the early and middle Pleistocene.

Following the Naro Moru Glaciation, loess with normal remanent magnetism was emplaced and weathered prior to the emplacement of Teleki Till (considered equivalent with isotope stage 6). This till is named from its type locality in Teleki Valley in the wetter Hagenia woodland; on the drier eastern side of the mountain, Teleki Till is less weathered and contains thinner surface paleosols. However, a detailed study of the relative weathering in the sand (250-63 μm) fractions of a section (TV23) comprising post-Teleki loess, Teleki Till, post-Naro Moru loess plus tephra, and Naro Moru till in Teleki Valley (see Chapter 5) showed that the post-Naro Moru paleosol was less highly weathered than the overlying post-Teleki unit (Vortisch, Mahaney and Fecher, 1987); the main differences in weathering were attributed to changes in paleoclimate and more aggressive leaching in the younger unit.

While relatively little is known about the time of onset of Liki I (Würm, Wisconsinan) Glaciation, considerably more information is available to document the beginning and the end of the Liki II and III substages. Liki II began just prior to 25,000 yr BP when slope instability led to burial of paleosols on slopes in lower Teleki Valley (Mahaney, 1987a). In overlying slope wash deposits a magnetic excursion ^{14}C dated at between ~21,000 and 24,000 yr BP, is possibly correlative with the Mono Lake excursion of North America (Barendregt and Mahaney, 1988). Moraines deposited during the Liki Glaciation extend to ~3200 m, but are best displayed on the western and eastern slopes of the mountain. Because they lack datable organic materials it is impossible to determine their maximum ages; however, ^{14}C dates on bog bottom sediments in lower Teleki Valley place the end of the Liki II substage at ~15,000 yr BP (Mahaney, 1988b). In all valleys ice receded relatively rapidly from 3200 m to *circa* 4000 m, from *circa* 15,000 yr BP to *circa* 12,500 yr BP (Mahaney, 1985, 1987a); thereafter during the late glacial several recessional moraines were constructed during recessional stillstand events (Mahaney, 1987b, c). These glaciers either disappeared completely or retreated to their present positions in cirques by the end of the Pleistocene.

The Holocene on Mount Kenya was comparatively warmer and wetter than today. Contrary to the opinions of some workers (Karlén, 1985, Johansson and Holmgren, 1986; Perrott, 1982), there is no evidence for ice advances in the early and middle Holocene (Mahaney, 1985, 1987c,d). Towards the end of the Holocene, during the Neoglacial, ice advances as documented by the Tyndall (~1000 yr BP) and Lewis advances (~100 yr BP) (Mahaney, 1985) show that ice reached *circa* 4400 m and remained in local cirque basins.

In later sections of this chapter paleoclimatic reconstructions that are possible from equilibrium line altitudes, glacial termini, and pollen recovered from paleosols, will be discussed.

11.2 Clay Minerals and Paleoclimate

Variations in amounts of halloysite-metahalloysite and gibbsite with decreasing amounts of Si and increased leaching are related to variable Quaternary paleoclimatic intensity accompanied by low to high available soil water. Clay mineral and weathering product assemblages can be used to reconstruct the intensity of paleoclimate from early to late Quaternary time on Mount Kenya. The available data tend to support a reconstruction of high intensity paleoclimate and increased wetness in the first two interglacials (Gorges-Lake Ellis; Lake Ellis-Naro Moru), followed by reduced precipitation in the Naro Moru-Teleki interglaciation (lower Brunhes), and ending with increased paleoclimatic intensity during the Teleki-Liki interglaciation (isotope stage 5). A true low intensity interglacial producing only halloysite-metahalloysite, without substantial removal of Si and genesis of gibbsite, may only occur at the end of the present postglacial period.

Quaternary paleosols older than the last glaciation (>100,000 yr) on Mount Kenya, East Africa (Figure 11.1) often show three distinct clay mineral trends that may be related to high, medium and low intensity leaching controlled by paleoclimate and variable degrees of available soil water. These paleosols, comprising aeolian, fluviatile, glacial and volcaniclastic sediments, were emplaced, for the most part, during periods of instability (*i.e.* glacials); weathering occurred primarily during periods of stability (*i.e.* interglacials). The model of clay mineral genesis proposed here may not operate in older paleosols that were exposed to several interglacial paleoclimates (Figure 11.1) while simultaneously receiving fresher aeolian sediments during each younger glacial episode.

At present the mean annual temperature at ~3000 m approximates 10°C. The estimated temperature depression during the last glaciation (Liki = Würm) is 7°C (Coetzee, 1967) giving a mean annual temperature of 3.0°C at ~3000 m during the Liki II glaciation (equivalent to the present temperature at ~4200 m near the terminus of the Lewis Glacier (Mahaney, 1989a). The estimated mean annual precipitation at ~3000 m is ~1200 mm (Coetzee, 1987). The estimated glacial temperature depression of 7°C might reduce mean annual precipitation to between 700 and 800 mm (similar to the present mean annual precipitation at ~4000 m elevation in the Central Peaks area).

On Mount Kenya two distinct assemblages of weathering products are known in pre-Pleniglacial paleosols (Vortisch, Mahaney and Fecher, 1987). Crystalline weathering products

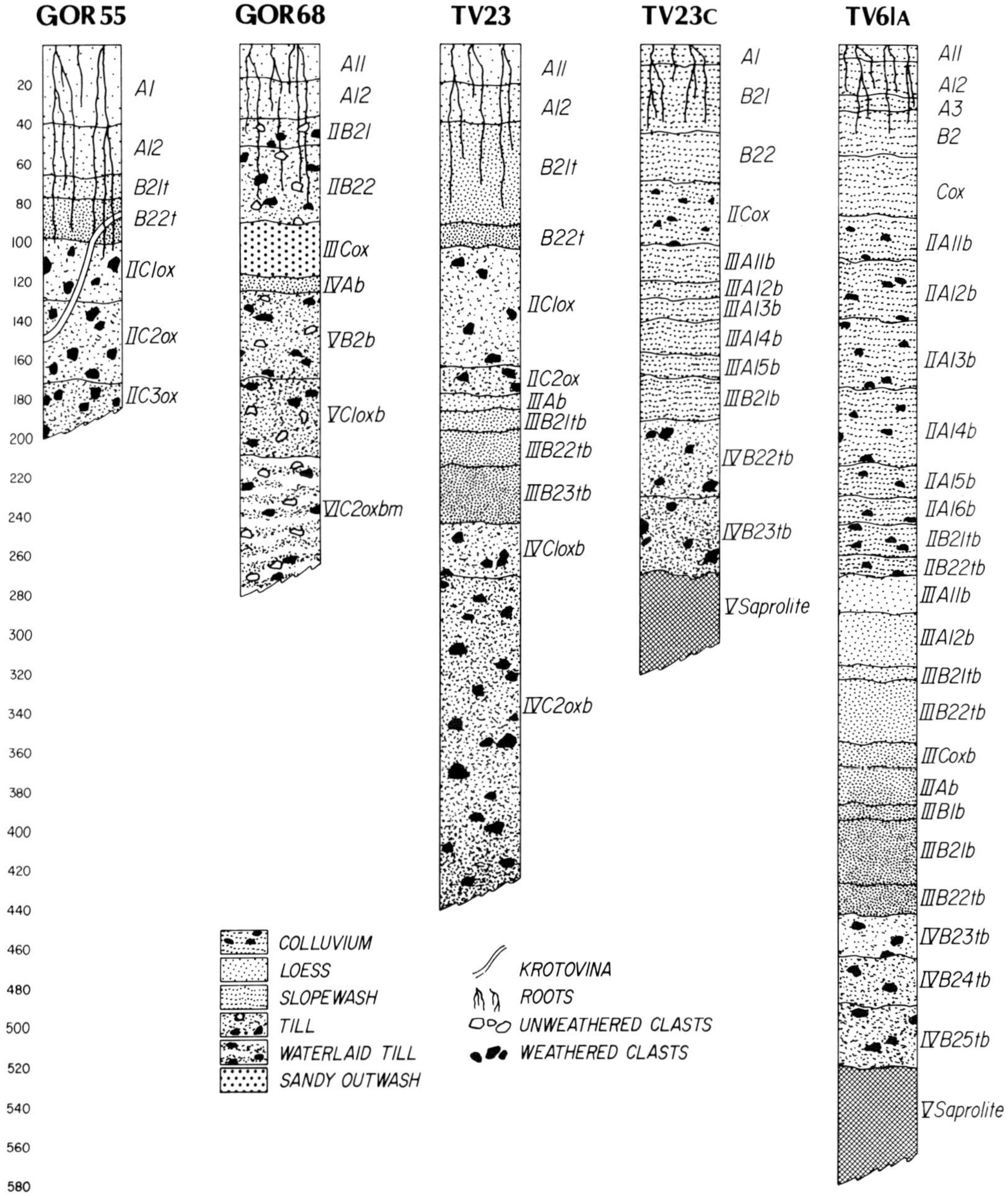

Figure 11.1: Stratigraphy of five Pleistocene sections showing positions of buried paleosol and sediments weathered during interglacial paleoclimates.

consisting of halloysite-metahalloysite and gibbsite dominate in over fifty compound paleosols studied in several different drainages (Mahaney, 1982a; Mahaney, Barendregt and Vortisch, 1989). Trace to minor amounts of goethite, lepidocrocite, and hematite occur in older paleosols, but they are considered less sensitive to leaching effects and changes in intensity of interglacial paleoclimates. Overall, about eight hundred paleosol samples have been studied (Mahaney, 1979a, 1981, 1982a, 1984a, 1985, 1989a; Vortisch, Mahaney and Fecher, 1987).

The distributions of halloysite-metahalloysite were analyzed to determine if they are related to the age or mineral composition of individual paleosols. In general, halloysite-metahalloysite clay minerals are found only in paleosols predating the last glaciation on Mount Kenya (Mahaney, 1989a). As shown in Figure 11.2 the amount of these clay minerals increases under low to medium paleoclimate intensity indicating that they tend to form under low to moderate leaching conditions, but disappear under aggressive leaching with high amounts of available soil water. Because the preservation of halloysite/metahalloysite is dependent on temperature changes during collection and storage, the samples were protected from sunlight and temperature changes en route to the laboratory.

The distribution of gibbsite, as shown in Figure 11.2 tends to increase from trace or small amounts in low to moderate intensity paleosols to higher amounts in high intensity paleosols. The pattern of its distribution suggests its genesis is related to more intensive weathering and leaching conditions in the soil microenvironment. Indeed, very intensively leached paleosols may undergo almost complete destruction of halloysite/metahalloysite clay minerals as gibbsite formation increases (Vortisch, Mahaney & Fecher, 1987).

Both halloysite/metahalloysite and gibbsite minerals are usually observed on altered volcanic glass particles and occasionally on feldspar crystals (Vortisch, Mahaney and Fecher, 1987). Because gibbsite dominates in intensively leached paleosols, and halloysite-metahalloysite is more prevalent in medium to low intensity paleosols, high amounts of both minerals might indicate more intensive high energy leaching (paleoclimate) followed by medium to low intensity leaching (paleoclimate) over periods of 100×10^3 yr. This polygenetic sequence might occur, for example, in paleosols formed in the early Quaternary under more intensive interglacial paleoclimate, followed by less intensive paleoclimate in the middle to late Quaternary. In this case a paleosol would develop most of its gibbsite concentation under high energy leaching conditions operating under high intensity paleoclimatic (wetter-warmer) conditions. Halloysite-metahalloysite, stripped of its Si during high intensity (very aggressive) leaching would tend to produce gibbsite [$A1 (OH)_3$]. In paleosol systems (relict profiles) open to the ingress of Si from aeolian deposition of fresh sediments, it is possible that, under lower paleoclimatic intensity, renewed Si might assist in the formation of halloysite-metahalloysite from gibbsite. In profiles open to the atmosphere, where both minerals are found, it may be that gibbsitic and halloysitic/metahalloysitic elements formed precisely in this way (see later section where both gibbsite and metahalloysite coatings are found on the same grain of volcanic glass).

Other buried paleosols offer the opportunity to study clay mineral formation during short (100 to 200×10^3 yrs) intervals of time followed by burial under aeolian, glacial, and volcaniclastic sediments. Several profiles on Mount Kenya (especially TV23, TV23C, TV61A, and GOR68) provide excellent examples of high intensity weathering followed by burial which tends to preserve nearly intact the clay mineral record of the paleoclimate that prevailed during a particular interglacial period.

Representative SEM photomicrographs of gibbsite and halloysite-metahalloysite minerals are presented in Figures 11.3-11.5. These illustrate weathering effects in progressively stronger grades of interglacial paleosols. The most important difference between these paleosols is the progressive increase in gibbsite with increased soil water and higher leaching intensity.

Both halloysite-metahalloysite and gibbsite form on glass particles (and on occasional feldspar grains) in various ratios. Figure 11.3 illustrates the occurrence of gibbsite in a profile

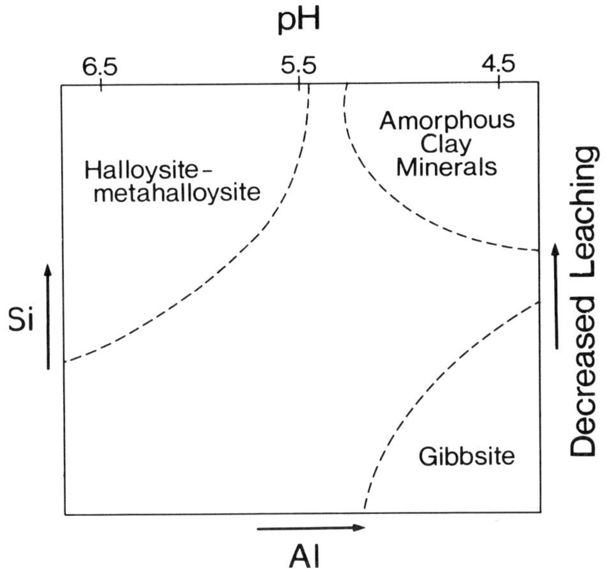

Figure 11.2: The relationships of gibbsite to increasing Al and metahalloysite to increasing Si in Mount Kenya paleosols.

[GOR58 (Figure 4.17)] open to the ingress of aeolian sediment since the Olduvai subchron (e.g. ~1.8 my). In this case, gibbsite may have formed by transformation of metahalloysite in a more aggressive leaching microenvironment. Figure 11.3B illustrates the nodular form of gibbsite common in many older paleosols in Gorges Valley. The high ratio of Al to Si (2:1) illustrates aggressive leaching conditions common in Gorges-Lake Ellis Interglacial paleosols only slightly modified by subsequent weathering during later interglacials.

Younger paleosols formed during the Naro Moru-Teleki Interglaciation (Figure 11.4 and 11.5) show predominantly halloysite-metahalloysite (with some Fe-oxides) illustrating less intensive paleoclimate compared with the older Gorges-Lake Ellis Interglacial paleosols.

Two main paleosol units are common on Mount Kenya. The first type, as for example GOR55, attained its assemblage of weathering products during the strongest interglacial episode and continually received input from aeolian sources. The second type (GOR68; TV23, TV23C and TV61A; Figure 11.1) formed during interglacials following periods of instability. Once the major weathered zone was in place the unit was covered with a veneer of new and relatively fresh sediment, thus preserving the older weathered sediment.

The weathering intensity of subaerially exposed sediments depends on the paleoclimatic intensity which ranges from low to high, and on the amount of volcanic glass which is the main primary mineral source of weathering (Vortisch, Mahaney and Fecher, 1987). As with halloysite-metahalloysite the chemistry of pore solutions and drainage/leaching in individual profiles are the controlling factors for gibbsite formation (see Dejou *et al.*, 1977; and Vortisch, Mahaney and Fecher, 1987). Therefore particle size distributions (Figure 11.6) are primarily important in controlling permeability and porosity. Under extreme conditions, paleosols with high clay (either of pedogenic or volcanic origin) might inhibit drainage and therefore provide negative feedback from the positive effects of high intensity paleoclimate. Higher contents of volcanic glass in some of the paleosols might possibly assist in the transformation of gibbsite to halloysite-metahalloysite or prevent the formation of gibbsite (see profile TV23) (Curtis and

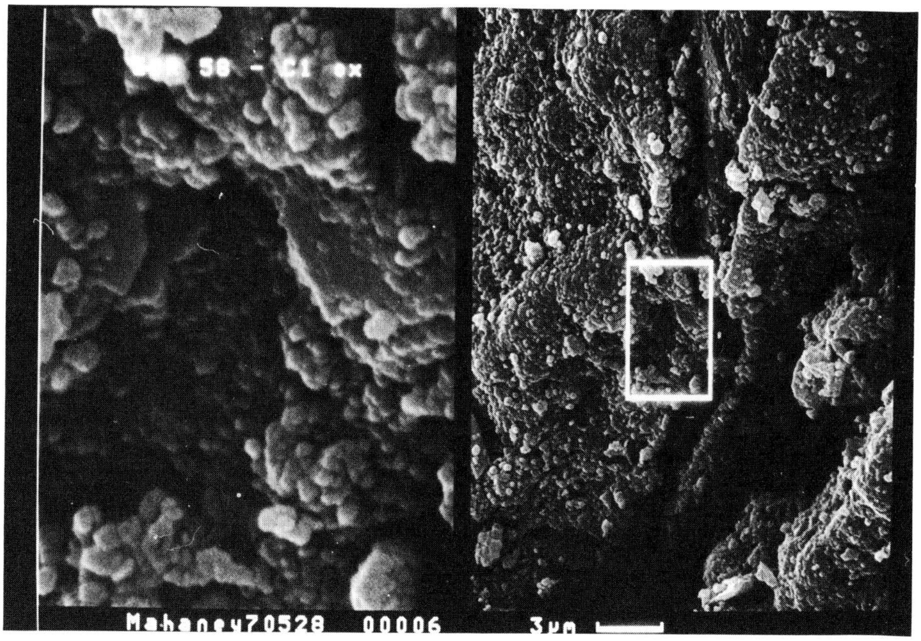

A.

B.

Figure 11.3: A, Highly etched feldspar surface in the IIClox horizon of profile GOR58 showing high concentration of gibbsite (nodular form) in enlargement; B, Nodular form of gibbsite in profile TV61A on volcanic glass particle.

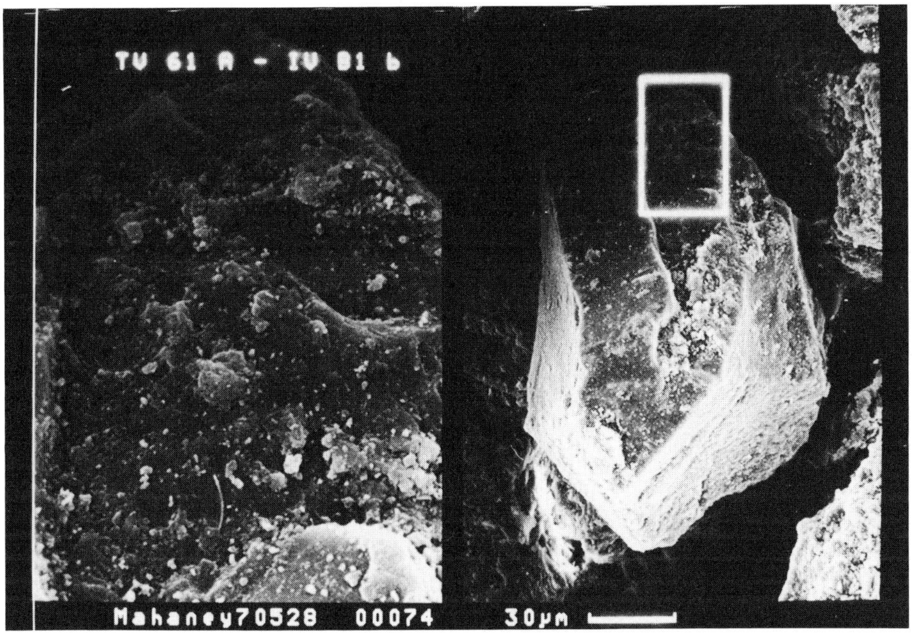

Figure 11.4: Etched and coated quartz grain showing conchoidal fractures and minor glacial-crushing features. Coating consists of clay minerals with Si:Al ratio 1:1 (metahalloysite) in post-Naro Moru paleosol.

Figure 11.5: Glass particle transformed to neoformed clay minerals with Si:Al ratio 1:1. Fe-oxides are also present.

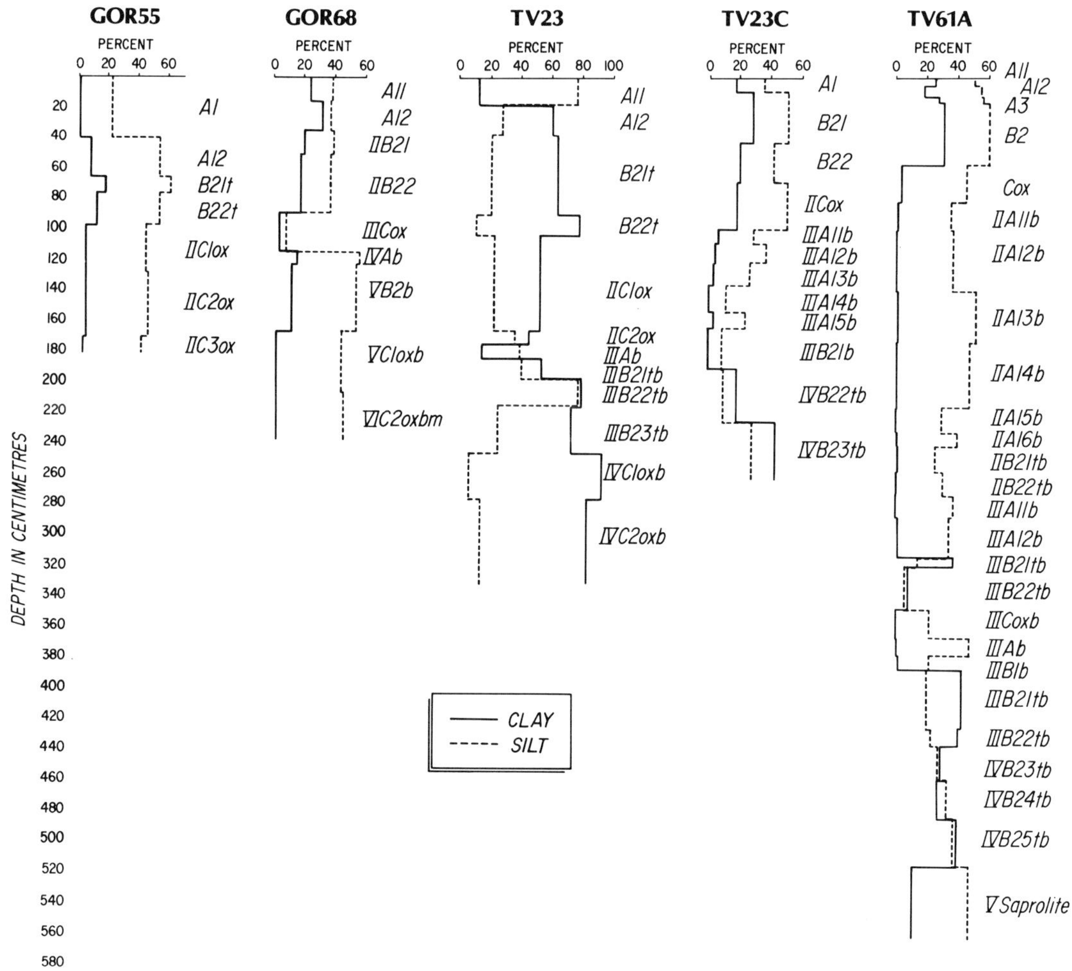

Figure 11.6: Particle size depth distributions in five paleosols in Gorges and Teleki valleys.

Spears, 1971). The possibility of the silicification of gibbsite (Dejou, 1977) is likely in sediments with low (< 50%) clay, high clast composition, and sufficient porosity to facilitate leaching.

Periods of instability on Mount Kenya involved glacial erosion and deposition, accompanied and followed by deposition of aeolian, volcaniclastic and fluvial sediments. In most cases soil-forming activity occurred during interglacial periods (Morrison, 1978, for discussion of soil forming intervals) which are known to have varied in terms of available water, wetting depth, and degree of weathering.

Increases in the amount of available moisture and in depth of wetting in individual paleosols can only occur during interglacial periods when forest cover rises on the mountain in response to higher temperatures and increased precipitation (Mahaney, Harmsen and Spence, 1990).

The paleoclimatic model shown in Figure 11.7) is based on the XRD analysis of over fifty paleosol profiles (Mahaney, 1979a, 1981, 1982b, 1985, 1988b; Vortisch, Mahaney and Fecher, 1987) and SEM analyses of 40 profiles. Five of these profiles are described in detail (Figure 11.6)

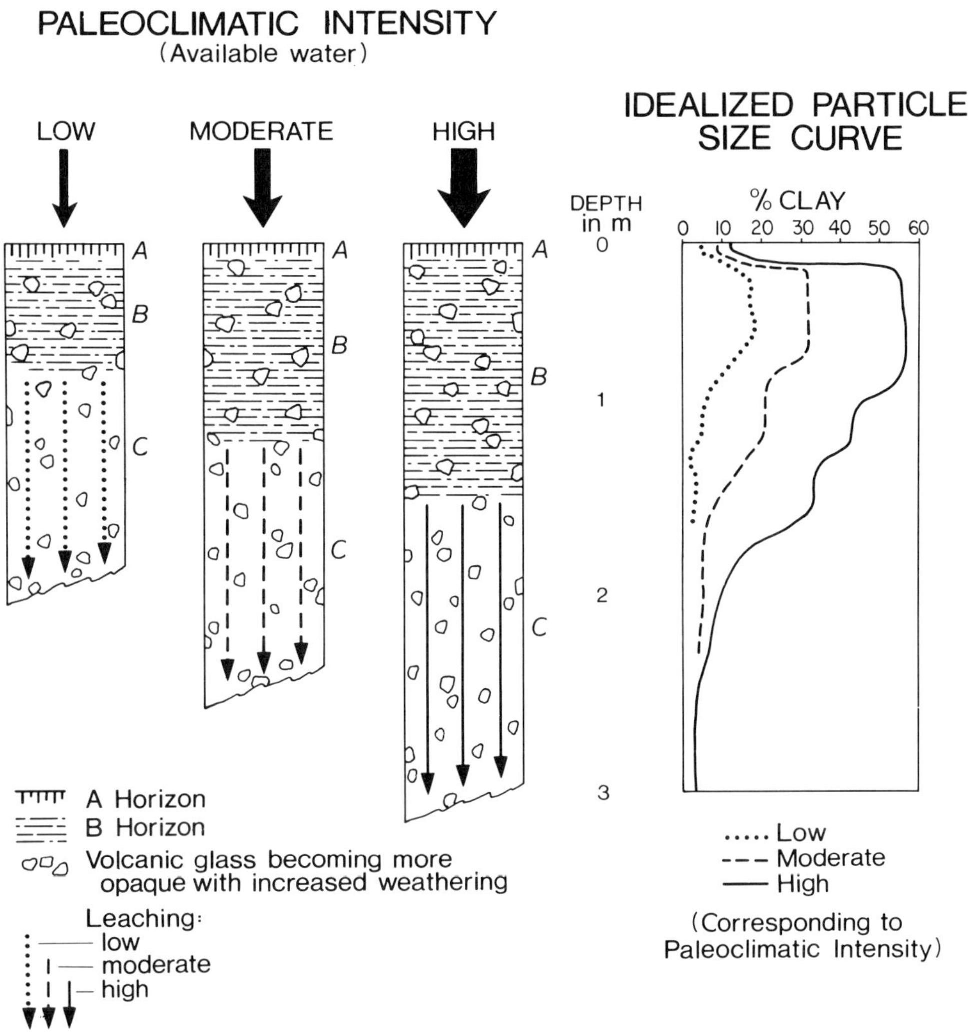

Figure 11.7: Paleoclimate model showing relationships between weathering intensity, available soil water, and particle size distributions.

giving the clay and silt trends with depth. The data show that all profiles (except lower TV23) have sufficient porosity, and hence drainage conditions, for the conversion of halloysite-metahalloysite to gibbsite and *vice versa*. While feldspar grains sometimes weather to form clay minerals and other weathering products, volcanic glass is the prime contributor to weathering products of various kinds (Vortisch, Mahaney and Fecher, 1987).

A gradient from low to high paleoclimatic intensity would tend to produce increasing available water for leaching in the soil systems. A low intensity paleoclimate would have the power to produce only halloysite-metahalloysite concentrations mainly in the B horizon complexes. An increase in available water to moderate paleoclimatic intensity tends to yield a mix of halloysite-metahalloysite clay minerals along with gibbsite (with the former dominating). A

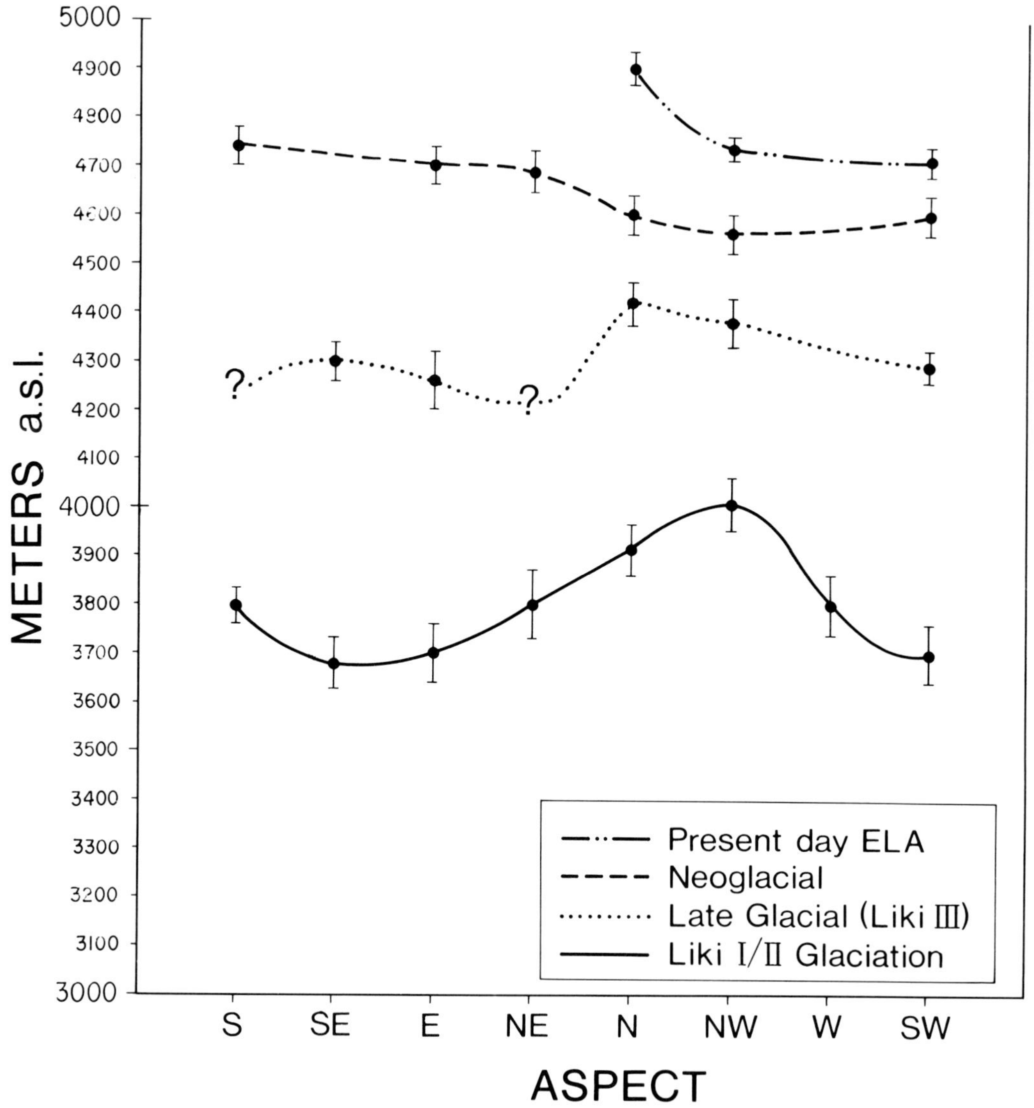

Figure 11.8: Equilibrium line altitudes for Liki I/II, Liki III and Neoglacial glaciers on Mount Kenya.

further increase in available water and onset of aggressive leaching conditions results in the loss of +90% of the halloysite-metahalloysite clay minerals and assists in the formation of abundant amounts of gibbsite.

XRD and SEM analyses of several Quaternary paleosols show that gibbsite and halloysite-metahalloysite dominate as weathering products in pre-Pleniglacial weathering environments. Volcanic glass is the dominant source material for weathering. Paleodrainage and leaching conditions, controlled largely by paleoclimate and available soil moisture (as well as by particle

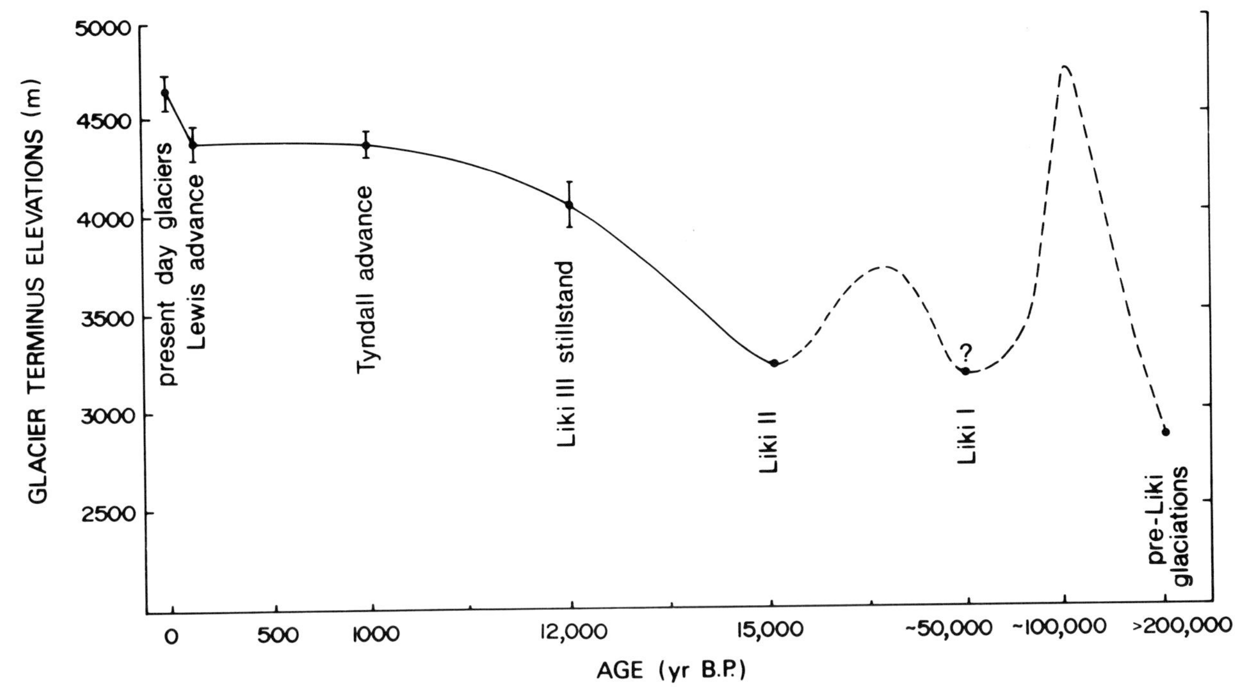

Figure 11.9: Glacier termini positions for different Neoglacial and Pleistocene glaciations on Mount Kenya.

size), appear to determine the increase in the ratio of gibbsite/halloysite-metahalloysite. Using the relative abundances of these minerals it appears possible to reconstruct the intensity of paleoclimates from the early to the late Quaternary.

11.3 Equilibrium Line Altitudes

Paleoclimate can be reconstructed from the mean altitudes of lateral moraines (Figure 11.8) of various ages and from termini positions (Figure 11.9). For present-day glaciers, ice fronts are 4650 ± 100 m, which is about 150 m higher than the frontal positions during the two-stage Neoglacial. From the landform evidence (Mahaney, 1985), the earliest Neoglacial moraines, emplaced during the Tyndall advance lie farther out on the cirque floors, and approximately at the same altitude as moraines emplaced during the Lewis advance (classical Little Ice Age). Because the older Neoglacial advance occurred only on the southwestern flank of the mountain, it is possible that microclimatic changes, induced primarily by changes in localized cloud cover (perhaps produced by local anabatic winds), might explain the local advances. The paleoclimatic change that produced the Lewis advance was widespread, affecting every cirque on the mountain, including three cirques on the eastern flank (presently unoccupied by ice).

During the late glacial the annual snowline was at $\sim 4100 \pm 150$ m (Figure 11.8) as determined from lateral moraine positions. Even though some of these moraines are considered to be of Holocene age, there is no evidence to support moraine emplacement over preweathered surfaces, nor is there any dating evidence from the moraines in support of a middle Holocene age (Mahaney, 1987a). The spatial extent of late glacial (Liki III) ice as shown in Figure 7.1 is considered to result from a halt in the recession and decay of the main Liki valley glaciers. The time of this recessional stillstand is fixed at ~12,500 ± 1000 yr BP (1987a,b). Paleoclimate as reconstructed from pollen evidence (Coetzee, 1967; Mahaney, 1985) confirms a colder and drier paleoclimate during the late glacial, although not as cold as during the Liki II Glaciation.

Liki II snowline positions are difficult to work out for all valleys owing to the lack of well defined lateral moraines. Taking the termini (Figure 11.9) of several major glaciers into account, the annual snowline was undoubtedly higher than 3200 m (mean snout positions), and probably at ~3700 m. Liki I termini, while slightly lower than Liki II end moraines, are considered to have had *circa* the same equilibrium line altitudes. The available pollen evidence from Sacred Lake, on the northeastern flank of Mount Kenya (Coetzee, 1967), shows colder and drier paleoclimate during the Liki II period and warmer paleoclimate during the deposition of older deposits considered equivalent in age with the Kalambo Interstade (> 25,000 yr BP; Coetzee, 1967).

Pre-Liki glaciations produced termini as low as 2850 m (somewhat lower on the eastern flanks of the mountain) that reflect the major source of moisture derived from the Indian Ocean monsoon. During the Naro Moru Glaciation ice descended at least 175 m lower on the western than on the eastern side, possibly reflecting a moisture source from the Atlantic monsoon. Just as the northern and northeastern sides of the mountain are dry today, during the pre-Liki glaciations very little drift was emplaced on the northern and northeastern slopes. Even during the Liki Glaciation ice termini in the Sirimon Drainage reached only to ~3400 m and ice in the Kazita Valley was no lower than ~3500 m.

The lowering of equilibrium line altitudes (ELA's) for Holocene glaciers on Mount Kenya, as described by Johansson and Holmgren (1985), is not supported by the ELA evidence presented by Mahaney (1985, 1987a, 1989a). The ELA boundary of 4600 m shown in Figure 6 in Johansson and Holmgren (1985) is, however, a reasonable approximation for late glacial ice in Teleki valley (see Figure 11.8). During the Neoglacial on Mount Kenya, beginning approximately 1000 yr BP (Mahaney 1982b, 1984b), ELA's were lower than today by about 100 m (Mahaney 1987a), which is sufficient to fill existing cirques with ice, but would not lead to ice advances out of major cirques.

11.4 Asymmetry of Mount Kenya Icecaps

The equilibrium line altitudes (ELA's) for Mount Kenya, calculated from the mean altitudes of lateral moraines, show a striking asymmetrical pattern when the main Liki and Neoglacial snowlines are compared (Figure 11.8). The Liki I/II icecaps appear to have had a pronounced asymmetry with the longest and largest glaciers on the eastern, southern and western quadrants, and somewhat smaller glaciers on the northern flank of the mountain. This pattern appears out-of-phase with Neoglacial and late glacial ELA's which were higher on the eastern flank of the mountain and somewhat lower on other slopes.

Considering the paleoclimatic implications of this data it is important to note that the Neoglacial glaciers on Mount Kenya were influenced greatly by shade and microclimatic factors in the Central Peaks area. These climatic variables give rise to ELA differences of *circa* 300 m. With respect to Liki III and the main Liki ELA's, the relation to major source areas of precipitation and temperature distributions was perhaps a more important controlling factor. The Neoglacial and present day ELA's tend to descend in areas with pronounced cloud cover (west and southwest flanks) and rise in areas that have less cloud cover and less precipitation (east-north). During the full glacial the situation was considerably different with the largest glaciers on the eastern and western flanks presumably responding to classical and Atlantic monsoonal effects, as well as to anabatic wind systems. Drying effects from the Saharan anticyclone, which persist over Mount Kenya from January to March, may have lasted longer each year influencing the amount of precipitation and cloud cover on the northern slopes. Glaciers on the southern slopes were intermediate in size during the last glaciation.

11.5 Holocene and Liki Interstadial Paleoclimates

In many ways the palynological evidence for Liki Interstadial/Stadial paleoclimate parallels the changes that occurred after the end of the Liki Glaciation. Overall temperature and precipitation increased and timberline ascended the mountain slopes in response to this paleoclimatic amelioration. A general description of Liki interstades is possible from the pollen data in Figure 11.10 (section TV61).

Section TV61, of 560 cm depth, in the Met Station Catena in Teleki Valley contains a sequence of weathered till, loess, colluvium and slopewash. Vegetation at the site is representative of the bamboo zone/montane forest ecotone. Relatively high pollen concentrations (Figure 11.10) were recorded only in the upper 340 cm of the sequence, showing marked variations in pollen content. The lowest pollen zone, A, is found in a loess deposit (unit III), dominated by grass pollen similar to that found at higher altitudes in the Afroalpine Belt, suggesting that vegetation might have been emplaced at the beginning of the Liki Glaciation (or earlier). The zone indicates relatively cool conditions. Two contradictory radiocarbon dates for pollen zone A, ca. 23,200 and 27,600 yr BP, probably reflect contamination with younger organics. The second pollen zone, B, corresponds with the lower section of colluvium, unit II, overlying the loess. More *Podocarpus* and Ericaceae pollen suggest that the lower part of the Ericaceous Belt was present at the site during deposition, implying cool conditions, but not as cool as those suggested for zone A. A radiocarbon date of ca. 25,400 yr BP for this section is also possibly contaminated. The third pollen zone, C, corresponds with the upper section of the colluvium, and is again dominated by grass, possibly from the Afroalpine Belt. Radiocarbon dates of ca. 21,530 and 25,170 have been recorded for the lower part of this pollen zone, suggesting that the greatest part of it was deposited during the Last Glacial Maximum. The fourth pollen zone, D, in younger slopewash, is possibly representative of the Holocene (although it gives a geomagnetic reversal dated at 21-24 x 10^3 yr BP), and shows high *Podocarpus* and other tree pollen associated with dry montane forest.

A marked similarity between the main trend of this pollen sequence, on the western side of Mount Kenya, and that from Sacred Lake, at 2400 m on its northeastern side (Coetzee, 1967), suggests that related environmental events occurred on both sides of the mountain. The pollen study tentatively associates the sediment sequence in the Teleki Valley with these events.

The detailed pollen and paleoenvironmental records described by Coetzee (1967) are similar to the landform-soil record described in Figure 7.19 (Mahaney, 1985). Pollen Zone Ui accompanied the first amelioration following the full glacial (Liki II Glaciation) ice cover on the mountain. During this period, the timberline approached 2400 m and receding glaciers in the major valleys halted at about 4000 m to build (Liki III) end moraines. At about 3000 m mass movements buried thick soil under slopewash deposits (Mahaney, 1985, 1986). Warmer periods ensued after *circa* 12,500 yr BP, culminating in a dense timberline (pollen zone V), and disappearance of the main valley glaciers. From the landform evidence (Figure 11.11), the period between about 12,500 and 8000 years ago was either dry or variable wet and dry, conflicting somewhat with the pollen record (Figure 7.19). Also, the landform evidence indicates only minor slope instability, streams at or near equilibrium (neither aggrading nor degrading their channels), and glaciers at equilibrium (Mahaney, 1985) during the early Holocene.

Paleoclimate during the middle Holocene appears to have been wet (Figure 11.11), bringing first a humid montane forest with warmer temperatures, and later a lower timberline at 2400 m with cooler temperatures (Figure 7.19). In the Afroalpine belt this time period is characterized by widespread slope instability, alluvial fan deposition, and minor stream degradation (Mahaney, 1984b, 1985, 1987b). The pollen evidence (Figure 7.19) indicates that maximum humid forest developed under higher temperatures and precipitation, a condition that continued to the end of the middle Holocene, at lower elevations. In the Afroalpine, widespread slope instability that led to emplacement of the Darwin Formation (see Mahaney, 1985, for discussion) (Figure 11.11) also

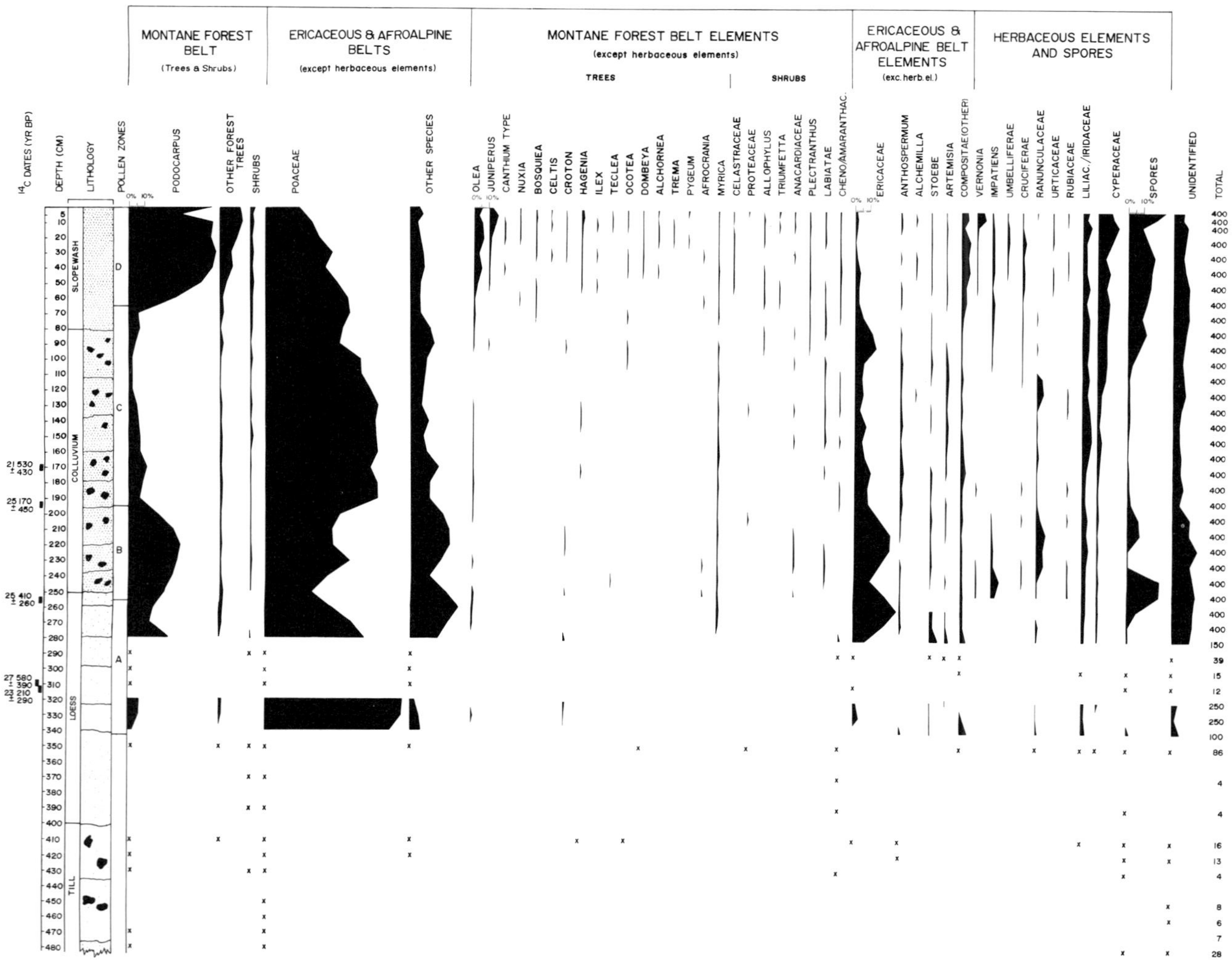

Figure 11.10: Pollen distributions in profile TV61 (colluvium = Liki II Stadial deposit; slope wash = Liki II Stadial deposit).

coincided with wetter and generally warmer paleoclimatic conditions inferrred from the pollen evidence. Pollen zone Y persisted to 3400 yrs BP, with the upper portion marked by isolated colluvial and alluvial events. Pollen zone Z began with cooler and drier paleoclimatic conditions accompanied by dune formation (Mahaney and Spence, 1986) and ended with Neoglaciation, which began *circa* 1000 yr BP, coinciding with the onset of dry montane forest at 2400 m.

Differences between the landform-soil evidence and the pollen chronology appear to have resulted from variations in climatic changes between Sacred Lake at 2400 m (Coetzee, 1967) and the Afroalpine zone (above about 4000 m) (Mahaney, 1985), or from somewhat different timing in the two areas for events of similar magnitude.

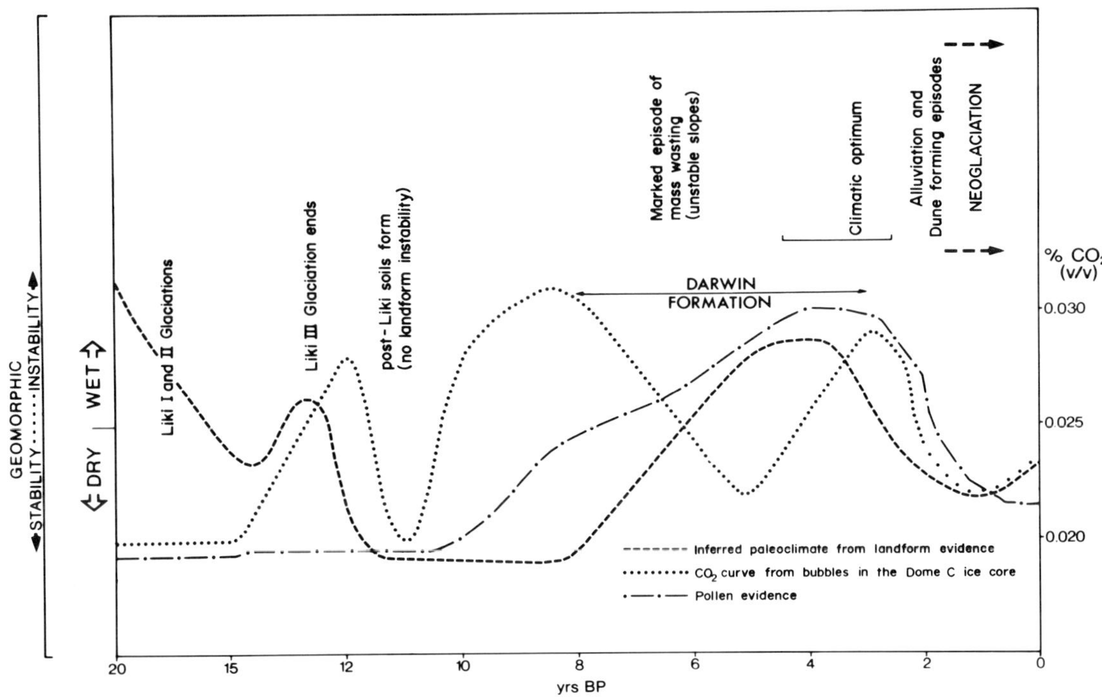

Figure 11.11: Inferred paleoclimate from landform evidence compared with the pollen sequence from Sacred Lake and CO_2 curve from Dome C ice core.

11.6 Will the Eleven Remaining Glaciers Survive?

Of all the East African mountains only the three highest have ice covers at present: Kilimanjaro, Ruwenzori and Mount Kenya. Of these three, Mount Kenya has eleven small glaciers with a total area of less than 1.0 km^2. Elsewhere in the tropical mountains, the Andes of Ecuador have a total ice cover in excess of 200 km^2 and in Irian Jaya ice covers approximately 7 km^2 (Hastenrath, 1984). In all of these areas documented ice recession since the 19th Century has been continuous. In Irian Jaya and Ecuador ice recession began in the middle of the last century and continues today. The East African ice recession was brought about by a sharp decrease in precipitation and decrease in cloudiness over the last two decades of the 19th Century, which also affected other tropical mountains. These changes induced a slight warming of ~0.5°C and an albedo increase of 2% (Hastenrath, 1984) with most change occurring in the 1920's and 1930's.

The study of glaciers on Mount Kenya is important if we are to understand major environmental changes in the tropical mountains. If the present drying and warming conditions continue, all eleven glaciers may disappear by the end of this century. At present all glaciers show negative mass balances. Repeated visits to these glaciers over the last fifteen years show they are continually thinning and their terminal positions are receding. Indeed some glaciers, like the Krapf, Heim, Forel, and Diamond are barely able to exist given their small size and absence of new snow cover. The Heim and Forel glaciers, in their present state, may slide onto the surface of the Tyndall Glacier at any time.

Given the tendency for worldwide warming to occur over the next couple of decades, there is little hope of ice surviving on Mount Kenya except in areas protected by cloud and steeper rock walls, *i.e.* the Tyndall and Lewis glaciers.

CHAPTER 12

ECOLOGICAL AND GEOMORPHOLOGICAL PROCESSES

12.1 Introduction

Pleistocene glaciations and two Neoglacial advances are documented on Mount Kenya. The three oldest glaciations occurred prior to the Matuyama/Brunhes boundary (0.73-0.79 my) and extended to ~2850 m in the Gorges Valley on the eastern side of the mountain. Deposits emplaced during various glaciations form the abiotic platforms on which the Afroalpine ecosystem evolved. The formation of an ecosystem is highly dependent on the potential source floras and faunas, on the dispersal of biotic elements among isolated Afroalpine regions, and on the ability of taxa to survive in the Afroalpine ecosystem from glacials to interglacials (Harmsen, Spence and Mahaney, 1990). The current Afroalpine ecosystem on Mount Kenya contains a number of endemic subspecies and species, but *few* endemic genera. This indicates a high degree of recent isolation, and relatively young age. The combination of isolation and extreme changes in paleoclimate is mostly responsible for the low number of species and also for the lack of structural complexity in most Afroalpine ecosystems. The question of the origin of Afroalpine biotic elements on Mount Kenya (from lower altitudinal ecosystems or more distant alpine systems) still is subject to considerable controversy.

Various moraines dating from the last glacial period have undergone sufficient levels of weathering and soil formation to currently support fairly mature and stable Afroalpine plant communities. The recent recession of Little Ice Age glaciers led to the exposure of raw unweathered tills allowing the study of primary invasion of biotic elements of the Afroalpine ecosystem. Thus, it is possible to discuss Neoglaciation insofar as it affected the restructuring of the Afroalpine ecosystem on Mount Kenya, and to discuss the role of glacial-interglacial cycles in this restructuring.

12.2 Abiotic Considerations

Those abiotic elements most important in understanding the restructuring of the Afroalpine are soils (paleosols) and paleoclimate. Both the ages and nutrient reserves are important in assessing the paleosols as "habitats" in which various plants (and animals) of the Afroalpine live and continue to evolve.

Paleosols dating from the Liki I/II and Liki III substages have been defined by Mahaney (1982b). Both groups of deposits have Inceptisols with A/B2/Cox/Cu profiles developed to depths of ~50-100 cm. In these profiles B2 horizons qualify as cambic horizons (Soil Survey Staff, 1975), and all have well-defined loess caps 20-30 cm thick (Mahaney, 1982b). Structure grades from granular in the A horizons, and weak blocky in the B2 horizons, to massive in the subsoil (C horizons). Leaching, and the formation of thick B horizons, are restricted because of the low microbial numbers (Mahaney and Boyer, 1986a), low CO_2, and hence restricted movement of pore solutions. Even with colors of 10YR 2/1 (black) (Oyama and Takehara, 1970) in the A horizons, these soils routinely give 20-40% organic carbon. Hence, some of them qualify as 01 or 02 horizons (Mahaney 1982b). In general these late Pleistocene soil sequences are somewhat similar to comparable field settings in the Rocky Mountains of the western U.S., with the important distinction that the Mount Kenya paleosols are thinner. Weathering products in these systems are restricted to minor amounts of amorphous clay minerals and small concentrations of goethite; halloysite/metahalloysite and gibbsite are limited to paleosols older than 100,000 yr (Vortisch, Mahaney and Fecher, 1987) and form mainly from the weathering of volcanic glass. Feldspar dissolution does not appear to yield any amount of neoformed clay minerals.

At these older, late Pleistocene sites, paleosols vary in their morphology as a response to topographic setting (catena, which controls drainage; Mahaney & Boyer, 1987), to elevations (~3200-~4350 m), and to development of bogs (or mires, controlling peat production). In general, catenas of full glacial to late glacial age yield important data on movement of organic carbon, fine-grained clasts, and Fe-oxides from high to low member positions. While not quite as dynamic as middle-latitude alpine catenas (Mahaney & Sanmugadas, 1983), these systems, on Mount Kenya, can be considered to control soil morphologic expression to a high degree. Variations in soil-profile expression from the lower limit of glaciation (~3200 m) to the high moraines at ~4200 m is difficult to calculate precisely because *time* and *paleoclimate* combine to control paleosol expression. The lower paleosols are older and formed in an environment that was warmer and wetter than the Afroalpine. To compound matters part of the ericaceous and lower Afroalpine belt have large areas of bog (especially prominent on the western flank of the mountain). The vast bulk of this area has developed on surfaces exposed following the recession of Liki II ice, and here soils are largely Histosols with thick (1.0 to 0.5 m) peat mats overlying subsoil C horizons on bedrock.

Holocene paleosols which formed between 8000 and 2000 yr BP have been documented by Mahaney (1984b, 1985). In general these paleosols are important in reconstructing alluvial fan histories; in most cases they are multi-story units of Inceptisols and Entisols, the latter consisting of A/Cox profiles. Collectively they are most important in paleoenvironmental reconstruction and in some cases they have been used to document the time required for the formation of A1 horizons in the upper Afroalpine climate (~160 yr; Mahaney and Boyer, 1986a). One profile in particular (TV4; Mahaney, 1984a) contains the youngest dated profile (~2000 yr) on the mountain with a B horizon (*i.e.* an Inceptisol).

Profiles formed in Neoglacial substrates above 4300 m are all Entisols with A/Cox/Cu profiles, thin depths (~45 cm), sandy and pebbly sandy loam and loamy sand textures. They are, for the most part, exceedingly well drained. For a complete discussion of these soils and their ^{14}C chronology see Mahaney and Spence (1989).

Paleoclimatic considerations discussed in the previous chapters show that ice termini positions, as determined by equilibrium line altitudes (mean annual snowlines), are important in assessing ecological processes on Mount Kenya. For present-day glaciers ice fronts are at 4650 ± 100 m, which is about 250 m higher than the frontal positions during the two stage Neoglacial. From the landform evidence the earliest Neoglacial moraines, emplaced during the Tyndall advance, were deposited farther out on the cirque floors and still approximately at the same altitude as moraines emplaced during the Lewis advance (Classical Little Ice Age) (Mahaney, 1985). Because the older Neoglacial advance occurred only on the southwestern flank of the mountain, it is possible that microclimatic changes, induced primarily by changes in localized cloud cover (perhaps produced by anabatic winds), might explain the local advances. The paleoclimatic changes that produced the Lewis advances were widespread, affected every cirque on the mountain, and most importantly they produced new substrates for plant colonization and succession.

During the late glacial the ice front was at *circa* 4100 ± 150 m as determined from recessional moraine positions. The spatial extent of late glacial (Liki III) ice as discussed in Chapter 7 is considered to result from a halt in the recession and decay of the main Liki valley glaciers. The time of this recessional stillstand is fixed at *circa* 12,500 yr BP (Mahaney, 1987a,b), plus or minus approximately 1000 yr. Paleoclimate as reconstructed from pollen evidence (Coetzee, 1967; Mahaney, 1985) confirms colder and drier conditions during the late glacial, although not as cold as during the Liki II Glaciation. Because it is obvious that the Afroalpine biotic communities must have been severely affected by the recorded shifts in Holocene climate, as corroborated by pollen analysis, it is important to assess the degree to which these climatic shifts may have impoverished the diversity of the Afroalpine biota (R. Harmsen, personal communication, 1987). This question will be taken up in a later section.

Liki II snowline positions are difficult to establish for all valleys owing to the lack of well defined lateral moraines. Taking the termini positions of several major glaciers into account the annual snowline was undoubtedly higher than 3200 m (mean snout position as discussed in Chapter 11), and probably at ~3700 ±150 m. Liki I termini, while slightly lower than Liki II end moraines, are considered to have had *about* the same equilibrium line altitudes, and hence termini positions. The available pollen evidence from Sacred Lake, on the northeastern flank of Mount Kenya (Coetzee, 1967), shows colder and drier paleoclimate during the Liki II period and cooler paleoclimate in older deposits considered equivalent in age with the Kalambo Interstadial (>25,000 yr BP; Coetzee, 1967).

Evidence for a considerable (~700 m) depression of vegetation belts, during the Liki II stadial, comes from a "moorland-like" paleosol, in slopewash and colluvial deposits, found in the Naro Moru drainage between 2500 and 3200 m. Several radiocarbon dates from buried paleosols underlying prominent slopewash deposits indicate that burial occurred *circa* 25,000 yr BP or at about the end of the Kalambo Interstadial (Coetzee, 1967). The pollen in the slopewash sediments indicates an increase in ericaceous taxa and a decrease in forest species (J. A. Coetzee and E. M. van Zinderen Bakker, personal communication, 1987), which suggests a major shift in vegetation belts.

Pre-Liki glaciations produced termini as low as 2850 m (somewhat lower on the eastern than on the western flanks of the mountain) that reflect the major source of moisture derived from the Indian Ocean monsoon. Just as the northern and northeastern sides of the mountain are dry today, during the pre-Liki glaciations very little drift was emplaced on the northern flank. Even during the Liki glaciation, ice termini in the Sirimon Drainage reached only to ~3400 m and ice in the Kazita Valley was no lower than ~3500 m. The buried paleosols of site TV61 and others indicate the existence of a well developed Afroalpine ecosystem following the Kalambo Interstadial. Similar Afroalpine ecosystems must have existed throughout the Pleistocene. At this state, however, it is possible only to present the outline of the structure of these ecosystems and their evolution over time. There is considerable evidence for an ice cap on Mount Kenya during the Liki Glaciation including polished bedrock pavement on Point Lenana, extensive lateral moraines to ~4100 m on the eastern flank, as well as striae on bedrock interfluves at ~3600-3800 m between Gorges and Hobley valleys. Because the Two Tarn area (4550 m.a.s.l.) in the upper Burguret Drainage was ice free during the Liki Glaciation the Teleki and Burguret glaciers appear to have been separated so that they operated more or less as independent valley glaciers. This gives rise to the hypothesis that the ice cap flowed mainly to the north, east and south with the strongest flow to the east and southeast – rather thicker on the eastern and southeastern sides, somewhat thinner on the southwestern flank, and thinnest on the northern side.

Glacial-interglacial cycles on Mount Kenya have been reconstructed using data discussed in several chapters. Six Pleistocene glaciations are known on Mount Kenya starting in the early Quaternary and occurring intermittently until the last (Liki) glaciation terminated ~15,000 yr BP in the lower valleys. Ice recession to ~4000 m.a.s.l. occurred within ~2500 yr, and produced prominent late glacial moraines that range in age from 11,000 to 13,000 yr BP. The abiotic portion of the Afroalpine ecosystem presumably came into quasi-equilibrium within a few thousand years. At the very least the Inceptisols that dominate in the Afroalpine belt attained their major properties in the first few thousand years. Prominent Neoglacial advances disrupted this equilibrium above ~4300 m.a.s.l. starting about ~1000 yr BP. These Neoglacial sequences provide an enormous body of information on the early stages of soil morphogenesis and plant succession above ~4300 m. Even in the Afroalpine environment with a diurnal temperature regime and relatively moist climate, the initial development of Inceptisols (A/B/C profile) takes ~2000 yr; and plant and lichen succession appear slow on most substrates studied.

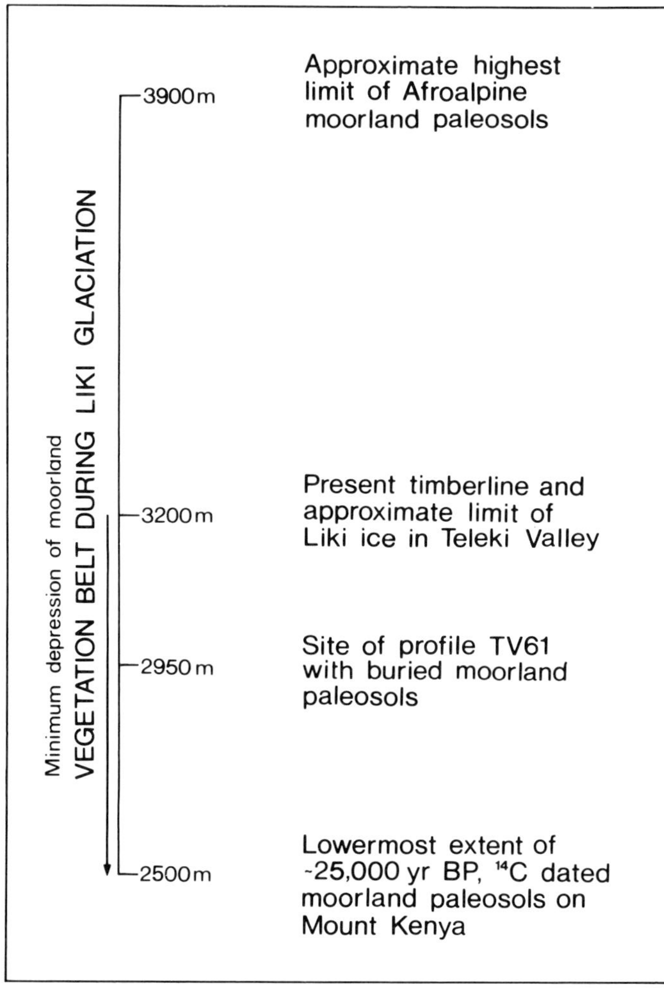

Figure 12.1: Diagrammatic illustration of vegetation depression on Mount Kenya deduced from the presence of the ^{14}C dated "moorland-like" buried paleosol in surficial sediments.

12.3 Restructuring of the Afroalpine Ecosystem

The unique climate of the Afroalpine, characterized by a high diurnal range of temperature, crossing the freezing point every 24 hr, is probably the main determining factor that has shaped the character of the ecosystem (Hedberg, 1957 1964; Harmsen, Spence and Mahaney, 1989). However, beside this climatic uniqueness, it is also important to assess the potential source floras and faunas, and the time available for succession to produce mature biocommunities. The changes that occurred during the Upper Tertiary, involving volcanic mountain building processes and continental rifting which pushed the land surface above 3000 m are discussed in Chapter 2. Mount Kenya, for instance, has had an Afroalpine ecosystem uninterruptedly (but with spatial expansions and contractions during each glacial and interglacial) for the entire period since the late Pliocene (Axelrod and Raven, 1978). The minimum physical constraints to maintain such an ecosystem (given present-day climate) is an area over 3000 m altitude. Maintenance of these conditions for several thousand years, given what is known from Neoglacial succession (Coetzee, 1967), should theoretically allow for the invasion, adaptation, and maturation of plants and animals into quasi-stable biocommunities. On the other hand, frequent climatic changes may have introduced enough disequilibrium to cause perpetual altitudinal shifts of most taxa. Accelerated intermountain dispersal may have occurred during glacials, when the lower Afroalpine belt was at its widest extent. Conversely, major contraction of the ecosystem (in space) during interglacials when the biotic elements were restricted to smaller areas in front of existing glaciers would have resulted in local extinctions. Because Mount Kenya is one of the oldest of the montane "islands", it is likely that some of its biotic elements have invaded other younger "islands" in the "biogeographic archipelago", especially the nearby Aberdares (Hedberg, 1970; Harmsen, Spence and Mahaney, 1990). Mount Kenya no doubt represents for all East African Mountains, the near-optimum Afroalpine biotic community (as discussed later on).

This biotic "archipelago effect" (Harmsen, 1989) of species dispersal among mountains throughout the Quaternary has undoubtedly led to multiple invasions by individual taxa followed by adaptive radiation. As Harmsen *et al.*, (1990) point out the main adaptive problem for invaders of tropical origin is one of coping with freezing temperatures. For instance, of two species of flightless beetles (*Parasystates*), one adapted by developing supercooling mechanisms, and the other by becoming frost hardy (Sømme and Zachariassen, 1981).

Endemism on the East African mountains in all taxa studied is a measure of the degree of biotic isolation and age (Hedberg, 1969; Harmsen *et al.*, 1990). The Ruwenzori Mountains, perhaps the oldest, least explored, and the most isolated summits, also have the highest degree of endemism – 14% of vascular plant species (R. Harmsen, personal communication, 1987) (compared with 3.5% on Mount Kenya and 2.8% on the Aberdares). Mount Elgon, for example, younger than the Ruwenzori or Mount Kenya, but also strongly isolated, has 7% endemic taxa.

With each onset of full glacial conditions on Mount Kenya and development of an icecap, vegetation belts shifted dramatically (Figure 12.1) and the Afroalpine biotic enclaves increased in size shifting downslope. Not all Afroalpine assemblages would have had to shift to lower elevations – many could have survived on interfluves (= refugia) which were largely unglaciated. Nevertheless the Afroalpine belt expanded in size and sustained a certain amount of disequilibrium as species had to survive lower temperatures (and probably lower precipitation) or face extinction. Because the East African mountains are relatively dispersed, it is quite unlikely that Afroalpine vegetation belts ever coalesced during maximum cold periods, as similar communities did in the Andes and in New Guinea (Simpson, 1975; Hope 1986).

Adaptation to the Afroalpine conditions by species originating from African lowland or montane taxa probably occurred during Plio-Pleistocene time. Somewhat comparable patterns of Quaternary speciation have occurred elsewhere in the paramo floras of New Guinea (Smith, 1986) and in the Andes (Prance, 1982; Smith and Cleef, 1988). On Mount Wilhelm, for example, approximately 73% of the alpine flora is endemic to the Malasian area with many species originating from lowland Malasian or Gondwanaland elements (19%) (Smith, 1977, 1984). More recently, Smith and Cleef (1988) showed that tropicalpine floras have large disjunct elements with ties to north (Africa/Asia) or to south (Andes/Asia). In addition, they believe the main tropicalpine areas are surprisingly different in plant evolution and dispersal patterns. Overall, when compared with the Andes and Indonesia, African tropicalpine areas are comparatively low in endemic species.

Many Afroalpine species have wide elevation (eurythermic) tolerances and share their habitat with elements of the Afromontane zone (Hedberg, 1969). Those that have produced special adaptations to high altitude climate are considered *bona fide Afroalpine endemics* (Harmsen, *et al.*, 1990). Hedberg (1965) believed 42% of the Afroalpine vascular plants have evolved from taxa in the montane forest or from other Afroalpine zone species. Among these the genera *Dendrosenecio, Oreophyton, Haplosciadium* and *Stoebe* are the most important. To this list Mabberley (1973, 1975a, b, 1976) added the giant *Lobelia* and many species of *Alchemilla* and *Helichrysum*. Other taxa, originating beyond the African montane biochoria, are called "disjuncts" or "cosmopolitan" species (Harmsen, *et al.*, 1989). Because their exact origins are often unknown, it is impossible at this stage to say exactly how far or from where they have migrated. However, from what is known of the moss flora (Spence and Pócs, 1989), it seems that long distance spore transport is possible. Within the vascular plants alone about 20% are from the Northern Hemisphere with many originating in the Mediterranean mountains (Hedberg, 1964). Because some species are undifferentiated from their Northern Hemisphere populations (*e.g. Arabis alpina, Cardamine hirsuta*, and most mosses), while others show considerable differentiation (many vascular plants), it is likely that transport and dispersal from palearctic areas have occurred over a long time frame (Spence and Mahaney, 1988b). It is possible that dispersal success has been aided by the presence of fresh substrates at the end of each glaciation. Indeed, even the occurrence of minor climatic upheavals, like the Little Ice Age may have resulted in an increase in this type of dispersal. This is hardly surprising when one considers documented Northern Hemisphere dispersal into the tropical areas of New Guinea and the Andes (Simpson, 1975; Smith, 1977).

Because the Southern African land mass is relatively small, north-south dispersal from the Southern Hemisphere sources into the Afroalpine has provided far fewer invaders than dispersal from the Northern Hemisphere. Approximately 4% of vascular plants and 1% of mosses *appear* to be derived from the Southern Hemisphere (Harmsen *et al.*, 1990).

When considering the fauna of the East African mountains there is virtually no endemism among the birds, while among the insects (Harmsen, 1989), and small mammals (Coe, 1989), the percentage is low.

The development of the Afroalpine ecosystem involved three distinct and overlapping processes: establishment of substrates, succession through several stages, and evolution or adaptation. These processes not only overlap one another, they influence each other by extensive development of feedback mechanisms (Harmsen *et al.*, 1990). Suffice it to say that, in simplistic terms, from the least diverse biocommunities found near existing glaciers, to the most diverse groups near timberline, the present flora and fauna represent the sum total (Figure 12.2) of all ecosystem-forming processes that have acted along the present climatic gradient from ~3200-~4500 m. This succession, as described briefly in Chapter 2, does not take into account fully the paramount effects of paleoclimate, differences in substrate and aspect, isolation, and remote evolutionary history.

The progressive and increasingly complex biota of the older deposits is probably not the result of biotic succession, but may be explained by lower extremes in daily temperature cycles and overall milder climate. The events described by Harmsen *et al.*, (1990) show the Neoglacial deposits and their sparse flora may be an analog of biotic events (and processes) that unfolded each time a glaciation occurred during the Pleistocene. Colonizations following earlier glacial retreats would have been slower because a pre-adapted flora would not have been available to aid dispersal, and the mountain was higher with somewhat more extreme and lower temperatures. With each new glacial episode the twin processes of speciation and dispersal between montane islands likely had an "additive" (Harmsen *et al.* 1989) effect. Volcanic events probably achieved similar effects by creating new and unoccupied outcrop upon which plants could become established, enter into succession with other plants, and evolve into greater or lesser important components of the Afroalpine ecosystem.

With the rise of the East African mountains and glaciations that followed, an altitudinal sequence of environments was created with distinct differences from either lower level Afromontane and distantly related Paleoarctic locales. This means that invasion and establishment by a

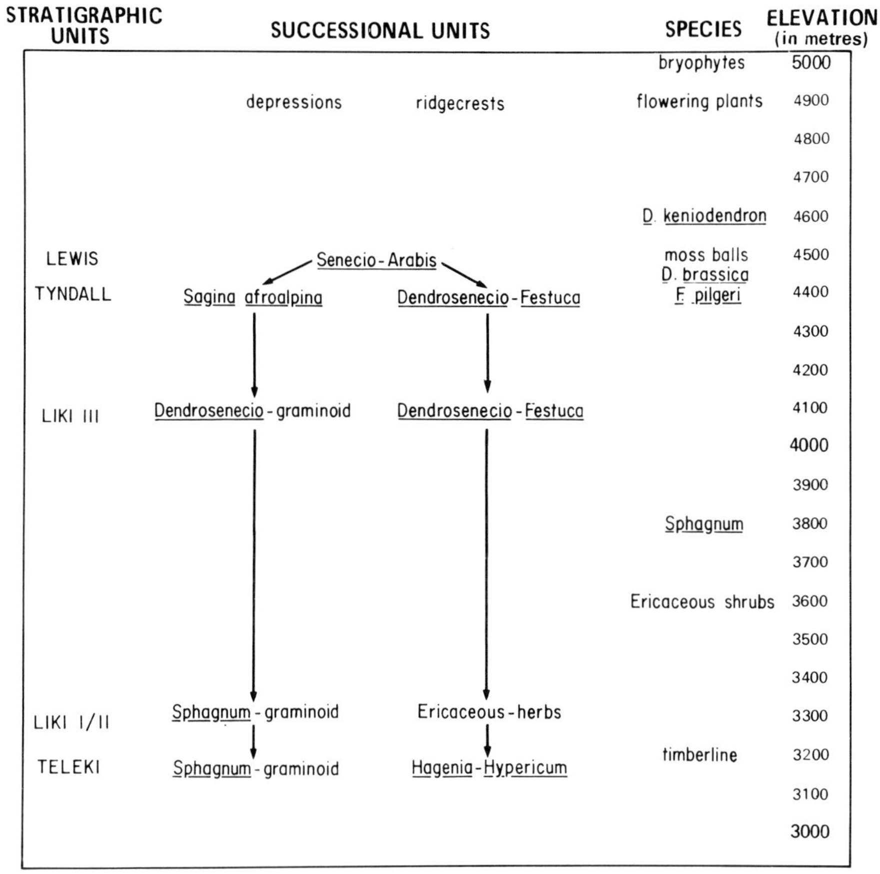

Figure 12.2: Quaternary stratigraphic units on Mount Kenya showing the successional sequence and species change over time at different elevations (After Harmsen, Spence and Mahaney, 1990).

taxon was in all cases hampered by major environmental barriers. Because of the relatively high level of endemism we can conclude that such invasions, despite these barriers have been frequently successful. Overall, species new to the Afroalpine had to develop morphological, physiological and reproductive changes that would eventually lead to isolation (both physical and genetic) and hence, to speciation. These evolutionary processes are still in progress and many Afroalpine species have not diverged completely so as to develop complete genetic isolation (Mabberley 1973, 1975a, Hedberg, 1964, Harmsen, 1989).

The Afroalpine ecosystem, despite its unique assemblage of giant groundsels and other exotic plants, appears to be a rather young biochore. Many species of plants and animals have well developed adaptations to the rigors of the Afroalpine environment (Hedberg, 1964), but there is little evidence of "co-evolved species complexes" (Harmsen *et al.*, 1990). Neither component of species diversity (richness or evenness) are high in Afroalpine communities, trophic networks are simple, and complex parasitic, symbiotic or commensal interactions are rare. All this indicates a low level of community structure (Harmsen, 1981). The unstable Quaternary Period, with attendant volcanism and episodic glaciation produced a series of quasi-climatic equilibria, each lasting at most ~ 100,000 yr or less. Such periods of time appear insufficient to achieve a high degree of interactive evolution. With the current tendency towards world wide warming it will be interesting to see how Afroalpine organisms meet the challenge of warmer temperatures and increased precipitation that may exceed even the levels reached during recent interglacials.

12.4 Microflora Distributions

Microorganisms consisting of bacteria, fungi, and actinomycetes have been identified and cataloged in buried (Ab), and subsurface horizons, of fourteen paleosols examined so far on Mount Kenya. A detailed microbiological survey (Mahaney and Boyer, 1986a) showed that microorganisms found in buried paleosols were also present in ground soils, and that microorganism communities developed quickly in ground soil horizons. Microorganism counts produced three distinct distributions or trends with depth, in surface and buried paleosols, that are related to biogeochemical contamination and its affect on radiocarbon dates. The most frequently encountered distribution, termed *sequential decline* (Mahaney and Boyer, 1986a), is found in paleosols with decreasing counts *vs.* depth. This distribution is associated with internally consistent and accurate radiocarbon dates. Paleosol profiles with *disjunct* or *bimodal* microflora distributions, at various depths, often yield internally *inconsistent* sequences of radiocarbon dated buried substrates.

A number of samples from sites in sections previously dated by radiocarbon (Chapter 9) covering the full range of ^{14}C dating were selected and analyzed for microbial counts. Both Holocene and Pleistocene paleosols, some of which are beyond the range of radiocarbon were studied. Sites from the higher Afroalpine (~ 4000 m) and ericaceous (~ 3500 m) vegetation belts, to the bamboo forest (~ 2700-3000 m) ecosystem, on a variety of deposits ranging from outwash, slope wash and till, to loess were selected for study.

Numbers of microorganisms per gram of sample proved consistently lower in the profiles studied than were usually observed in paleosols of temperate and arctic regions (Mahaney & Boyer, 1986a). While, in general, numbers of microorganisms studied *decrease sequentially* with depth (profiles TV34, LE3 and LE4), distributions were also found to be *disjunct* (profile TV23) or *bimodal* (profiles TV4a and TV33) (Figures 3A and B). Indeed, if wood or buried soil were present in the parent materials of profiles with sequentially declining numbers, it would be expected that such samples would produce valid ^{14}C dates.

Variability in numbers and distributions of microorganisms are considered a function of paleosol evolution, diagenetic history, and unconsolidated nature. In general, numbers of bacteria and fungi, in buried A horizons, have concentrations similar to or less than those in adjacent horizons. For example, horizons TV4a-Ab (Figure 12.3A) and LE4-IIIAb (Figure 12.3B) contain fewer organisms than their corresponding C horizons. These data suggest that, in most cases, residual organic matter (Table 12.1) is not a sufficient energy source to stimulate microbial activity as might be expected if it provided a usable form of energy. Even where Ab horizon microbial counts are higher than in adjacent Bb and Cb horizons (*e.g.* profiles LE3 and TV23), fungi of ~ $1.5 \times 10^3 g^{-1}$ and bacteria as high as $3 \times 10^4 g^{-1}$ appear to have little or no effect on radiocarbon ages (Figures 12.3A and B).

Because phosphorus might be an important limiting factor for the growth of both fungi and bacteria, available P was analyzed in all paleosols. Low (< 1 ppm) values in several older Ab horizons may well preclude the growth of microorganisms. Even at the wetter sites (Table 12.1) values for P only increase to 4.87 ppm. The range in pH from 4.7 to 5.7 (Table 12.1) is not considered a limiting factor for the growth of fungi and bacteria.

Profile TV33 (Figure 12.3), formed in a complex of alluvial fan materials, exhibits a disjunct distribution where the Ab horizon at 109-120 cm is apparently free of microorganisms. The relatively high microbial counts determined for the upper A/C horizon, close to the fibrous root systems of the surface grass-sedge plant community (Figure 12.3A), suggest a possible source of nutrients (including radiocarbon) that might periodically move downwards to the population of microorganisms below by soil moisture percolation processes.

The lowermost paleosol in profile TV33 (Figure 12.3A) contains relatively high microbial counts suggesting contamination by ground water influx. During sampling constant discharges of ground water occurred in the lowermost paleosol, as well as from other nearby sites in alluvial fans. This suggests that fans act as rechargeable valley aquifers. Ground water contaminated with spores and new organic carbon may produce high fungal counts and inconsistent sequences of radiocarbon dates (Figure 12.3A). Three of the most numerous species in TV33, *Mortierella parvispora* Linnemann, *Cladosporium cladosporioides* (Fres.) de Vries and *Gymnoascus roseus* (Raillo) Apinis are species of wide temperature amplitude and high reproductive capacity, which often produce spores transported by groundwater influx (Mahaney and Boyer, 1986a).

Those radiocarbon dates determined from organic matter in the presence of low or nil counts of microorganisms (*e.g.* TV33-Ab, 109-120 cm: Figure 12.3A) are considered to yield reliable ages. Within acceptable limits radiocarbon dates appear to be compatible with thermoluminescence dates (*e.g.* profile TV4a-Ab horizon; TL age of 1170 ±270 yr) (Figure 12.3A). In addition, relative age indicators such as topographic position, soil stratigraphic differences, and amino acid racemization dates (previously discussed) support the radiocarbon dates reported here. Where microorganism counts are high ($>1.5 \times 10^3 g^{-1}$ for fungi and $> \sim 3.0 \times 10^4 g^{-1}$ for bacteria), radiocarbon dates become inconsistent within paleosols producing erroneous ages, which are considered too young.

Assuming mean residence times (n = 3) for ^{14}C in the Afroalpine of ~240 yr, estimates of biomass (Burgess, 1963; Lockwood, 1981; Widden, 1977), and total microbial numbers approximating those determined here, the conclusion that the microflora is inefficient both at utilizing residual organic matter as well as adding substantially to its mass (except under the circumstances cited) seems unavoidable. The results obtained on Mount Kenya suggest that the distribution of microflora counts can be used to determine if biogeochemical contamination occurred in buried paleosol horizons. This is especially important where regional stratigraphy and paleoclimatic interpretations rest on known radiocarbon dates, where dated sections are known to be internally inconsistent, and as a means of assessing the overall impact of ecological processes on the suitability of samples for radiocarbon dating.

12.5 Frost-Patterned Soils and Microorganisms

Frost-patterned soils are a common phenomenon in arctic and alpine areas (Denny, 1940; Zeuner, 1949; Gradwell, 1957; Soons, 1967; Mackay and Mathews, 1974; Pérez, 1987a, b) and in the Afroalpine zone (Hastenrath, 1984). How they form in the various structures known by their appearance as soil stripes, nets, polygons, or pipkrakes still invites considerable conjecture (Washburn, 1973; Troll, 1958; Outcalt, 1971a,b). Among the most widely distributed patterns on bare ground are needle ice features (piprakes), so named because they have a raked appearance (Figure 12.4). Columnar ice crystals form (Figure 12.4) in these soils during the freezing cycle. Elongated shaped crystals are tapered at the surface and blunt at the basal end. They usually support small "soil caps" of fine-grained material which collapses with melting of the ice columns. Mackay and Mathews (1974) conclude that since stripes often form at right angles to incoming solar radiation the exposed side melts first giving rise to the pattern of ridges and troughs usually observed. On the other hand, Troll (1958) thought nocturnal winds might be implicated in striped soil formation.

The occurrence of needle ice, formed by diurnal temperature changes on Mount Kenya, was studied to determine its origin and role in assisting with the development of different species of fungi and bacteria. Previous work by Boyer and Mahaney (1986a) established the microfloral

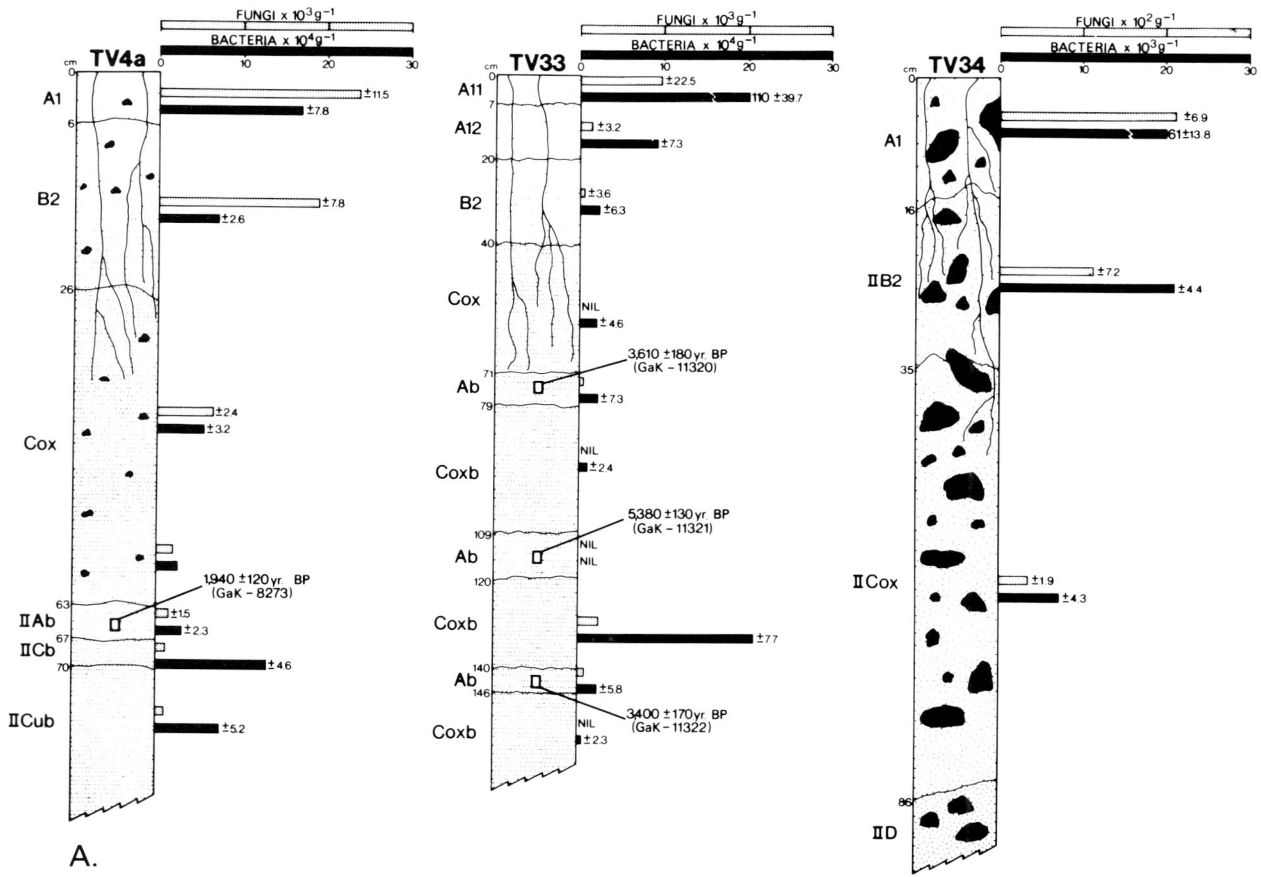

Figure 12.3A: Quaternary paleosol sequence and lithostratigraphy on Mount Kenya: A, profiles TV4a, TV33 and TV34 in the upper Afroalpine belt; For microbial counts, horizontal bars represent the means of three replicates ± one standard deviation. The 14C dates are from whole-soil samples taken from buried organic-rich horizons. Thermoluminescence age determinations for buried horizons are: TV4a-Ab: 1170±270 yr; TV33-Ab (lower: 6400±1700 yr; TV23-Ab: 62,000±20,000 yr). Roman numerals refer to lithologic discontinuities as follows: TV4a, I = outwash, II = fine-grained alluvium; TV33 = slope wash throughout; TV34: I = loess mixed with till, II = till; LE3: I = loess, II = outwash, III = loess, IV = till, V = lacustrine sediment; LE4: I = loess, II = alluvium, III = loess, IV = till, R = basalt; TV23: I = loess, II = till, III = loess + tephra, IV = till.

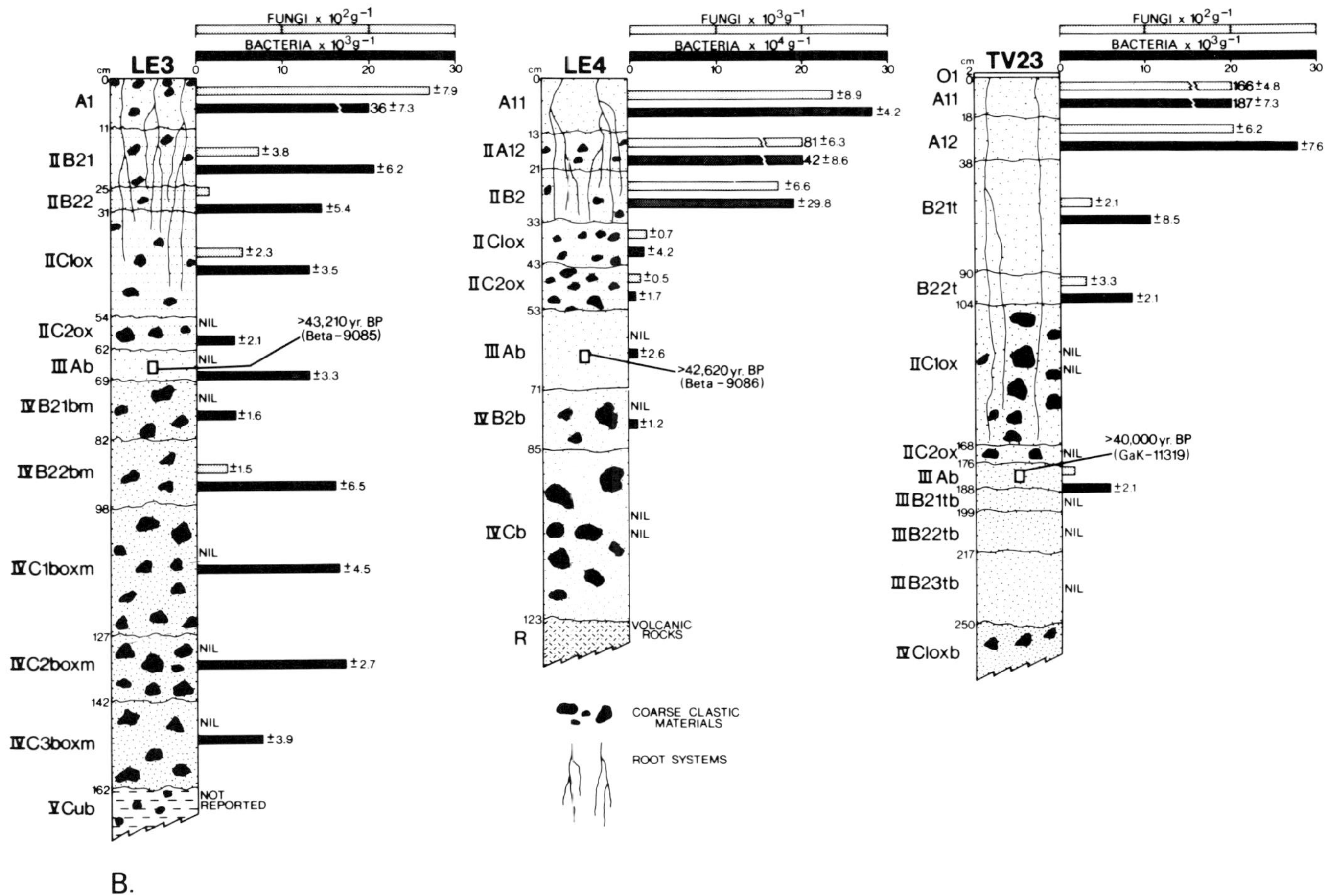

B.

Figure 12.3: B, profiles LE3 and LE4 in the ericaceous belt, and profile TV23 in the bamboo forest/*Hagenia* woodland belt. (See Figure 12.3: A.)

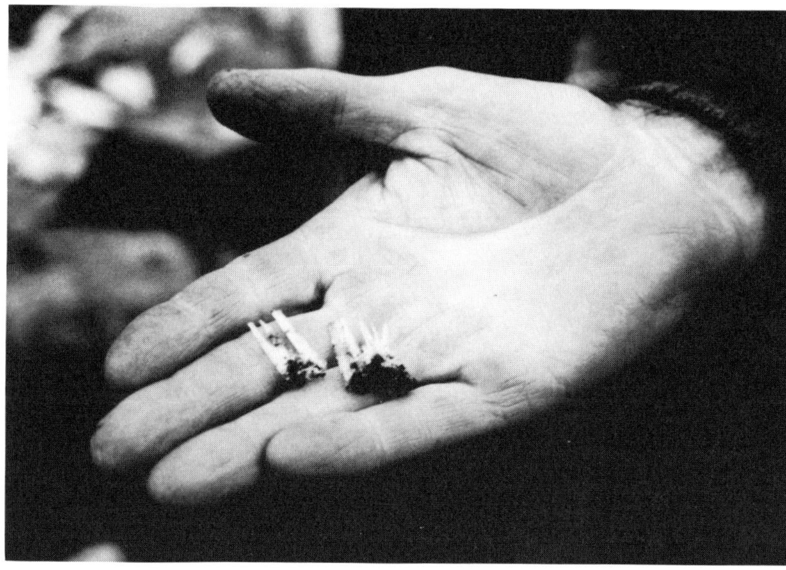

Figure 12.4: Needle ice about 4 cm long.

composition of surface soil horizons not affected by needle ice. Since these surface soils expand and contract with the formation and dissipation of needle ice nearly every night, it is possible that such activity might affect the growth and development of certain species of microflora. Eight surface soil samples, collected from frost-churned A horizons in the surface of a prominent moraine of Liki III age below Simba Col (Figure 12.5) in Upper Gorges Valley, were studied to determine their physical characteristics and microfloral composition.

The resistence of fungi to freezing and thawing or avoidance probably originates from establishment in and adaptation to the Afroalpine (or arctic and alpine) environment. Morphological

Table 12.1
Data for organic carbon (O.C.) and available phosphorous (P) in buried and relict paleosols on Mount Kenya

Site	Horizon	O.C. (%)	P (ppm)	pH (1:1)
TV4a	Ab	7.59	-	5.6
TV33	Ab	3.66	1.00	4.7
	Ab	1.66	2.73	4.9
	Ab	8.13	4.87	5.0
TV34	A1	13.72	3.70	5.2
LE3	Ab	6.94	0.53	5.7
LE4	Ab	8.52	0.63	5.6
TV23	Ab	0.07	0.28	4.7

attributes, mycelial morphology and size, as well as spore form may be significant in the prevention of intracellular freezing and hence may contribute to frost hardiness or supercooling. For most fungi and other microorganisms, intracellular characteristics are probably of more significance to freezing resistance. These assume a variety of structural and chemical forms (Iniss and Ingraham, 1978), which resolve themselves into those preventing intracellular-freezing, desiccation or both.

Table 12.2
Percent water, field capacities, permanent wilting points, and available moisture for soils below Simba Col, Gorges Valley, Mount Kenya Afroalpine area.

Sample	Pw (%H_2O)	1/3 Bar (Field Capacity) (%)	15 Bar (Permanent Wilting Point) (%)	Available Water (FC-PWP) (%)
N1S1	30.40	46.01	15.93	30.08
N1S2	29.74	44.44	15.06	29.38
N1S3	29.62	41.39	14.88	26.51
N1S4	31.46	43.46	19.29	24.17
N1S5	32.80	46.22	19.97	28.25
N1S6	30.37	41.96	16.35	25.61
N1S7	19.30	28.66	9.49	19.17
N1S8	23.58	33.72	14.09	19.63

Fungi from surface soils presumed to be either tolerant of (supercooled), or resistant to freezing, were examined for their growth responses following low temperature exposure. Because of the close relationship between freezing injury and moisture stress the response of these fungi to media of increasing water potential (Table 12.2) was also examined.

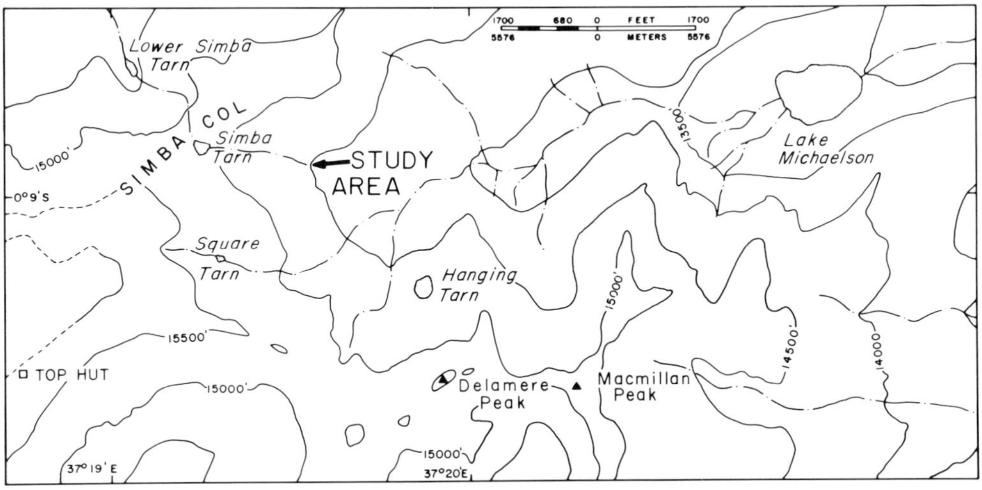

Figure 12.5: Site near Simba Col where needle ice and surface A horizon material was collected.

The soil formed in the moraine is of Liki III age (12,500 yrs BP), 89 cm in depth (Table 12.1), and has a pebbly sandy loam texture throughout. Usually profiles of this age are formed from two parent materials, till overlain with loess (Mahaney, 1984a), but in this case no loess was identified. Areas stripped of vegetation cover appear to provide the proper setting for the development of needle ice.

Crests and gentle slopes of end moraines and valley sides exposed to wind and water erosion are most susceptible. Human and animal activity, which is becoming more of a problem on Mount Kenya (Mahaney, 1985), may result in removal of vegetation cover and exposure of the upper parts of thin A horizons to freezing processes. Needle ice formed in stripped areas seemed to be more prevalent on the eastern side of the mountain where cloud cover is less frequent and nighttime temperatures lower (Mahaney, 1981).

Coarse grained pebbly surfaces are not conducive to the formation of needle ice. Because freezing leads to the congealing of coarse-grained sediment, rather than to the segregation of ice, the slow growth of needle ice from the soil surface is confined to finer sediments.

The details on particle size characteristics for samples NIS 1-8 are shown on Figure 12.6. The position of the curves less than 1000 μm in diameter is nearly linear, which is typical of the partially weathered and unweathered glacial sediments on Mount Kenya and elsewhere (see Mahaney, 1984b). These distribution curves correlate closely with particle size characteristics of

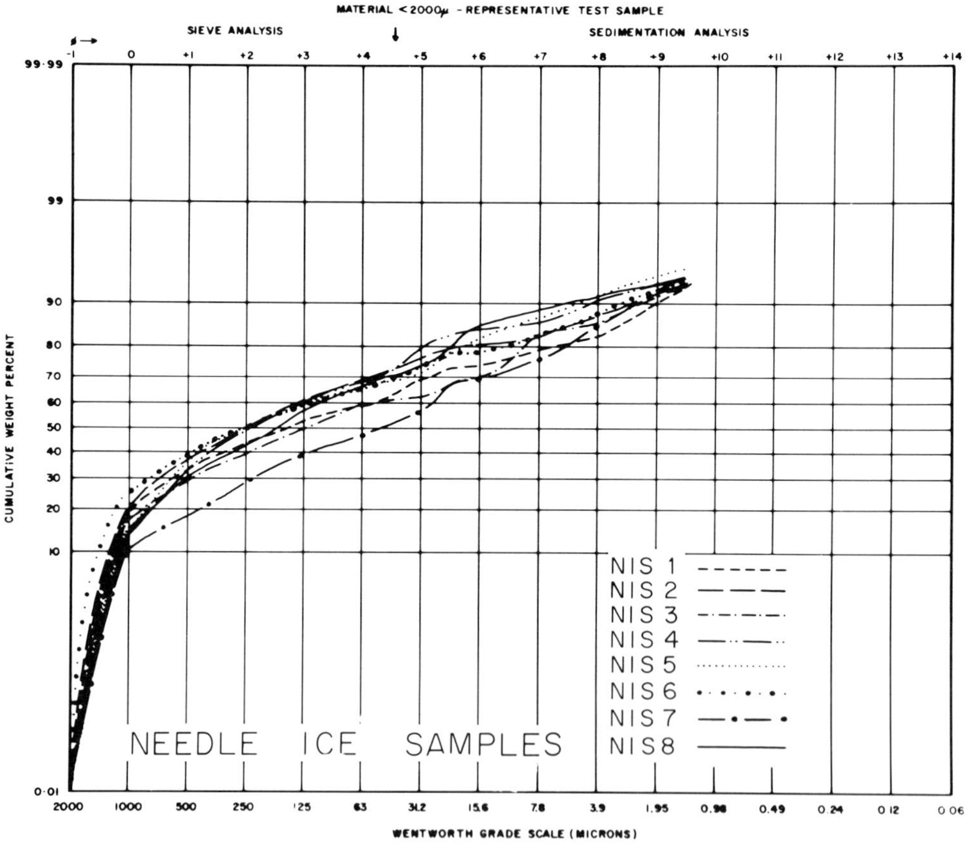

Figure 12.6: Particle size distributions for eight samples of ground paleosol in which needle ice formed.

soils studied in other areas (Embleton and King, 1971; Williams, 1957), and in another locality on Mount Kenya (Zeuner, 1949) (Figure 12.7). Samples NIS 1-6 and 8 show little variation in particle size; sample NIS 7 contains less sand and more silt (~20 percent). However, needle ice formed in sample NIS 7 in the same way and with the same strength as in the other samples. Mean phi calculations from the particle size data (Figure 12.7) suggest that samples with values between 2.5 and 4.5 are capable of forming needle ice, and field observations do not suggest that needle ice activity is more advanced in samples with higher mean phi (heavier textured).

Water retention (percent water, field capacity and permanent wilting point) was calculated to determine the range in available moisture among the samples (Table 12.2). Available water ranged from ~19 to 30 percent in samples NIS 1-6 and 8. Sample NIS 7 had 19.17 percent available water reflecting its lower sand and higher silt content. However, no discernible differences between needle ice forms in sediments with variations in particle sizes were observed, suggesting that the range of sandy loam soils is adequate for pipkrake formation. Mineralogical determinations of the <2 μm grade size material revealed that no crystalline clay minerals are present in the samples. Primary minerals include minor to moderate amounts of quartz, plagioclase feldspar,

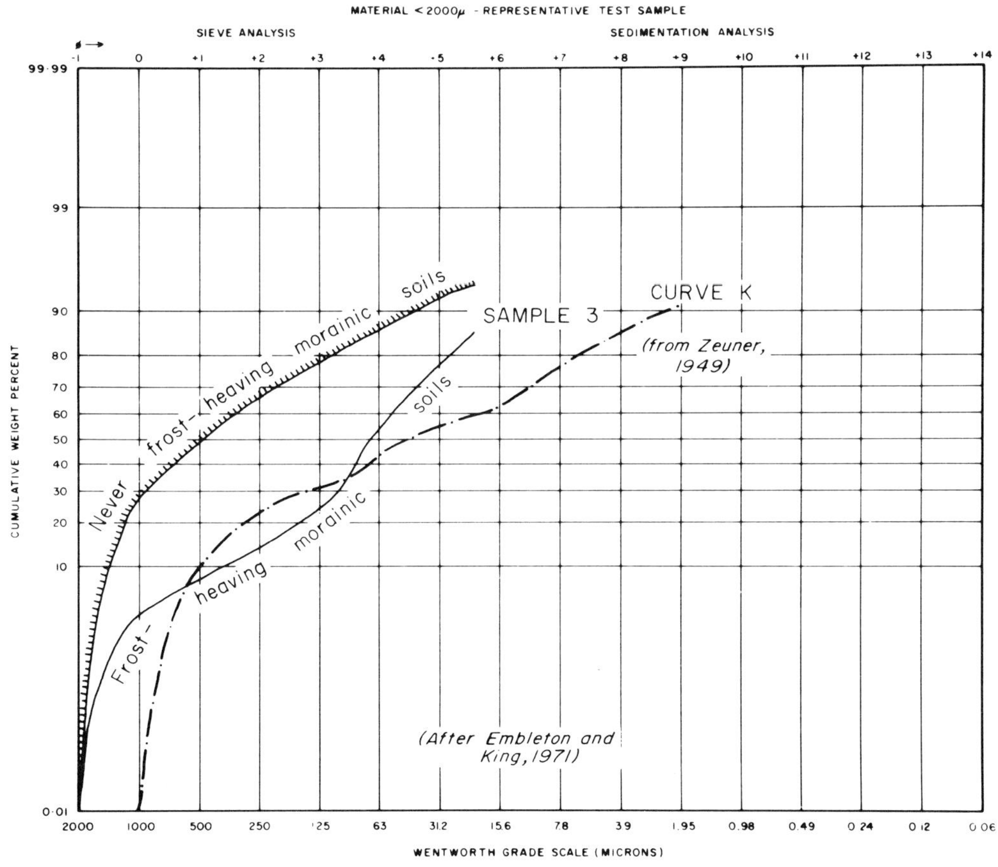

Figure 12.7: Particle size distribution limits for needle ice formation, curve K (from Zeuner, 1949) and sample 3 from Embleton and King (1971).

magnetite, hematite and pyroxene (Mahaney, 1985). The absence of crystalline clay minerals indicates that they are not essential for needle ice development.

Counts of fungi and bacteria show few if any obvious relationships to the physical parameters previously discussed. The highest counts of bacteria and fungi were made in sample NIS7 which had lower Pw and available water (Table 12.2). Microbial counts of bacteria and fungi varied independently of water content, but the numbers were lower than in surface soils not subject to freezing effects (Boyer and Mahaney, 1985). This indicates that diurnal freeze thaw or lack of plant cover inhibit the growth and survival of soil microflora.

The most common fungi isolated both from soils subject to frost heaving and from soils not subject to frost heaving is *Mortierella*, and in particular *M. parvispora* (Table 12.3). Other

Table 12.3
Counts of microorganisms in needle ice soils

Sample Site	Bacteria x 103	Fungi x 102
1	12.2	26.2
2	21.6	4.1
3	11.9	7.5
4	67.7	71.3
5	8.4	12.1
6	5.6	6.8
7	17.6	36.2
8	12.5	14.1
	Σ = 157.5	178.3
	$\bar{x}$ = 19.61	22.2

mucoraceous genera from Temperate areas, including Mucor, Absida or Rhizopus were not observed in the Mount Kenya samples. The next common fungi in soils subject to frost action are Hyphomycetes. In this group *Penicillium*, *Trichoderma*, *Fusarium* and *Paecilomyces* were isolated form cultures in declining amounts (Mahaney and Boyer, 1986c). Because the dominant Afroalpine plant forms are tussock grasses and sedges, it is not surprising that all microbial genera observed belong to grassland biomes (Christensen, 1981).

The growth characteristics of these fungi are summarized in Tables 12.4 and 12.5; Figures 12.8-12.11). Collectively, the mean growth rates (Table 12.4) at 23 oC and 15 oC did not reveal any differences between the ten fungi in frost-affected and non-frost affected soils. Both fungi groups are composed of mesothermic species that yield higher growth rates at 23 oC (Deverall, 1968).

Growth rates for fungi in frozen soils averages 0.87 cm/day, whereas those in unfrozen soils average 0.86 cm/day (Table 12.4). Because temperature has a similar effect on both sample groups, it is likely that fungi from frozen soils do not have lower optimum growth temperatures resulting from exposure to diurnal freezing effects. The hypothesis that some microorganisms may undergo rapid growth as an adaptive mechanism may be affected by a negative correlation between moisture content and resistance to freezing (Mazur, 1968; Mahaney and Boyer, 1986c).

Table 12.4
Growth characteristics of fungi isolated from frozen soils (left) and unfrozen soils (right)

Species	*Growth cm/day 23oC	15oC
Aspergillus janus Raper and Thom	0.32	0.28
Fusarium semitectum Berkeley and Rafcnel	0.76	0.66
F. sambucinum Fuckel	0.51	0.25
Mortierella gemmifera M. Ellis	0.80	0.73
M. parvispora Linneman	0.97	0.80
Paecilomyces elegans (Corda) Mason and Hughes	0.33	0.21
P. marquandii (Marsee) Hughes	0.66	0.57
Penicillium charlesii Smith	0.27	0.19
Trichoderma harzianum Rifai aggr.	2.08	0.85
T. viride Pers ex S.F. Gray aggr.	2.01	1.36
Aureobasidium pullulans (de Bary)Arnaud	0.52	0.38
Cephalosporium acremonium Corda	0.40	0.22
Mortierella exigua Linneman	0.97	0.78
M. humilis Linneman	1.02	0.63
M. marburgensis Linneman	1.50	0.95
M. zonata Linneman	0.82	0.64
Paecilomyces parvus Brown et Smith	0.43	0.28
Penicillium pulvillorum Turfitt	0.36	0.30
P. waksmanii Zaleski	0.33	0.18
Trichoderma harzianum Rifai aggr.	2.23	1.45

*Mean of 3 plates. Inoculum 5.0 mm diameter.

Fungi grown on agar media at 15- and 90-bar water potential and calculated as a percent of maximum growth reveal important differences between both groups (Table 12.5, Figures 12.10 and 12.ll). Growth of fungi in frozen soil amounted to 63.3 percent of the maximum of 15-bar and 28.4 percent at 90-bar, whereas fungi in unfrozen soil grew at a mean rate of 48.6 and 15.5 percent of the maximum. These results indicate that the ability of fungi to resist desiccation in frozen soils may result from variations in the osmotic potential of the mycelium.

Recovery of growth following exposure to freezing temperatures for 24 hours and subsequent freeze thaw cycles was determined by calculating linear growth after 24 hours at 25 oC (Table 12.6, Figures 12.10 and 12.11). The results, given as percentages of the controls, show the growth rate of fungi in frozen soil at 84.4 percent, whereas fungi in unfrozen soil grew at 34.8 percent. Clearly there is a greater resistance to freezing exhibited by fungi in frozen soil. At least two groups of fungi are present in both frozen and unfrozen soils. The first group shows low growth rates and resistance to high osmotic potentials and includes species of *Aspergillus*, *Fusarium*, and *Penicillium* with growth rates exceeding 95 percent of the controls. The second group consists of species with low osmotic potentials and rapid growth rates, including species of *Mortierella*. The first group is not limited to frozen soils because *P. pulvillorum*, *P. waksmanii*, *P. charlesii*, and *A. janus* showed fast growth after freezing (Table 12.6, Figures 12.10, 12.11). Species of *Mortierella*

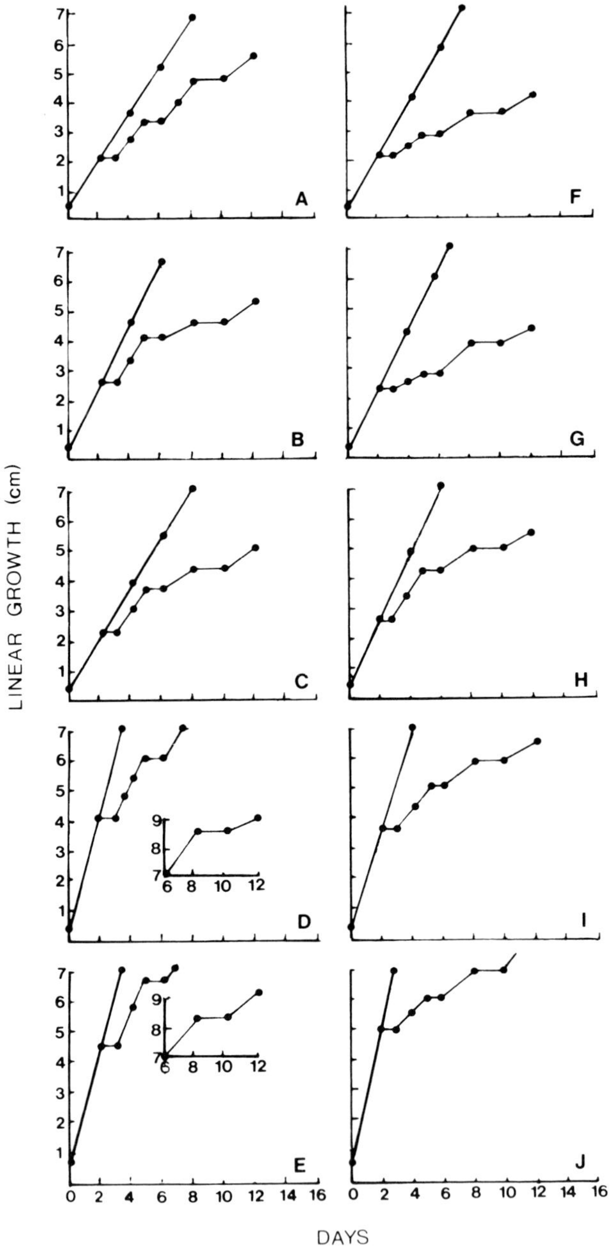

Figure 12.8 (A-J): Growth rates of fungi from frozen soils (A-E) and unfrozen soils (F-J) on malt agar. Constant temperatures (23°C) are linear. Alternating temperatures (-10°C) (23°C) are stepped.

A = *T. harizianum*	D = *M. gemmifera*	G = *M. humilis*
B = *T. viride*	E = *F. semitectum*	H = *P. parvus*
C = *M. parvispora*	F = *T. harzianum*	J = *P. pulvillorum*

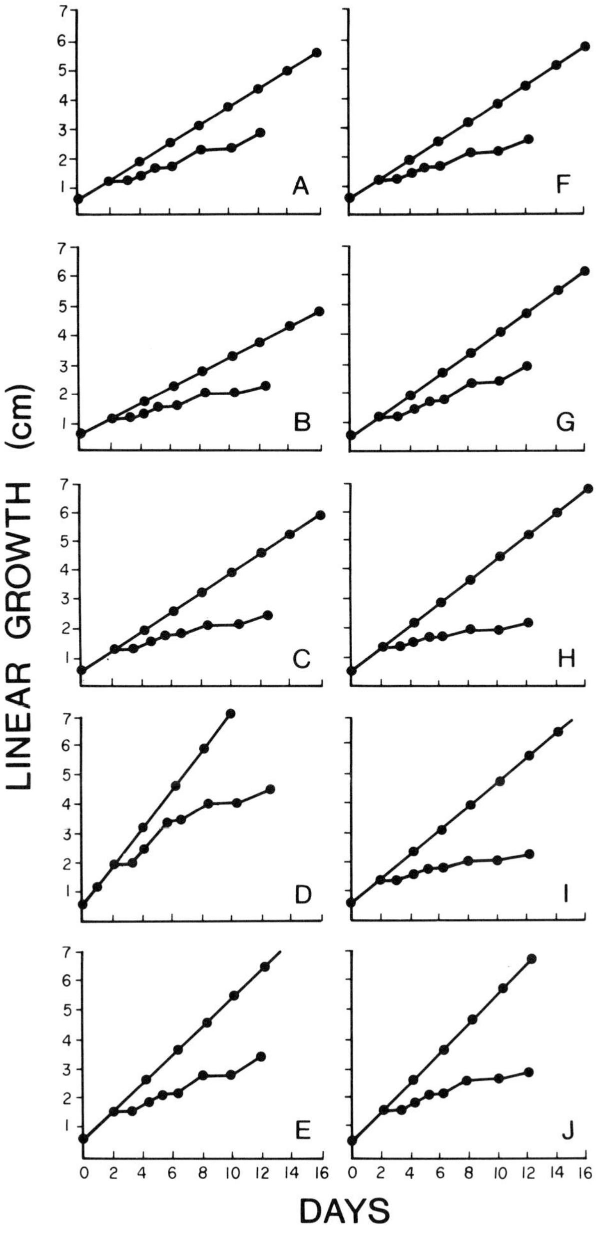

Figure 12.9 (A-J): The growth rates of fungi from frozen soils (A-E) and unfrozen soils (F-J) on malt agar. Constant temperatures (23°C) are linear. Alternating temperatures (-10°C) (23°C) are stepped.

A = *P. charlesii*
B = *F. sambucinum*
C = *A. janus*
D = *P. marquandii*
E = *P. elegans*
F = *A. pullulans*
G = *C. acremonium*
H = *M. humilis*
I = *M. marburgensis*
J = *P. waksmanii*

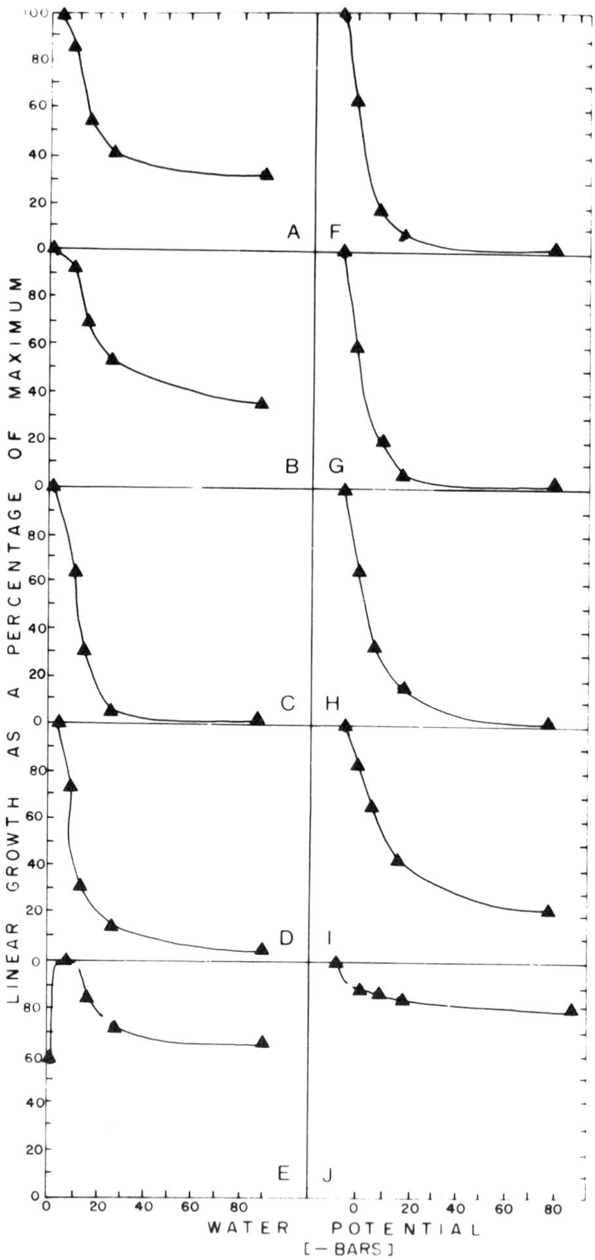

Figure 12.10: (A-J): Growth of fungi at 23°C from frozen soils (A-E) and non-frozen soils (F-J) on media of varying water potential. Growth is expressed as a percent of the maximum. See text for further details.

A = *T. harzianum*
B = *T. viride*
C = *M. parvispora*
D = *M. gemmifera*
E = *F. sambucinum*
F = *M. zonata*
G = *C. marburgensis*
H = *M. humilis*
I = *M. harzianum*
J = *P. pulvillorum*

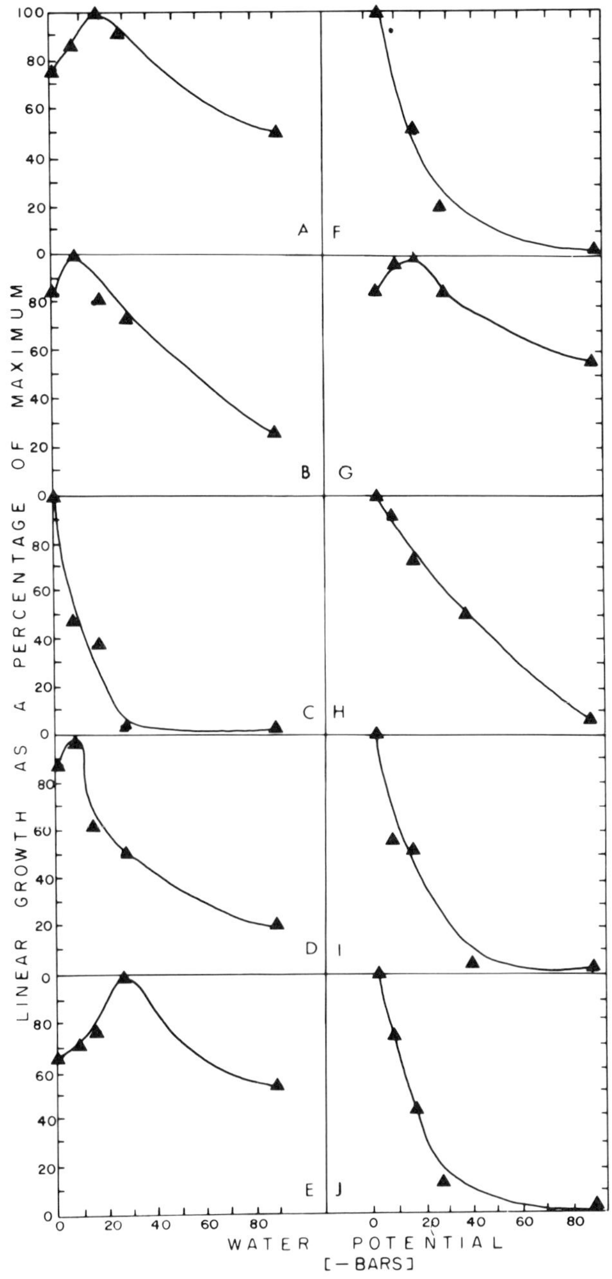

Figure 12.11: (A-J): Growth of fungi at 23°C from frozen soils (A-E) and non-frozen soils (F-J) on media of varying water potential. Growth is expressed as a percent of the maximum..

A = *F. semitectum*
B = *A. janus*
C = *P. elegans*
D = *P. marquandii*
E = *P. charlesii*
F = *A. pullulans*
G = *P. waksmanii*
H = *C. acremonium*
I = *P. parvus*
J = *M. humilis*

Table 12.5
Growth of fungi from frozen soils (left) and unfrozen soils (right) on media of varying water potential

Species	Percent of maximum growth at 23oC (Table 4) in media at - 15 bar	- 90 bar
Aspergillus janus	89	25
Fusarium semitectum	97	51
F. sambucinum	83	64
Mortierella gemmifera	28	5
M. parvispora	30	0
Paecilomyces elegans	31	0
P. marquandii	62	18
Penicillium charlesii	78	52
Trichoderma harzianum	63	32
T. viride	72	37
Aurebasidium pullulans	58	5
Cephalosporium acremonium	51	0
Mortierella exigua	17	0
M. humilis	44	0
M. marburgensis	5	0
M. zonata	7	0
Paecilomyces parvus	52	52
Penicillium pulvillorum	83	83
P. waksmanii	97	97
Trichoderma harzianum	42	42

in the two soils were easily distinguished. All four species in unfrozen soils, while capable of reproducible responses, were clearly sensitive to freezing (Table 12.6). Some species, including *M. gemmifera* and *M. parvispora* showed a high resistance to freezing injury (Mahaney and Boyer, 1986c). As suggested by Iniss and Ingraham (1978), the presence of intercellular and intracellular fats in these species may help to make them frost resistant.

After three cycles of freezing and thawing fungi growth may have been curtailed by lack of nutrients. Even though substantial nutrition was evident from fungi growth, those in frozen soil grew at 93 percent of the controls on media frozen three plus times. In unfrozen soils the mean growth was 58.7 percent (Table 12.6). All species of fungi survived freezing; however, four species of fungi in unfrozen soil including *C. acremonium*, *M. exigua*, *M. marburgensis A. pullulans* showed suppressed marginal growth and reinitiation of growth from the interior of the colony.

While the conditions were not identical, microscopic observations on the spores of fungi to freezing and thawing point out some of the ways by which deleterious effects of freezing may be avoided. In slide cultures of *Mortierella*, regardless of soil origin the growing points were invariably injured by freezing. Regrowth following freezing was always initiated from branches remote

Table 12.6
Post-freezing growth following one 24 hour cycle of freezing and thawing for fungi from frozen soils (left) and unfrozen soils (right)

Species	Growth as percent of control 1 cycle	Species	Growth as percent of control 1 cycle
Aspergillus janus	100	*Aureobasidium pullulans*	48
Fusarium semitectum	97	*Cephalosprorium acremonium*	50
F. sambucinum	94	*Mortierella exigua*	16
Mortierella gemmifera	98	*M. humilis*	78
M. parvispora	89	*M. maburgensis*	30
Paecilomyces elegans	75	*M. zonata*	39
P. marquandii	77	*Paecilomyces parvus*	17
Penicillium charlesii	97	*Penicillium pulvillorum*	95
Trichoderma harzianum	97	*P. waksmanii*	90
T. viride	60	*Trichoderma harzianum*	20
	834		483
	$\bar{x}_f$ 83.4		$\bar{x}_u$ 48.3

from the tips. The distinction between species of frozen and unfrozen soils appeared to be related to the extent of freezing injury rather than to the kind. In no case was recovery associated specifically with asexual or sexual spores, either sporangia, gemmae or zygospores (Mahaney and Boyer, 1987). Although these may be initiated in proximity to the hyphal tips of some species. Among the Hyphomycetes particularly *Aspergillus* and *Penicillium* species as well as *Fusarium*, hyphal tips appeared to be largely unaffected by freezing whereas more distal reisolated hyphae were then inversibly altered, the large vacuoles disappearing and the cytoplasm becoming granular. Conidia of the species of *Penicillium* and *Aspergillus* produced in the immediate vicinity of the growing hyphae were susceptible to freezing injury and did not germinate whereas more mature conidia were not affected by low temperature. It is concluded from this that growth of the colonies of these fungi was reinitiated following freezing from existing hyphal tips. In the case of *Trichoderma* species no other explanation of the differences in response to freezing could be discerned. The vegetative hyphae of *T. harzianum* from unfrozen soils are extensively altered by freezing, the normal vacuolate mycelium appearing granular and unstructured throughout. *T. harzianum* from frozen soils was also altered but less extensively. Regrowth of both fungi arises from branches of distal hyphae (Mahaney and Boyer, 1986c). In the case of *T. viride*, unlike *T. harzianum*, phialospores are produced immediately behind the hyphal tips. While mycelium of *T. viride* appears to be severely damaged by freezing, regrowth from conidia was observed and may have been responsible for the further extensive growth of the colony observed in Petri dish experiments.

Needle ice processes on Mount Kenya do not appear to be assisted by crystalline clay minerals which might adsorb water in greater quantity. Surface paleosols with mean phi values ranging from 2.4 to 4.4 produced needle ice and showed no difference between ensuing size and shape. Particle size curves for the Mount Kenya samples are similar to those reported by other workers on Mount Kenya and elsewhere.

Fungi isolated from needle ice samples are not site specific and are widely distributed species common in temperate and other alpine regions. While the environmental conditions imposed by diurnal freezing and thawing under Afroalpine conditions are rigorous (+ 10°C to 5°C on average) they are not extreme, and evidence is lacking that *specific* physiological charcteristics are a requirement for species survival. Certain physiological attributes including greater resistance to freezing, more rapid recovery following freezing injury, and greater tolerance to water stress appear more marked in the predominant fungi of soils affected by needle ice. Perhaps three types of fungi may be identified as capable of resisting the deleterious effects of needle ice. The first type is characterized by species of *Penicillium, Aspergillus* and *Fusarium* fungi with high osmotic potentials, slow growth rates, and rapid spore initiation and maturation. A high osmotic potential may be of great importance to the survival of hyphae providing protection against internal freezing and resistance to osmotic water loss after freezing of the substrate and during thawing. The second type is represented by fungi such as *Mortierella* spp. and also *T. harzianum*. They are characterized by low osmotic potentials, rapid growth rates, and delayed spore production. Freezing causes extensive and apparently irreversible alteration of the vegetative mycelium. The distal hyphae of fungi in both genera are often killed by freezing, with continued growth arising from branches, or perhaps from resistant reproductive bodies such as chlamydospores. The extensive presence of fats in species of *Mortierella* raise the question as to their possible role in providing resistance to freezing injury (Iniss and Ingraham, 1978). While on initial isolation both *M. gemmifera* and *M. parvispora* seemed to grow faster than other species studied, this could not be assessed quantitatively by these methods. A final group of fungi among those examined possessed characteristically, low growth rates and low osmotic potentials but high rates of spore production. *A. pullulans, C. acremonium, T. viride* and species of *Paecilomyces* exhibited extensive mycelial mortality based on microscopic observations and on observed recovery rates but proved persistent to competition from most other fungi, as a consequence of the production of many spores with greater resistance to freezing injury.

12.6 Mud Polygon Genesis and Small Mammal Grazing Activity

Mud polygons, formed in a valley train deposit in Teleki Valley on Mount Kenya, were studied with respect to their physical, mineralogical, chemical, and biological characteristics. Developing in fine-grained alluvium of postglacial age, these polygonal systems are composed of numerous, nearly isomorphous units, located on the banks of existing drainages, in areas stripped of vegetation cover.

The formation of soil frost phenomena on Mount Kenya has been described by many investigators (Troll, 1958; Hedberg, 1964; Zeuner, 1949; Hastenrath, 1973, 1984), but no experimental work has been carried out to determine more about the physical setting and processes by which mud polygons form in the Afroalpine area. Reconnaissance mapping in all the major glaciated drainages shows that mud polygon development in outwash deposits tends to increase in drainages with large populations of hyrax. In Teleki Valley (Naro Moru Drainage, Figure 12.12) a mud polygonal system at site TV37 was studied along with the local soil flora (Site TV4) and fauna to determine the factors controlling the development of these geomorphic features. A "cake" of mud polygon sediment was excavated and transported to the Geomorphology and Pedology Laboratory at York University for laboratory analysis.

The mud polygon site was chosen for study because it is representative of many sites on valley train deposits in upper Teleki Valley (Figure 12.12). The occurrence of mud polygon development appears higher in Teleki Valley than on similar deposits in other drainages on Mount Kenya. During the course of investigations small herbivorous mammals were observed removing vegetation, followed by erosion through overbank flooding (Figure 12.13) or deflation by wind activity,

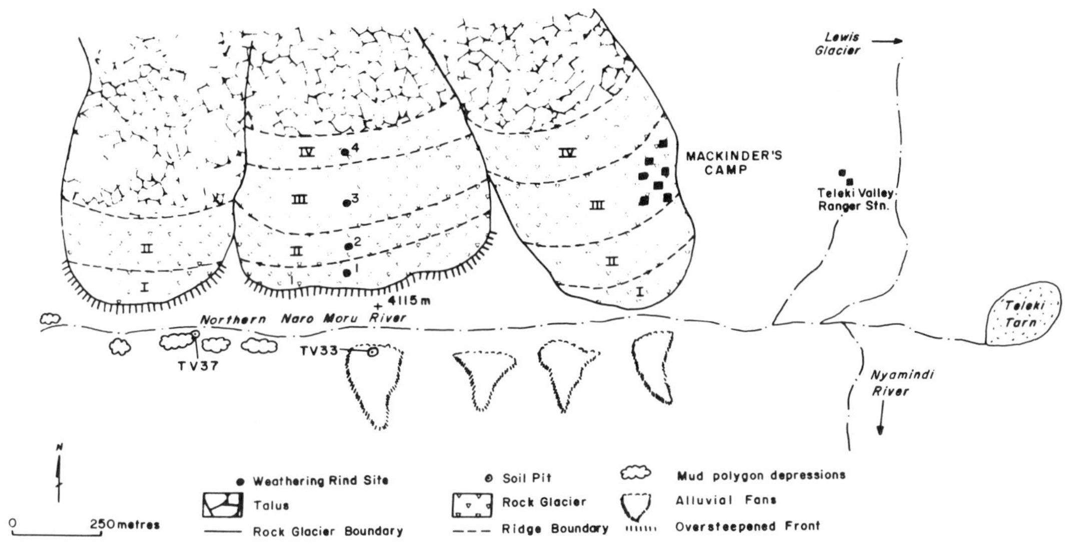

Figure 12.12: Location of site TV37 in Upper Teleki Valley where "the cake" was collected in 1984.

so that surface depressions could form. In these depressions (depth of 30 to 10 cm) polygons formed in drainages above 3700 m (Figure 12.14).

The mud polygon site (TV37) shown in Figure 12.12 grades into undisturbed turf-covered soil (Site TV4; Figure 7.18). This compound paleosol has a complex history and two mutually reinforcing dates, derived from thermoluminescence (TL) and radiocarbon, are available to place the ground soil at less than 2000 years old. A ^{14}C date of 1940 ± 120 yr BP (GaK-8273) on the Ab horizon is substantiated by a TL date of 1170 ± 270 yr (Berger and Mahaney, 1989). The ^{14}C date suggests that the buried soil in clay loam alluvium formed prior to 1940 years ago, and was subsequently covered by fresh pebbly sandy clayey sediment. The ground soil formed in this outwash developed into an Inceptisol that is the youngest radiometrically dated soil on Mount Kenya with a cambic (color B) horizon (Mahaney, 1982b, 1984b, and 1985).

Particle size data for the compound paleosol revealed variations in sand, silt and clay resulting from original stratification in the deposit (Table 12.4). The mix of textures (Table 12.4) reflects rather broad ranges in sand, silt, and clay ratios. The clay loam texture in the Cox horizon appears representative for the formation of mud polygons shown in Figure 12.14.

The shapes and sizes of the mud polygons (Figure 12.14) are related to position; the largest entities are located in the center of each mud polygon zone. Towards the periphery of the depression the polygons are reduced in size, the smallest measuring approximately 5 cm x 5 cm. The size of individual polygons does not appear to affect shape, as some large polygons also have only four sides. While it is impossible to determine the rate at which large polygons form at the expense of smaller ones, it was obvious from field observations that expansion and contraction around the periphery of the mud polygon depressions added new organic materials to the fine clay-rich material there.

The mineralogical composition in the $< 2\mu m$ fraction of material in profile TV4 (site TV4a is 1 m west of site TV4) showed only trace amounts of illite and metahalloysite with moderate to

Figure 12.13: Drainage near site TV4 with hyrax-cropped surfaces caving into the water.

Figure 12.14: Mud polygons near Naro Moru River, upper Teleki Valley. The knife is 30 cm long.

small amounts of albite, anorthoclase, olivine, and quartz reflecting the nature of the country rock in the transported regolith from which this soil formed. Outside of a trace amount of illite nothing in the clay mineral composition suggests that expansion and contraction after wetting and drying is assisted by expandable minerals.

The chemistry of site TV4a (Tables 9.1 and 9.2) yielded important information on the paleosol in which mud polygons formed. The uniform pH with depth and exchangeable cation data indicate that this surface is poorly drained and not well leached. In fact, the presence of Na in such high quantitites (2.95 meq/100g in the Cox horizon) suggests that the most mobile cations tend to increase in the subsurface horizon where mud polygons form after the surface cover is stripped. The cation data reflect higher Na_2O, CaO, and MgO in most of the local rock, whereas the cation exchange capacity at depth reflects the textural changes shown in Table 12.7. Data for the ground paleosol suggest only minor translocation of organic matter downward in the pedon; this pattern is complicated by the presence of the buried paleosol which tends to contribute some carbon and nitrogen for downward movement into the C1b horizon. The data for total salts show higher amounts in the A1 and B horizons followed by uniform distributions below. Base saturation, which is slightly higher in the surface, drops off in the lower horizons of the ground paleosol, and in the paleosolum of the buried soil, to rise again with depth. The values of "free" Fe_2O_3 in these soils are within the ranges expected for soils of this age in the Afroalpine belt on Mount Kenya (Mahaney, 1984a and b).

Table 12.7
Particle size data for the TV4a soil profile

Site	Horizon	Depth (cm)	Sand % (2mm-63μm)	Silt % (63μm-2μm)	Clay % (<2μ m)
TV4a	A1	0-8	16.7	64.8	18.5
	B2	8-29	86.5	7.0	6.5
	Cox	29-59	25.1	48.4	26.5
	IIAb	59-62	37.4	45.6	17.0
	IIC1b	62-78	26.5	54.5	19.0
	IIC2b	78-85	90.0	6.1	3.9
	IICub	85+	67.3	24.2	8.5

To determine how these mud polygons react to saturation the site was flooded with water. Within 30 minutes the polygons expanded and most were fully closed (Figure 12.15). Within ninety minutes the surface began drying out and cracking, along the already predefined planes of weakness (Figure 12.16). While there is no instrumental meteorological data for sites on Mount Kenya, limited observations of nighttime temperatures (Coe, 1967; Coetzee, 1967; and Mahaney, 1980) suggest that surface temperatures routinely drop to -5° nearly every night rising to between $+8^{o}$ and $+15^{o}$C everyday. Following a period of desiccation (3 hours) and a nighttime air temperatures of -6°C, no appreciable change was observed in mud polygon form, *i.e.* they remained open although some needle ice had formed in places. From these observations the conclusion that diurnal freezing and thawing achieves no expansion and contraction seems most plausible. Needle ice forms quite well in these finer sediments as well as in other coarser-grained sediments on Mount Kenya.

The vegetation around mud polgyon sites is characteristic of wet-mesic sites of the upper Afroalpine region of Mount Kenya at approximately 4000 m (Coe, 1967; Hedberg, 1964). The valley walls, covered with stands of *Dendrosenecio keniodendron* or *Dendrosenecio keniensis* with an understory of *Alchemilla argyrophylla* and *Festuca pilgeri*, merge on the valley floor with a flora of grasses, sedges and forbs in which acaulescent cusion-like forms predominate (Hedberg, 1964; Mahaney and Boyer, 1989). The important grasses and sedges include *F. pilgeri, Agrostis trachyphylla* and *Carex monostachya. F. pilgeri* is greatly reduced in size compared with the more typical form observed on drier sites. Others of rarer occurrence include *Luzula abyssinica, Poa schimperiana* and *Juncus* (spp). Within the forbs the predominant species are *Ranunculus oreophytus, Haplocarpha rueppellii* and much more rarely *Carduus platyphyllus, Haplosciadium abyssinicum* and juvenile forms of *D. keniensis*. Only one species of shrub is present, *Alchemilla johnstonii*, a form adapted to wet soils and solifluction terraces and exhibiting a nearly prostrate habit at the base of cushions or reduced tufts of grasses or sedges. Within the polygon depressions many seedlings of *Subularia monticola* are observed. Islands of vegetation (Figure 12.14) are dominated by grasses mostly *F. pilgeri* and *A. trachyphylla. C. monostachya* is less frequently observed.

A total of two hundred meters of transects assigned randomly on both sides of the polygon depressions were sampled for species cover (Table 12.8). The predominant species are grasses and sedges accounting for approximately 90 percent of the plant cover. The three dominant species: *A. trachyphylla, F. pilgeri* and *C. monostachya* are cropped intensively by colonies of hyrax and rodents (Figures 12.17 and 12.18).

The concentration of polygon depressions in this area led to an attempt to explain their origin. They are located in juxtaposition to the rock glaciers described previously by Mahaney (1980) (Figure 12.12). Because this site is a favored habitat of colonies of the Mount Kenya hyrax (Figures 12.17 and 12.18) the surrounding bottomland vegetation around the polygon site had been heavily cropped as a result of their feeding activities (Figures 12.17 and 12.18). *F. pilgeri, A. trachyphylla* and *C. monostachya* were selectively grazed. Burrows and trails of the groove-toothed rat (*Otomys orestes orestes*) are also exceptionally common among the cushions of grasses and sedges, the trails worn bare and covered with fecal pellets. Both skulls and bone fragments of these two species were collected on site (Mahaney and Boyer, 1989).

Table 12.8
Percent cover for principle species of the valley floor in proximity to polygon depressions.[a]

	Total Cover %
Agrostis trachyphylla	60
Festuca pilgeri	14
Carex monostachya	10
Alchemilla johnstonii	8
Ranunculus oreophytus	7
Haplocarpha rueppellii	3
Luzula abyssinica	2
Carduus platphyllus	2
Poa schimperiana	2
Dendrosenecio brassica (seedlings)	1
Dendrosenecio keniodendron (seedlings)	1
Mosses	2
Bare ground	

a. Data from Mahaney and Boyer, 1989

Figure 12.15: Cracks between mud polygons filling with water during controlled flooding experiment. The knife is 30 cm long.

Figure 12.16: Cracks reappearing in mud polygons following desiccation during controlled experiment. The knife is 26 cm long.

Figure 12.17: Hyrax (*Procavia johnstoni mackinderi*) cropping *Agrostis* and *Festuca* down to the fibrous root systems.

Figure 12.18: Hyrax colony on Liki III recessional moraine near the rock glacier complex. Note the lawn-like uniformity of the cropped paleosol surface.

The patterns of movement and feeding habits of these herbivorous mammals are probably sufficient to initiate the changes leading to development of depressions, and eventually to mud polygon activity. Seedlings and fresh growth of *F. pilgera* and other grasses and sedges stimulated by grazing (Figure 12.19) are favored by the hyrax over coarse tussocks or cushions. As a result vegetation becomes patchy – a system of over-grazed grassland, barren soil, and scattered islands of uncropped tussocks (Figure 12.14). By feeding on the fibrous roots of grasses where exposed and further eroding the vegetation through their nocturnal activities, the groove-toothed rat probably accelerates the process. Population densities of both animals may be enhanced by the presence of refuse left by tourists.

Following suppression or removal of vegetation, enhanced needle ice activity may prevent reseeding or colonization. Erosive forces sufficient to create the depressions may arise from both wind and water. Periodic flooding of the terraces has been observed (Mahaney, 1979a, 1982a) and the annual influx of waterborne seeds of *Subularia monticola*, germinating on the polygons, supports water as an important agent of erosion (Mahaney and Boyer, 1989).

The uniformity of polygon depressions also leads to the conclusion that overcopping by herbivores contributes to their origin. The depressions appear to be of a generally uniform size. Tussocks of grasses, principally *F. pilgeri*, are always present in considerable numbers and they are of a size and distribution compatible with the grazing patterns described previously. The tussocks are extremely resistant to displacement; their fibrous root systems bind the soil tenaciously, and the exposed sides are often covered in mosses. Where undercutting is evident among tussocks it is not unidirectional contributing to the conclusion that once formed the system is probably quasi-stable over short periods of time.

Four stages (Figure 12.20) in the development of mud polygon complexes in the Mount Kenya Afroalpine area have been outlined by Mahaney and Boyer (1989). Stage 1 requires the suppression or elimination of vegetation by cropping so that the surface is bare and subject to needle ice activity. In Stage 2 the vegetation cover is removed, flooding and/or aeolian activity can proceed to erode surface sediments. Erosion leading to the development of small depressions (Stage 3) provides a quasi-stable environment in which mud polygons may form (Stage 4).

Evaporation of water between soil grains creates tension which stretches blocks of soil, creating near-uniformly distributed stress across exposed surfaces leading to contraction. The contraction force is created by surface tension between grains; as more water is lost the top surface of water pulls grains away from one another. Because some of these grains have electrically charged surfaces (where cation exchange capacities are high) they may be pulled together by electrical forces attracting clays, organic micelles, and cations. The contraction process places the topmost layer of mud under stress which is later relieved by strain producing the polygon. Since the cracks appear at random times, the tensile stress must not be uniformly distributed across the exposed soil surface, and therefore contraction centers are not uniformly spaced. Even though these mud polygon cracks do not intersect always at 120° angles, they often produce a combination of random and regular polygonal patterns.

Because these polygonal systems are three dimensional, it may be that mud polygon formation is the result of tensile and shearing stresses working horizontally and vertically. In the horizontal axis contraction occurring in response to tensile stress may place an adjoining area under shearing stress. Along the vertical axis, faces sliding past one another might create new shearing stresses, the combination of which might produce polygonal cracks intersecting between 90° and 120°.

No curling was observed on these mud polygons which may be related to the relatively slow rate of evaporation in the Afroalpine environment creating insufficient stress on the surfaces, or to the particle size distributions.

Air and ground temperature have been recorded in the Afroalpine belt by different workers (Hedberg, 1964; Coe, 1967; Mahaney, 1980). Of these Coe (1967) gives the most comprehensive

Figure 12.19: Hyrax cropping area in front of the Liki III rock glacier, upper Teleki Valley

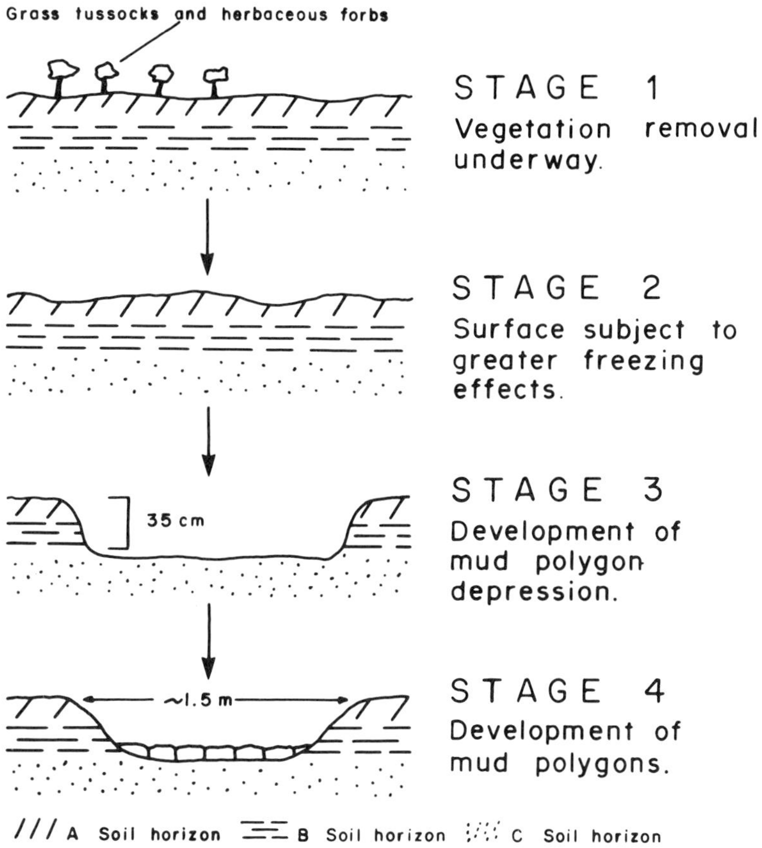

Figure 12.20: Stages in the development of mud polygons in upper Teleki Valley.

record (Figure 12.21) but without actual soil or sediment temperatures. To determine the thermal characteristics at different moisture states a mud polgyon "cake" (15 cm in thickness) x 35 cm x 35 cm (Figure 12.22A) was installed with thermistors at different depths (Figure 12.22B) in a freezer in the Geomorphology and Pedology Laboratory at York University. The ambient air temperature was cycled from ~ -7°C to ~ 12°C over 24-hour periods of 12 hours below freezing and twelve hours above freezing to duplicate as closely as possible conditions on the mountain. Even though nighttime air temperatures on Mount Kenya rarely drop below -6°C, the lower temperatures used in this experiment (~ -7°C) would serve to greatly stress the surface sediments.

The first freezing cycles (24-hour periods) were carried out under *dry* conditions (Figures 12.22A and B). The results show that the air, 2 and 6 cm temperatures fell below 0°C, while the 12 cm level dropped below 0°C for only 1 hr. Under *dry* conditions freezing activity appears to have the potential to affect mud polygon activity.

The second set of freezing cycles was carried out with the "cake" at field capacity (water content = 13 ml/cm^3) (Figure 12.23A, B). Contraction cracks are in essentially the same positions after the first freezing cycle. The data in Figure 12.23B clearly show that with water content at field capacity it is possible to freeze only the top 2 cm of the "cake" and only for a period of ~ 4 hr.

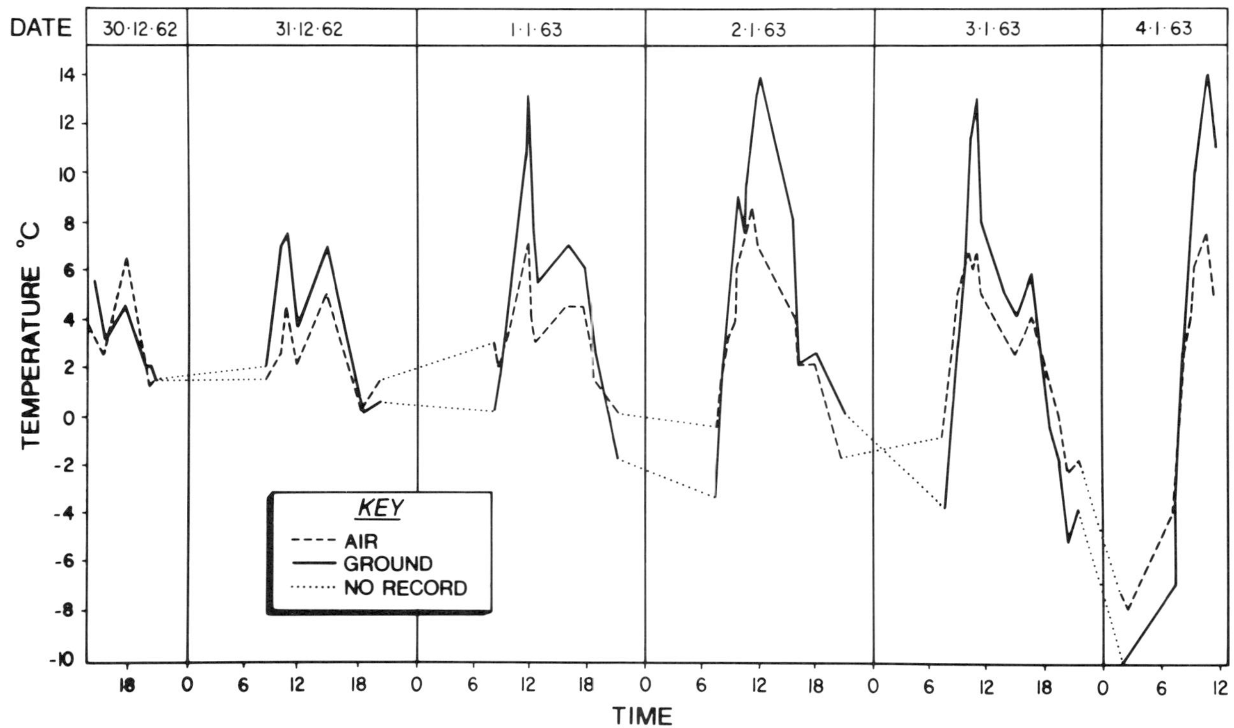

Figure 12.21: Air and ground interface temperatures for 6 days in 1962 and 1963 at 4175m in Mackinder Valley (from Coe, 1967, p. 59).

A.

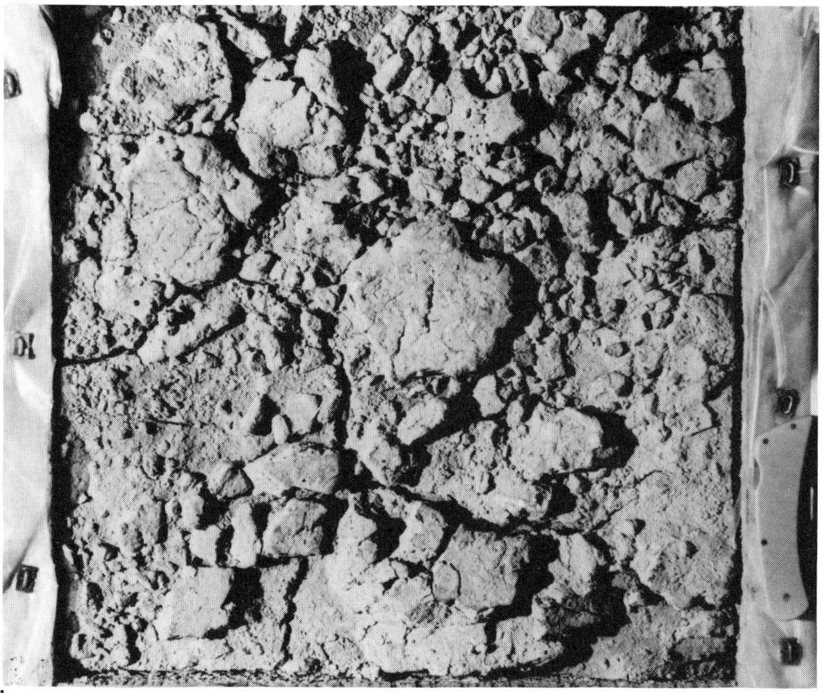

B.

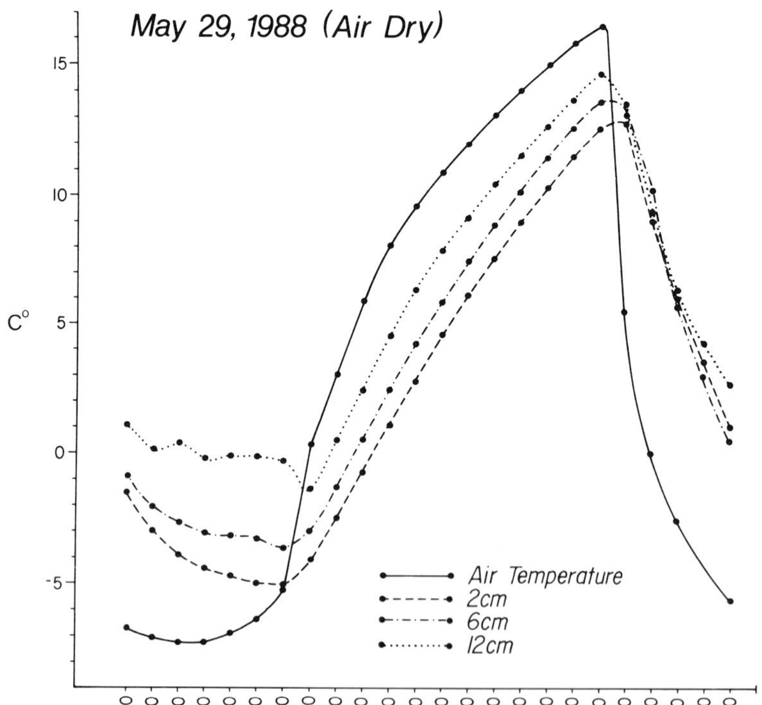

Figure 12.22: A, "Cake" of mud polygons in the dry state 29 May, 1988; B, 24-hour temperatures in ambient air and at different depths.

A.

B.

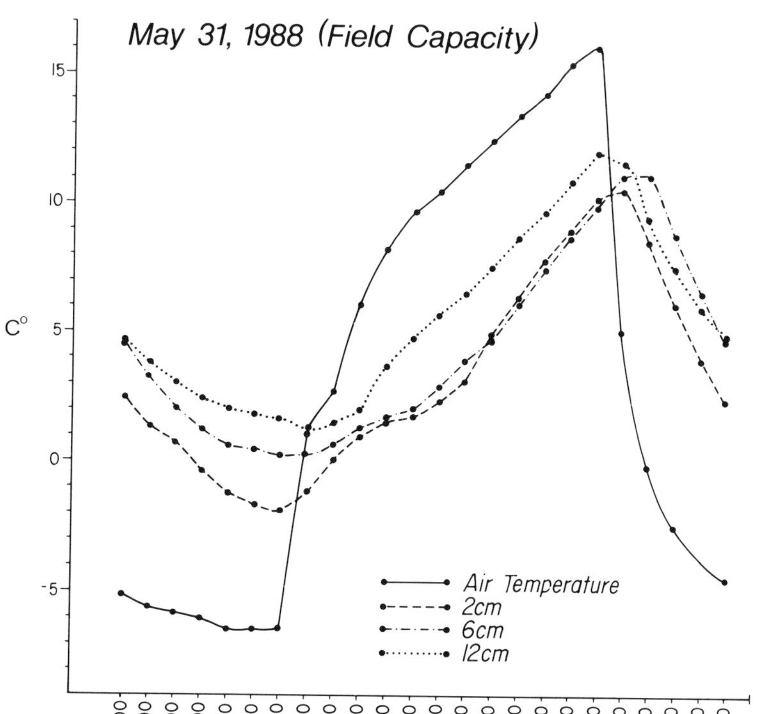

Figure 12.23: A, "Cake" of mud polygons at field capacity 31 May, 1988; B, 24-hour temperatures in ambient air and at different depths.

The third round of freezing cycles was carried out with the "cake" at saturation. The results at the highest moisture level show that the lowest temperatures at the 2 cm depth reach $+3^{o}C$ (Figure 12.24A, B). Because ground soils and paleosols in the Upper Afroalpine usually have moisture contents between field capacity and saturation (water content = 26 ml/cm^3) it seems unlikely that freezing contributes to the development of mud polygons. However, during the dry season (January-March) it is possible that freezing might penetrate to depth and contribute to the development of frost phenomena.

The grazing habits of Mount Kenya hyrax and groove-toothed rat are considered essential in the formation of mud polygons in Upper Teleki Valley on Mount Kenya. Through selective over-grazing of vegetation, leading to the loss of plant cover, the following succession occurs: bare ground→depression→mud polygon formation.

While particle size ratios are important in terms of controlling the water-holding capacity of surface paleosols, the absence of expandable clay minerals in surface sediments is not seen as a limiting factor in the development of mud polygons. Soil chemical data, especially organic matter relationships, show that organic constituents might be important in influencing water-holding capacity. Flooding of the polygons prove they automatically congeal after wetting and contract upon drying out, with the contraction cracks assuming nearly their pre-flooding shape. Wetting and drying of these samples showed that clay-rich material formed small polygonal shapes. Further laboratory work showed that some freezing at depth is possible but only under dry soil conditions. At moisture levels of saturation and field capacity, freezing temperatures do not penetrate to sufficient depth to contribute to mud polygon development.

12.7 Geophagic Behavior Among Buffalo

Soil mining and licking behavior (geophagia) of African buffalo (*Syncerus caffer caffer*) on Mount Kenya has been reported numerous times in the literature (Mahaney, 1987c). The hypothesis that this behavior is a method for the buffalo to acquire salt and possibly other nutrients is proposed and tested. Analysis of the chemical content of the paleosols involved (Tropudalfs and other compound Quaternary paleosols of similar composition) show conclusively that some factor other than salt must control this behavior. While the total salt (NO_3^-, Cl^- and SO_4^{-2} content) is negligible in these soils, free iron (dithionite extractable) is relatively high, and may impart the black hues frequently observed in buffalo dung. Further analyses of African buffalo dung for Fe content would tend to confirm or deny this relationship.

During eighteen months of Quaternary geological and pedological investigations on Mount Kenya (Mahaney, 1979a, 1982b, 1983a,b, 1984a, 1984b) African buffalo (Figure 12.25) were frequently observed using their horns to dislodge clods of soil from paleosols and engaging later in geophagia. Most frequently the paleosols affected were located in the bamboo forest zone and bamboo forest/*Hagenia-Hypericum* woodland ecotone, between 2800-3050 m.a.s.l., but buffalo were observed infrequently behaving similarly at the lower limit of forest ($\sim$2100 m) near sites previously documented by Mahaney (1979b). Moreover, buffalo were observed making paths through the timberline and lower Afroalpine zones (Mahaney 1979c). In these higher vegetation zones buffalo were not observed mining the soil (paleosol), possibly because they are Inceptisols dating from the last glaciation, and hence lack appreciable amounts of clay and iron.

The western approach to Mount Kenya along the Naro Moru Track contains many roadcuts bearing Alfisols – thick, reddish brown paleosols (see Chapter 9) that have formed following episodic emplacement of glacial, aeolian, and slope wash deposits. Below $\sim$3000 m.a.s.l. these soils become older, deeper, exhibit more reddish hues, and in general have higher contents of free and total iron. Buffalo roadcut mining activity has been observed *only* along the Naro Moru and Chogoria tracks. While it is not always possible to determine the sex of individuals engaged in roadcut mining, such activity appears to be carried out both by immature and mature buffalo

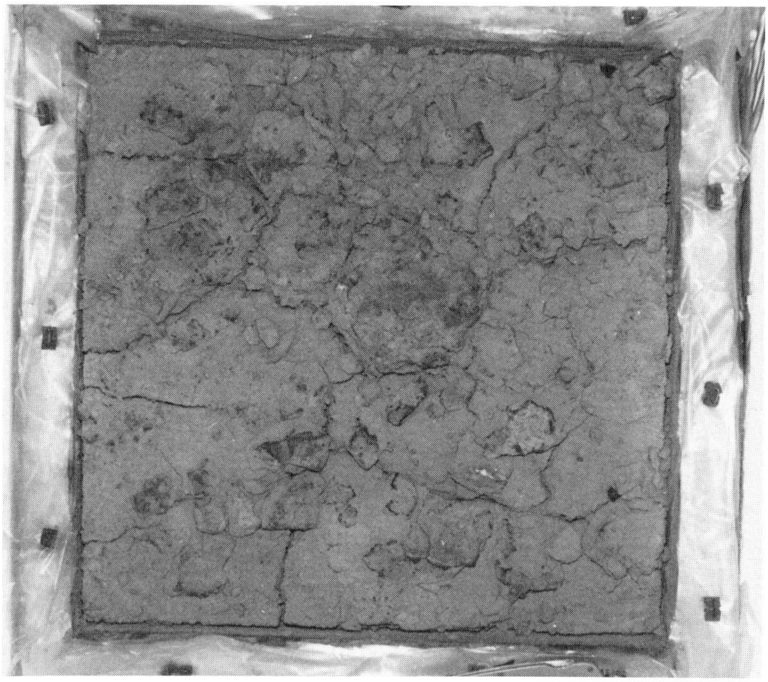

A.

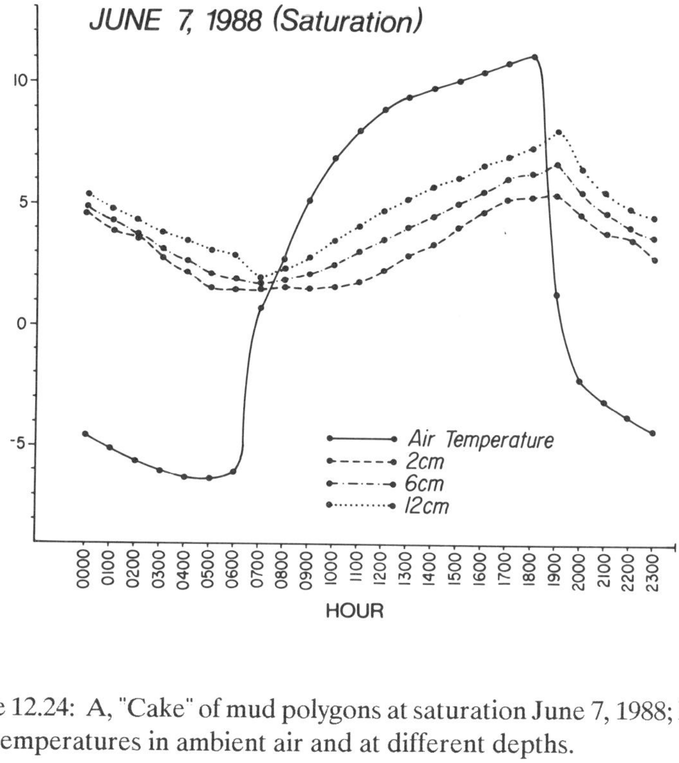

B.

Figure 12.24: A, "Cake" of mud polygons at saturation June 7, 1988; B, 24-hour temperatures in ambient air and at different depths.

Figure 12.25: African buffalo (*Syncerus caffer caffer*) in lower Teleki Valley montane forest zone.

operating *alone*. Stream cuts on Mount Kenya provide exposed bedrock and younger surficial materials that do not contain the reddish-brown Alfisols, and which apparently limited site selection by buffalo for mining.

The results of laboratory analyses on samples typical of the sites mined by buffalo activity are given in Chapter 5 (Tables 5.2 and 5.4).

Particle sizes of mined samples reveal that in general clay ($< 2\mu m$) materials tend to increase with depth, achieving a maximum in B and C horizons. It is from these lower horizons that African buffalo tend to mine material for licking. Clearly buffalo show a preference for clay-rich and reddish-brown soil as indicated from the colors determined in the field.

A number of chemical parameters were established in an effort to select possible causes which might explain the buffaloes' mining behavior. The soils are acid throughout, which results from high precipitation, estimated at ~ 1300 mm (52 in.) per year (Pedgley, 1966). Exchangeable cations are low throughout, which is a further demonstration of the ability of moisture to remove some of the products of weathering. With the exception of the surface A horizons where calcium is recycled by plants, Ca^{+2} is low throughout the profiles. The cation exchange capacities (CEC) are highest in the surface horizons reflecting the higher organic matter content; they then decrease somewhat in the B and C horizons as organic matter declines. The relatively high CEC in the lower horizons is maintained by high percentages of clay. Correspondingly base saturation is low in all cases, especially below the surface horizons. Total salts (NO_3^-, SO_4^{-2} and Cl^-) are found in moderate amounts in surface horizons as a result of input from newly fallen leaves, twigs and dung. Organic carbon and nitrogen distributions are remarkably uniform with depth among the profiles examined. Taken collectively there is nothing in these data to suggest a possible cause for the buffaloes' mining behavior.

Copper and cobalt, while important as nutritional supplements in bovid diets, are not considered to occur in any quantity in Mount Kenya paleosols. Phosphorus might serve as a dietary link (M. Mloszewski, personal communication, 1986), and hence a stimulus, although it is generally present in small amounts in areas of higher rainfall (*e.g.* western flank of Mount Kenya).

Licking by buffalo is described by Sinclair (1977) and Mloszewski (1983) as a response to salt encrustations on plants, presence of precipitated salts (mainly NaCl) near springs, and precipitated salts on other buffalo bodies. The hypothesis that buffalo "mine" roadcuts for salts, as suggested by P. M. Snyder (personal communication, 1976) is not supported by the data available for Mount Kenya paleosols. Nor is there any reason to suppose that geophagic behavior is merely a quest for lime. One chemical constituent that tends to be found in appreciable quantity, however, is dithionite-extractable Fe (sum of organically-bound, amorphous and crystalline Fe in paleosols), often referred to as "free" Fe_2O_3 in that it is the result of oxidation of part of the total Fe over elapsed time for weathering. The data available suggest it is the only common element that is sufficiently abundant in the mined soils to provide a possible stimulus for buffalo mining behavior in areas where roadcuts are available. Roadcut mining was always carried out by *individual* buffalo, and did not appear to be a herd-oriented activity (Mahaney, 1987c).

The altitude reached by buffalo on the mountain also appears to support the hypothesis that iron is the stimulus behind the licking behaviour. With increasing elevation the need for erythrocytes in blood increases bringing a corresponding need for iron (Stickney and van Liere, 1953). In humans, at such elevations (2800-3050 m.a.s.l.) the erythrocyte count is approximately 1.4 times that at 2100 m.a.s.l. (Matthews, 1954). Infrequent buffalo geophagic behavior at 2000-3000 m appears compatible with an erythrocyte deficiency because those elevations are above the normal height for the species in Kenya (see Mloszewski and Mahaney, 1989).

Even if elevation is not a factor contributing to buffalo geophagic behavior, other factors might merit further study, *e.g.* the sex of individuals carrying out geophagic activity on Mount Kenya. If bulls participate in this activity more than cows, it may be linked to semen motility, and hence have a cause centered in the reproductive function (Charanjitsingh and Gangwar, 1977).

As a result of random observations reported here, it is not possible to report the sex of individuals involved.

Some samples mined by buffalo have relatively higher concentrations of bromine. Because their dung is low in Br it may also serve as a stimulus for geophagic behavior (R. G. V. Hancock, personal communication, 1988).

Analyses of African buffalo dung and other geophagic sites have shown that the black hue of this material (link between Fe ingestion and black color as suggested by L. M. Mahaney, personal communication, 1986) is probably the result of the incorporation of Fe in its various forms. It appears that the licking behavior of these large bovids is unrelated to the presence of salts and lime.

12.8 Prehistoric Human Occupations

Despite all the excavations on Mount Kenya only a few sites in Lake Ellis and Gorges Valley were found to provide scattered evidence (Mahaney, Barendregt, Churcher & Spence, 1986) suggesting that prehistoric people might have lived and hunted on Mount Kenya at high elevations (above 3000 m). Earth mounds in Gorges Valley near the Nithi River, postulated to have a prehistoric origin (Hook, in Zeuner, 1948), are considered to be natural features (most are crevasse fillings on moraines; see Chapter 4). Rock shelters (Figures 12.26 & 12.27) in Gorges Valley have probably been occupied for the last few hundred years (Mahaney, Barendregt, Churcher and Spence, 1986). Pottery and stone artifacts from two areas, and a shaped cobble of Precambrian gneiss (Figure 12.28) found on the surface of moraines of early Liki age near Lake Ellis (Plate 2), suggest that prehistoric people occasionally visited various areas on the mountain up to 3500 m. This accords with the record of obsidian artifacts found above 3200 m (10,500 ft.) by Col. R. Meinherzhagen (Zeuner, 1948).

The mounds located between Mugi Hill and Lake Ellis and extending towards the timberline zone are of two types. The smaller, randomly distributed, and asymmetrically shaped mounds (~1.0 m high x ~2.0 m wide x ~3.0 m long) are possibly of wind-blown origin (Mahaney and Spence, 1986) but may have another natural origin as vegetation – protected erosional remnants (T. J. Charsley, personal communication, 1986).

These mounds date from pollen zone Z, following emplacement of the Darwin Formation on Mount Kenya (Mahaney, 1983a, 1984a,b, 1985), which is known to have had a cooler and drier paleoclimate. The second group of mounds appears symmetrical in cross-section, of somewhat larger size (~1.0 m high x ~4 m in diameter) and elliptical in shape. As reported by Mr. Raymond Hook (in Zeuner, 1948) and Baker (1967), they commonly have angular volcanic (basaltic) fragments in their centers. Unlike Zeuner and Baker, Mahaney and Charsley (1986) only observed mounds of the second type to the north of the Nithi River below "The Gates". While these features may be explained as cultivation areas designed by prehistoric cultures, additional work showed they are of natural origin (see Figure 4.14). It proved impossible to confirm "linearity" in mound arrangement as originally observed by Hook (Zeuner, 1948). However, it was possible to find stone artifacts of Middle Stone Age type (identified by Dr. S. Wandibba) within the Kech Tarn hollow just northwest of Mugi Hill.

In the area around "The Gates" (Plate 1) a number of caves and rock shelters (Figures 12.26 and 12.27) lie concealed close to a prominent waterfall which spills over the outermost moraine of Teleki age (Chapter 5) in lower Gorges Valley (Mahaney, Barendregt, Churcher and Spence, 1986). To find the caves and rock shelters one must hike (about 150 m) above the waterfall at "The Gates" and descend into a narrow gorge formed by the Nithi River. In 1984 a number of bone fragments of small mammals and birds, as well as wood, were uncovered in rock shelters above the cave network. In 1985, a fragment of pottery dated at 1100-1300 A.D. was found by G. Watson (personal communication, T. J. Charsley, 1986) in the scree downslope of rock overhangs

Figure 12.26: Rock shelters in Gorges Valley ~ 100 m above "The Gates".

Figure 12.27: Rock shelter in Gorges Valley from which wood and bones were recovered.

Figure 12.28: Cobble of biotite gneiss probably rounded for grinding activity by Middle Stone Age cultures inhabitating the eastern slopes of the mountain to ~3500 m.a.s.l. The knife is 9 cm long.

at "The Gates". Two artifacts of Middle Stone Age type were also found at this site at the same time (identifications of pottery and artifacts by Dr. S. Wandibba). It is possible that the caves and/or the rock shelters were used on a permanent or semipermanent basis by the people who left the artifacts in the general area.

Near Lake Ellis (Plate 2) a cobble (Figure 12.28) approximately 90 mm in diameter was found on the surface of a left lateral moraine of early Liki age. This moraine, situated along the north shore of Lake Ellis, is of early-Würm age (< 100,000 yr BP). Correlation with similar moraines in Hinde, and Kazita valleys can be established using relative dating indices of age, such as topographic position, soil profile expression, and weathering of surface boulders. In addition, buried Ab horizons in paleosols give high values (0.42) (Mahaney, Boyer and Rutter, 1986) for the D/L ratio of aspartic acid, suggesting an early Liki age. Radiocarbon dating provided an infinite age of > 42,000 yr BP for emplacement of surface outwash and loess. Near the site where the cobble was found, a large boulder about 0.9 m in diameter of biotite gneiss, was observed partly in the upper paleosol solum. This boulder and the cobble (Figure 12.28) had a 20 percent cover of *Parmelia* (spp.) suggesting that both samples had been *in situ* for at least a few hundred years (Mahaney, 1973 and 1974 for discussion of the effects of stone movement on lichen cover).

Examination of photographs of the cobble by Dr. M. D. Leakey suggests that it "is almost certainly an artifact" (personal communication from Dr. M. D. Leakey, 19 March, 1984).

The boulder and cobble are composed of migmatitic quartz-feldspathic gneiss (Mahaney and Charsley, 1986) containing variable but small amounts of biotite which give the rocks their poorly developed gneissosity. They are metamorphic rocks from the late Proterozoic Mozambiquian Orogenic Belt, typical of those widely outcropping in the area to the north and east of Mount Kenya. Similar rocks are presumed to underlie the whole Mount Kenya massif, and gneiss and quartzite fragments are relatively common in the area northeast of Lake Ellis towards Ithanguni (Plate 4) (Baker, 1967, p. 57). These fragments are presumed to be accidental clasts brought to the surface by explosive activity of the Ithanguni trachytic volcano. The occurrence of boulders and cobbles near to Lake Ellis can be explained either by their incorporation in morainic material derived from the west (a contrary direction to their present disposition in Liki till presumably brought from the Upper Hinde Valley), or by importation into the area by a human agent.

From the available evidence it appears that prehistoric people occasionally traversed and hunted in the lower moorlands (lower Afroalpine zone) and Hagenia Woodland, especially on the eastern and northeastern flanks of Mount Kenya. That at least some of this occupation occurred several centuries to perhaps hundreds of thousands of years ago, is suggested by the presence of a cobble artifact with a large-size lichen thallus, Middle Stone Age artifacts (hand axes), and sizeable rock shelters in Gorges Valley.

12.9 Historic Human Impact

When tropical forest soils are cleared and plowed the nature of the soil system is altered (often irreversibly) in a number of ways. Cultivation and plowing tend to change the soil and produce variations in physical and chemical properties. With prolonged plowing many soil properties are modified, the rate of which varies with the management system and soil type (Mahaney, 1979b).

On Mount Kenya Tropudalfs (Soil Survey Staff, 1975) that developed under indigeneous montane rainforest in Kenya's Central province have been disturbed by agricultural activity (Figure 12.29). Prolonged plowing and cultivation over a 4-year period (1983-87) led to deterioration of the soil system producing changes in solum thickness, soil structure, shear strength and texture. The most notable changes in chemical and organic properties occur with respect to salts (mainly NO_3), CEC (cation exchange capacity), some extractable cations (mainly K^+), organic carbon, percent nitrogen and C/N ratio. Sites were selected to determine the effect on the soil of forest cover removal, followed by plowing and cultivation of the soil at ~2300 m on the Naro Moru Track.

The cleared area is located on gently sloping (7-12°) land adjacent to the Naro Moru River (Figure 12.30) along the western flank of Mount Kenya in Central Province. These interfluves of early Quaternary age consist of residual regolith in phonolite outcrops that belong to the Mount Kenya volcanic suite (Chapter 2). The soils forming in these interfluves are probably ~2.0 my in age and are situated too low on the mountain to have been subjected to deforestation during Quaternary glaciations.

The macroclimate of the Mount Kenya area is controlled by the tropical easterly wind belt as well as the tropical monsoon (see Chapter 2 for details).

The montane rain forest is characterized by the presence of evergreen hardwood trees and a few conifers. Cedars (*Juniperus procera*) and *Podocarpus* (*P. gracilior; P. milanjianus*) dominate from its upper limit at ~2900 m to 2,000 m where the montane forest merges with savanna, pastoral, and cultivated lands. About 1900 A.D. the entire area as far as the Aberdare Mountains (100 km to the west) was covered with montane rain forest (Baker, 1967; Coe, 1967); between 1900 and 1930 the forest was cleared for agriculture and later replanted, in part, with pine and cypress, which was cut in 1983. Following clear cutting the area was planted (mainly by hand)

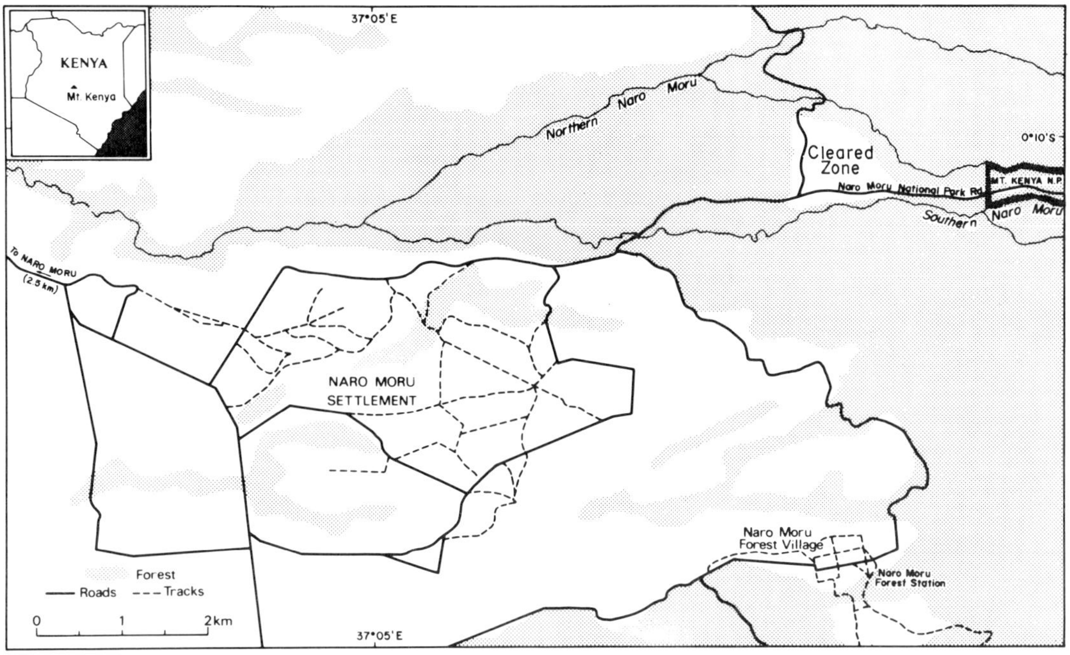

Figure 12.29: Cleared area of montane forest, Naro Moru River Valley.

with potatoes. The present forest boundary is continually under pressure from fire and agricultural activity (Mahaney, 1979c).

A Tropudalf representative of forest soils on interfluves in the field area was described and sampled to approximately 2.5 m (Figure 12.31). The soil in the montane forest was later cleared and subjected to agricultural activity over a four-year period. As a result of plowing the A horizon was completely obliterated to a depth of 25 cm (to the middle of the B22t horizon or upper part of the argillic horizon). Deterioration of grades of soil structure in the profile indicates severe disturbance in the plowed horizons. The textural classification (Soil Survey Staff, 1951) for the solum of the natural soil profile is clay. The plowed horizons tend to be somewhat lower in clay size material, which may be attributed to removal by wind and/or water processes.

Decreased thickness of the plowed A horizon parallels the findings of other workers who have documented lowering of the soil surface and thinning of the A horizon (Bouma, 1969; Kondrat'yev and Gritsenko, 1971; Levin, 1969; Mahaney and Ermuth, 1984; and Mahaney, 1989a). Deterioration of soil structure with tillage parallels the findings of Low (1972, 1973) and is not unexpected.

The plowed horizons showed increases in average shear strength (n = 10) from 1.1 to 2.0 kg/cm^2, which probably results form soil compaction due to the movement of farm machinery, people and animals. Overall, increases in shear strength with depth occur more slowly in the natural forest soil as a result of greater space and lower density.

Available soil H_2O calculated from field capacity (1/3 Bar) – permanent wilting point (PWP = 15 Bar) yielded values that were noticeably higher for the plowed horizons suggesting that higher silt content might be responsible for the increase in available moisture. The percent H_2O

Figure 12.30: Deforested part of the montane forest at ~ 2200 m.a.s.l. planted with potatoes. In four years these paleosols lost their nutrient reserve and showed effects of wind and water erosion.

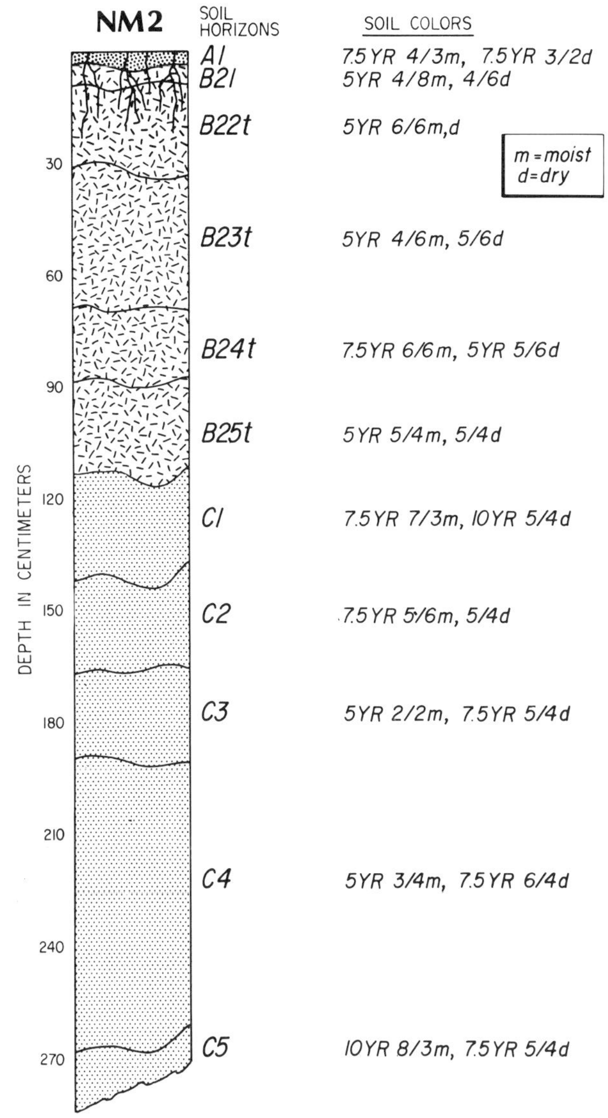

Figure 12.31: Paleosol NM2 formed in residual regolith in the Naro Moru River Valley.

at saturation is highest in the natural soil solum which may result from higher organic matter content.

Particle sizes were calculated in the plowed paleosol to determine the effect of clearing the land and plowing on local soil systems. The results show that sand remains fairly uniform or increases slightly with depth. Clay percentages are lower below the solum, and silt tends to increase slightly in the surface horizons. This suggests that clay plus silt may be added to the solum of the paleosol, possibly by wind or water deposition. A slight increase (~3-4 percent) in silt content in the plowed horizons (after only 4 years) may be due to comminution of sand by plowing. The tendency for plowing to comminute mineral material is documented by several workers, including Fiskell and Calvert (1975), and Soane and Pidgeon (1975). The clay texture of the surface soil horizons may also impede drainage and leaching.

Clearing the montane forest, plowing, and cultivation combine to affect some chemical characteristics of the paleosol. In general organic C and total N in the surface horizons of the natural soil are reasonably high (C = 5.3%; total N = 0.8-0.5%). The natural soil is acidic (pH = 5.3) and has low C/N ratios (6 to 7). In the solum of the plowed soil, organic C dropped off with depth (by 30 percent) in the B21t horizon and pH was ~6.0. The decline of organic C in the solum of the plowed soil is significant, and affects the cation exchange capacity (CEC), which also declined from 28.7 to <20 meq/100 g. With bare fallow and/or manuring the organic C reserve might be restored in a temperate latitude soil, but possibly *not* in a tropical soil (J. P. Watson, personal communication, 1982). In the natural forest soil increases in organic C are caused by surface litter being mixed with soil and by decomposition of roots. Because the age of the natural soil is ~2.0 my the organic C values are considered to represent steady-state conditions.

The distribution of total salts shows increases in the Al and lower C horizons which reflect the litter fall on the surface and depth of leaching in the paleosol. In the Al horizon of the plowed soil total salts decline to <0.20 (mmhos/cm at 25°C).

Rainfall on unprotected tropical soils may cause serious deterioration when large pores in the lower solum become filled with finer materials washed in from above (Mahaney, 1979b). Soil animals such as termites may also influence the number of large pores (Watson, 1970), although the field site is probably too high for termite activity. There may be fewer burrowing animals in the plowed soil. Rapid drainage through the lower solum and subsoil becomes progressively more difficult during and just after heavy tropical thunderstorms, which are common in the Naro Moru area. Increasing surface runoff, even on slopes of 5-10° can produce serious erosion of topsoil in cleared areas. Perhaps more important, the degradation of soil structure in plowed horizons by raindrop impact, movement of animals and people, and accelerated decomposition of organic matter make the soil more susceptible to erosion and decrease its water holding capacity at saturation.

The rapid deterioration of soil organic matter following removal of the forest, plowing, and cultivation of potatoes is the most important effect of plowing in this area. Ancient soils (paleosols) in tropical forests are in dynamic equilibrium with the environment and a balance ensues between the rapid breakdown of soil organic matter and the equally fast accumulation of fresh litter. Decomposition of soil organic matter is rapid even under shade (Cunningham, 1963) in tropical forests. Mahaney and Boyer (1986a,b) showed that the numbers of bacteria and fungi reach 10^6 and 10^5 per gram, respectively, in paleosols in the upper *Podocarpus-Hagenia* woodland vegetation belts on Mount Kenya which suggest microbial decomposition is nearly comparable with the middle latitudes. Nye and Greenland (1961) demonstrated that some movement of organic matter occurred even in cleared tropical soils, a condition not observed in this study. Although decomposition is rapid in shaded microclimates, organic material continually accumulates under existing forest cover. Measurements of litter fall/per acre (0.4 ha) per year are not available, but the values for organic matter in other areas of tropical Africa suggest it must range at least from 2,000-4,000 kg per acre (0.4 ha), per year (Nye, 1961), which is sufficient to maintain soil organic matter at a fairly high level (~9 percent). With the removal of the forest only surface litter and newly fallen organic matter are left to replenish soil organic matter, which together with accelerated decomposition, eventually lead to reduced fertility.

Compared with soil organic matter a similar pattern occurs with respect to total N, which decreases slightly under agricultural activity. Apparently, most food crops including potatoes respond well to N after a lapse in time following clearing of the land (Nye and Greenland, 1960). In this case lack of N will eventually limit production. Replenishment of total N requires fertilization or 7-12 year periods of bare fallow (Cunningham, 1963).

Food crops grown in the field area are probably affected by low total P. Usually organic P is 55 to 75 percent of total P in the solum (Cunningham, 1963). Judging from the distribution for P (2.9 to 8.2%) in the natural soil, it appears that P is marginal and will have to be replaced by fertilization if crop production continues.

The major clay mineral in these soils is metahalloysite with a lower CEC of ~10 meq/100g. Loss of organic matter over four years brought a reduction in CEC and a slight loss of extractable cations. In the cleared area, Ca^{+2}, Mg^{+2}, and Na^{+} decreased somewhat and K^{+} increased slightly (possibly from fertilization). In addition, increased soil acidification might produce even larger cation deficiencies which, if unchecked, could have serious implications for continued agriculture.

Various studies of shifting cultivation (DeSchlippe, 1956; Nye and Greenland, 1960) have shown that using concepts developed in the middle latitudes has led to an underestimation of the potential of tropical soil productivity. Similarities between temperate and tropical soil systems including their responses to agriculture involve changes such as structure deterioration, compaction, increased surface runoff, and accelerated erosion. In temperate latitudes total N and total organic matter may remain constant over several decades of cropping (Warren, 1956) or lose up to 40 percent of total organic matter and N (Mahaney and Ermuth, 1974). As shown in this study, severe depletion of organic constituents parallels the findings of Cunningham (1963) for tropical soils. Attempts to grow one crop repeatedly on tropical forest soils by eliminating bare fallow, and without fertilization, will result in rapid loss of organic matter and total N. Without additional research into the effects of clear cutting montane forest on tropical soils there is little hope for continued agricultural productivity in this area.

12.10 Conclusions

Ecological processes can best be understood and interpreted within the framework of Quaternary geologic and geomorphic processes. The six major Pleistocene glaciations and two minor late Holocene Neoglacial advances occurred as responses to Quaternary paleoclimatic changes. Major swings from dry and colder glacials to wet and warmer interglacials also affected weathering rates, paleosol evolution and ecosystem development. The Afroalpine and associated ecosystems on Mount Kenya, and on the other tropical mountains, owe their present form and structure to these paleoclimatic oscillations.

Restructuring of the Afroalpine ecosystem is still in progress with a number of endemic species having developed novel and useful adaptations to the rigors of the Afroalpine. Perhaps even new species that have established themselves on fresh Neoglacial substrates will eventually develop adaptive mechanisms which will allow them to disperse onto older Liki substrates and survive competition from endemic and well established taxa. Indeed, with developing structural complexity that is bound to accompany increased numbers of species in the future, the evolving biocommunities of the Afroalpine will likely be faced with their greatest challenge – to cope with changes in precipitation that will likely accompany disappearance of the existing glaciers.

CHAPTER 13

INTERMOUNTAIN CORRELATIONS AND THE DEEP SEA RECORD

13.1 Introduction

Geological investigations over the last century have provided a wealth of information on glacial/interglacial fluctuations in the tropical mountains. Diamicts have often been used as evidence for pre-Würm/Wisconsinan glaciations, but with a paucity of absolute dates, and in some cases uncertainty of a glacigenic origin (Mahaney *et al.*, 1989). On Mount Kenya, morainal evidence for five Pleistocene glaciations, and preliminary evidence, from clasts in lahars, for a sixth period of expanded ice cover during pre-Olduvai time, highlights the complexity of the landform sequence. Firm correlation of these earlier glaciations with other areas will have to await additional paleomagnetic surveys and the development of new isotopic dating methods.

Correlation (Plate 6) of Liki (late and later Pleistocene) and Holocene glacial advances on Mount Kenya with other areas is possible primarily because many deposits have been dated by radiocarbon. Although the time of onset of the Liki Glaciation and age of correlative deposits in other areas is still uncertain, it would appear to be ~100,000 yr or less. Thus, the time column in Plate 6 begins at ~100,000 yr BP and continues, with increasing numbers of radiometric dates, up to the present. A wealth of chronostratigraphic information is available for the period < 30,000 yr BP; however, a greater number of dates must become available before exact correlations can be made between different glaciated areas. Only on Mount Kenya, in New Guinea, and Peru are absolute ages available for Neoglacial deposits. However, from the available descriptions of these late Holocene deposits, they appear to be coeval with the Little Ice Age. In other areas, older Neoglacial deposits were either buried by younger glacial sediments or completely removed by glacial erosion. It is also possible in some cases that older deposits went unrecognized by various workers.

13.2 Pleistocene and Recent Glaciations

The abundance of geomorphic information available from Mount Kenya shows that Liki I/II moraines are found in the ~3200-3400 m.a.s.l. elevations on all slopes of the mountain. Although Liki I moraines have not been dated, owing to the scarcity of wood and absence of tephra, they most certainly are equivalent to isotope stage 4. Liki II moraines, which are more prevalent on Mount Kenya, were deposited between ~25,000 and ~15,000 yr BP. These may be considered equivalent to the classical Würm/Wisconsinan moraines of Europe and North America. The only direct chronostratigraphic information on Liki II moraines concerns limiting minimum ages of ~15,000 yr BP (Mahaney, 1988b). The time for the onset of the Liki II stade, and concurrent end of the Liki interstadial paleoclimate, is fixed at ~25,000 yr BP by ^{14}C dating of periglacial sediments in the Met Station Catena (see Chapter 5). The retreat of Liki ice from ~3200 m.a.s.l. to ~4000 m.a.s.l., in about 3000 yr, brought emplacement of the Liki III moraines as a result of recessional stillstand during the late glacial (Mahaney, 1987a). Indeed, ice recession was so rapid that all major valleys on Mount Kenya were left free of recessional moraines between ~3300 and ~4000 m.a.s.l. (Figures 13.1, 13.2, 13.3 and 13.4). Late Holocene (Neoglacial) advances, culminating in the emplacement of Tyndall and Lewis moraines and outwash, began approximately 1000 yr BP (Spence and Mahaney, 1988a).

Pleistocene moraines deposited in three stages and differentiated mainly by topographic position, are known on other East African mountains, including Kilimanjaro, Mount Elgon, Ruwenzori, and the Aberdares (Hastenrath, 1984; Osmaston, 1989a,b, Hamilton, 1984; and

Figure 13.1: Upper Teleki Valley from 3850 m.a.s.l. showing a U-shaped cross-valley profile mantled with solifluction deposits. The Liki III recessional moraines are in the upper valley at ~ 4000 m.a.s.l..

Figure 13.2: Liki III recessional moraine (= arrow) in upper Teleki Valley near the Liki III rock glacier complex.

Figure 13.3: Small kame or crevasse filling behind moraine shown in Figure 13.2.

Figure 13.4: Mackinder Valley, on the northwestern slope of the mountain, showing a minor recessional moraine (= arrow) below the Gregory Glacier.

Hamilton and Perrott, 1978). During the last glacial maximum in East Africa the total ice extent is estimated at 800 km^2 (Hastenrath, 1984), of which the main ice centers (of approximately equal volume) were Mount Kenya, Kilimanjaro, and the Ruwenzori. Following the late glacial, ice was limited to these three mountains. During the Neoglacial, total ice cover on the three highest mountains amounted to <30 km^2, with Mount Kenya contributing only a small fraction of this total area.

The data on Plate 6 show early Würm/Wisconsinan age equivalent moraines only on Mount Kenya. Correlatives of the Liki II stade on Mount Kenya are found on Kilimanjaro, the Aberdares, Ruwenzori and Mount Elgon (the latter two mountains have geologic-climatic names for deposits; Plate 6). No Pleistocene tills are reported in the Virunga Mountains, but paleosols correlative with post-Liki I/II paleosols and postglacial pedons have been described (Mahaney, Hancock and Barendregt, 1989). Of the six highest East African mountains (Plate 6), only Mount Kenya has a well-documented and dated early and middle Holocene nonglacial sequence (see Chapter 7). The only documented early Holocene glaciation on Kilimanjaro is called the Little Glaciation (Osmaston, 1989a). On Kilimanjaro, moraines of the Little Glaciation appear similar in morphology to late glacial moraines described on Mounts Kenya and Elgon. Because there is no proof that they were emplaced following a glacial resurgence during the early Holocene, it is possible that their origin is related to recessional stillstand at the end of the Pleistocene.

The Neoglacial record on the East African summits of Mounts Kenya, Kilimanjaro and the Ruwenzori appears to have some similarities. Two advances have been recorded on Kilimanjaro and Mount Kenya with absolute ages of deposits associated with the Mount Kenya record. On Mount Kilimanjaro the ages of late Holocene moraines are based largely on topographic position (Osmaston, 1989a), there being no radiocarbon dates available. The Neoglacial moraines in front of existing glaciers in the Ruwenzori belong to the Lac Gris advances, which are considered to have culminated several hundred years ago (Osmaston, 1989b). It is likely the early Lac Gris advance occurred at the same time as the Tyndall advance on Mount Kenya. Whether or not these advances were in phase with the Recent Glaciation on Kilimanjaro is subject to considerable speculation. Mount Elgon and the Aberdares were ice free during the Neoglacial.

Major mountains in Ethiopia, including Mount Badda and the Simen Mountains were glaciated (to ~4000 m.a.s.l.) during the last major cold period (Hastenrath, 1977; Hurni, 1989). Periglacial landforms have been described as low as ~3500 m. Older deposits equivalent with Liki I on Mount Kenya and the early Fourth Glaciation on Kilimanjaro are not known in Ethiopia. The most extensive Pleistocene moraines in the Simen Mountains are found on north-facing slopes as low as 3750 m (Hurni, 1989). The Bale Mountains of Southern Ethiopia show evidence of more intensive glaciation during the Würm, with snowlines about 300 m lower (~3950 m.a.s.l.) than in northern Ethiopia (Hurni, 1989). Because there has not been any detailed chronostratigraphic investigations in the Bale Mountains, it is impossible to make definite correlations with the stratigraphy shown in Plate 6.

Older outwash gravels and diamict deposits in the mountains of New Guinea may correlate with early Würm/Wisconsinan – equivalent glacial advances (Peterson and Hope, 1972; Löffler, 1972). Moraines at ~3500 m probably correlate with the Liki II/Fourth Glaciation/Mahoma deposits on Mount Kenya, Kilimanjaro, and the Ruwenzori. In the upper valleys at ~4000 m.a.s.l., end moraines of late glacial age may have been deposited during recessional stillstand as they were on Mount Kenya. In any case, the early and middle Holocene appears to have been ice free in New Guinea. Although Neoglacial moraines on Irian Jaya (Plate 6) imply four separate advances (Peterson and Hope, 1972), only nivation deposits are known on Mount Wilhelm (Löffler, 1972).

Kinabalu, the only glaciated summit in the Indonesian Archipelago (Plate 6) (Koopmans and Stauffer, 1968), seems to have been ice-free during most of the Holocene. Moraines dating from the early Holocene (9180 ±120 yr BP; ^{14}C date from G. Jacobsen-G. Hope, personal communciation, 1989) may document an early Holocene advance or late glacial stillstand. Additional ^{14}C ages are required in order to substantiate the original interpretation of Koopmans and Stauffer (1968). Owing to the elevation of Kinabalu (4101 m.a.s.l.), no Neoglacial moraines or periglacial deposits have been described in the summit area, nor are any likely to be found.

The Mexican volcanoes have complex glacial sequences with the oldest documented moraines on Malinche Volcano of middle Würm/Wisconsinan age (MI group, ^{14}C dated at ~30,000-35,000 yr BP). There are no documented early Würm/Wisconsinan moraines. A complex sequence of MII and MIII moraines, of late glacial and early Holocene age, are found on Malinche Volcano (Heine, 1984). These deposits appear correlative with Liki III moraines on Mount Kenya. The Neoglacial moraine sequence – drifts MIV and MV – date from ~2000 yr BP, suggesting that the Neoglacial may have started earlier in Mexico than in East Africa. On Iztaccihuatl and Ajusco volcanoes, a somewhat different sequence is reported by White and Valastro (1984), who show earlier and later advances during the Diamantes Glaciation (Würm/Wisconsinan equivalent) that appear out-of-phase with Malinche Volcano. The Alcalican moraines of late glacial age are broadly equivalent with the MII and MIII drifts on Malinche. Fluctuation of the Iztaccihuatl ice cap appears to have occurred throughout the Holocene emplacing the Ayoloteplto moraines of varying ages. The MIII and IV moraines on Malinche probably equate with the late Holocene moraines in East Africa.

Moraines from the last glacial maximum (Plate 6) are known in the mountains of Costa Rica and Guatemala, with lower limits at ~3500 m (Hastenrath, personal communication, 1989). No older or younger tills have been recognized in these areas. The relatively low elevations (~3000-4000 m.a.s.l.) of most peaks, and their subtropical locations (8-15° N), suggest that they were below the paleosnowline during the late glacial and late Holocene.

In the Andes Mountains of South America fairly well-defined moraine sequences, defined by various workers, appear correlative with other tropical mountain areas. In the Colombian Andes, drifts equivalent with the "classical" Würm/Wisconsinan record of Western Europe and North America have been described by Gonzales *et al.* (1965). These Aduriameina moraines probably date to about ~15,000 yr BP, making them equivalent to Liki II drifts on Mount Kenya. In the higher valleys, a younger set of moraines, deposited during the Mamancanaca advance (or recessional stillstand?), bear a striking resemblance to Liki III moraines on Mount Kenya and Cirque Stage moraines on Mount Elgon (Plate 6). The early and middle Holocene in the Colombian Andes, and elsewhere in South America, was largely free of ice advances. Neoglaciation appears to be confined to the last ~1000 yr (called the Bolivariano advances) including the Little Ice Age.

In the Venezuelan Andes, tills equivalent with early and late Würm/Wisconsinan deposits (Schubert, 1972, 1974) have been known for the last century. Moraines between 2600 and 2800 m.a.s.l. that are more highly weathered and eroded, probably belong to the early Würm/Wisconsinan Glaciation. Deposits between 3000 and 3600 m.a.s.l., with fresher surfaces and lesser degrees of erosional modification, probably belong to the late Würm/Wisconsinan Glaciation. The youngest moraines are older than 13,000 ^{14}C yr BP (Salgado-Labouriau, Schubert, and Valastro, 1987), and probably were being deposited around 19,000 to 16,000 ^{14}C yr BP (Schubert and Rinaldi, 1987). Moraines described by Pérez (in preparation, 1989) also look very similar to late glacial moraines in the high valleys of the East African mountains. Evidence for climatic cooling during the Little Ice Age, beginning about 600 yr BP is known from recent palynological investigations (Rull, 1987; Rull *et. al.*, 1987). The ^{14}C dates for the latest Neoglacial event correlate closely with lichenometric dates on moraine substrates (Pérez, personal communication, 1989).

The last glaciation in Ecuador (Hastenrath, 1981) left a moraine sequence similar to that of Venezuela and Colombia. The "classical" Würm/Wisconsinan Glaciation emplaced tills as low as ~3000 m, and appears to have left multiple lateral moraines of late glacial age in the higher valleys (Clapperton, 1986, 1987). Moraines of early to middle Holocene age have not been described;

however, the Neoglacial which may have begun ~2000 yr BP, bears a striking resemblance to the late Holocene glacial sequences in East Africa, as well as to the sequence in Peru (Hastenrath, 1981; Clapperton, 1972; Birkeland, 1989).

Radiocarbon dates from Ecuador (Clapperton, 1987) show maximum limiting ages on the "classical" Würm/Wisconsinan moraines of ~36,000 yr BP. Younger late glacial moraines appear to have been emplaced about 11,000 yr BP or during the Younger Dryas (Clapperton, 1987).

The most precise documentation of the last glaciation in Peru comes from the Quelccaya ice cap (latitude 14°S) (Mercer and Palacios, 1977). Here the last glaciation ended approximately 14,000 yr BP. During an earlier (>28,000 yr BP), and undated Würm/Wisconsinan advance, glaciers were about twice the length of late Wisconsinan ice. A short-term readvance of the Quelccaya ice cap began about ~11,500 yr and ended 11,000 yr BP, or prior to the Younger Dryas stade in Europe. By 10,000 yr BP, the Quelccaya ice cap was approximately at its present size. Between 2700 and 1600 yr BP, the ice cap was smaller than today; a Little Ice Age advance occurred between 600 and 300 yr BP.

In the Cerros Cuchpanga of Central Peru, the regional snowline was probably 100 m lower than today during the Neoglacial (Wright, 1984). Outwash deltas covered with peat gave minimum ^{14}C ages (on the base of the peat layers) of 1100 ±70 and 430 ±70 yr BP, indicating that the culmination of the early Neoglacial advance is likely coeval with the early advance in East Africa. A readvance of cordilleran glaciers at the end of the Pleistocene is considered to have occurred between 10,000 and 12,000 yr BP (although available ^{14}C dates are considered "contradictory" (Wright, 1984).

In East-Central Peru, pollen stratigraphy from a 30-m core in Laguna Junín showed that from 12,000 to *ca.* 3000 yr BP, about 30% of recovered pollen originated in the Andean forest, which is about 600 m lower today. After 3000 yr BP the pollen input to Laguna Junin from the lower source decreased considerably, perhaps as a result of deforestation by humans. Between 12,000 and 24,000 yr BP pollen concentrations are dominated by grass taxa along with some shrub species (mainly Compositae). Silty material in this section of the core is considered to be outwash from piedmont glaciers advancing out onto the plain from the cordillera. An older (>39,000 yr BP) set of moraines and outwash fans probably was emplaced during the early stage of the last glaciation (Hansen, Wright and Bradbury, 1984). It was this older suite of moraines that dammed the Laguna Junín.

In northern Chile, tills from the last glaciation are limited to the "classical" Würm/Wisconsinan and the Tapado readvance (or recessional stillstand?) (Plate 6). The late glacial Tapado moraines in the high valleys, although poorly dated, probably correlate with the main/late-glacial moraines in Peru. The La Laguna moraines probably correlate with Little Ice Age deposits found elsewhere in South America.

A sequence of moraines dated by K/Ar and ^{14}C on Mauna Kea has been reported by Porter *et al.* (1977). Moraines in this sequence correlate with isotope stages 2, 4, 6 and 8 and provide probably the most detailed chronostratigraphy of any tropical mountain. On Mauna Kea, a succession of small ice caps formed on the highest summit in Hawaii, and the only known glaciated mountain in the Central Pacific. Unlike Mount Kenya, Mauna Kea volcano was active during times of glaciation and so tills are found interbedded with lava flows and pyroclastic deposits. Absolute age determinations of these volcaniclastic units make it possible to arrive at maximum and minimum ages for glacial deposits. The earliest Würm/Wisconsinan glacial deposit directly dated by K/Ar gives an age of ~55,000 yr. Only additional research in other areas will determine if the culmination of the early Würm/Wisconsinan advance was in or out-of-phase with the Hawaiian advances.

13.3 Liki II Paleosol Correlations

On Mount Kenya and in the Virunga Mountains of East Africa "moorland-like" buried paleosols are known from the early and late stades of the last glaciation (Mahaney, Harmsen and Spence, 1990). These paleosols formed under a drier and cooler paleoclimate (Coetzee, 1967) when, according to palynological evidence, vegetation belts were depressed at least 700 m from their present positions. These paleosols generally contain a pollen count that is high in ericaceous or lower Afroalpine taxa (Cooremans and Mahaney, 1989), but their most outstanding characteristic is the thick Ahb horizon (0.3-0.7 m) that forms the peaty epipedons in paleosols found today above timberline. Below the lower limit of Liki Glaciation on Mount Kenya, these paleosols have been ^{14}C dated at between ~25,000 and 12,500 yr BP. In three localities they show evidence of a geomagnetic reversal that probably equates with the Mono Lake excursion of western North America (Barendregt and Mahaney, 1988).

At some sites in the Virunga Mountains, as on Mount Kenya, the Liki paleosol merges at the ground surface with the modern Ah or Al horizon. In such cases, which are very common on the eastern and western slopes of Mount Kenya and on the slopes of Visoke in the Virunga Volcanoes, a Pleistocene "relict" feature becomes part of the modern soil-forming process. Because the microbial count is so low (see Chapter 12) in these paleosols, decomposition is slow and the dark black (10YR 1/1,2/1) colors still persist. Even the increased microbial numbers that must have accompanied the upward migration of vegetation belts at the end of the last glaciation have not had sufficient time to remove much of the peat accumulation which occurred during the Liki Glaciation. Indeed the presence of this paleosol as low as 2500 m.a.s.l. on Mount Kenya, and to at least 2700 m.a.s.l. in the Virunga Mountains, suggests that microbial populations of paleosols, both in the Hagenia woodland and bamboo forest are sufficiently low as to maintain the existence of peaty, buried A horizons.

13.4 Correlations with Nearby Mountains and Western Europe

Late Quaternary chronostratigraphic correlations between Sacred Lake (2440 m.a.s.l.) and Lake Rutundu (3410 m.a.s.l.) on the northeastern flank of Mount Kenya (Coetzee, 1987), and the main massif itself (Plate 7), are possible given the prevalence of recovered core material, degree of landform mapping, and number of radiocarbon dates (Plate 7).

The landform record alone shows three tills from the last glaciation, identified as Liki I, II and III stades. Only the last two Pleistocene stades (Liki II and III) are well dated from sediments associated with moraines (not from the moraines directly). As discussed previously (Chapter 7) there are no data to support the hypothesis of early or middle Holocene glacial advances (Johansson and Holmgren, 1985). There is, however, a growing body of geomorphic-paleosol information in support of a middle Holocene warmer and wetter interval from ~8000-~4000 yr BP (see Chapter 11). Deposits emplaced during this interval (Darwin Formation, Plate 7) are prevalent in almost all drainages on Mount Kenya and in the Karisoke area of the Virunga Mountains.

The Neoglacial record on Mount Kenya comprises two advances that began at least ~1000 yr BP (see Chapter 8), and possibly as long ago as ~2000 yr BP. Even though the pollen record (Coetzee, 1967) calls for cooler paleoclimate in the middle Holocene, this apparently was not cool enough to support the expansion of existing glaciers.

Recovered cores from Sacred Lake, Lake Rutundu (Coetzee, 1967), and a number of other mires and bogs (Mahaney, 1987a), show a near-synchronous sequence of events beginning with pollen zone R (Liki I time). Pollen zone R (Sacred Lake) was accompanied by colder paleoclimate, with timberline at ~2400 m, and glaciers descending to ~3200 m. Liki interstadial paleoclimate (pollen zone S; Sacred Lake-Kalambo Interstade) persisted for at least 5000 yr, ending about ~26,000 yr BP (Coetzee, 1967), with the beginning of the Mount Kenya Hypothermal (pollen zone T) that lasted until ~15,000 yr BP. The late glacial period, from ~15,000 yr to ~10,500 yr BP, shows alternating warmer (pollen zone Ui) and colder intervals (pollen zone Uii). Following the late glacial period, paleoclimate became warmer and wetter (with one cooler episode (~5000-6300 yr BP, comprising pollen zones V, W and X, Sacred Lake) (Coetzee, 1967). This was followed by the climatic optimum (~5000-~2500 yr BP) (pollen zone Y, Sacred Lake). It is interesting to note that Neoglaciation, which occurred during pollen zone Z, was presumably drier, but with a similar temperature as in "cooler" middle Holocene time. Perhaps Neoglaciation eradicated all evidence of earlier Holocene advances in local cirque basins.

The data from Lake Rutundu, ~700 m higher up in the ericaceous belt (3140 m.a.s.l.), show only three pollen zones for the postglacial period (Plate 7). The early and middle Holocene are dominated by warmer and wetter conditions that persisted until ~2200 yr BP. No middle Holocene "cooler" period (pollen zone X at Sacred Lake) was detected in the Lake Rutundu core material (probably due to the altitude). The Lake Rutundu core shows essentially the same distribution of ericaceous belt pollen in the upper levels, but with possible cooler temperatures (Coetzee, 1967).

A recovered core from the Cherangani Hills mire (2900 m) shows essentially the same stratigraphy as at Sacred Lake (Coetzee, 1967), but with a higher degree of differentiation in late glacial time. Two warm and two cold periods are known between ~14,000 and ~10,000 yr BP from the recovered core material. These colder periods may equate with late glacial (recessional) moraine positions near the rock glaciers in Teleki Valley (Figure 12.12), and below Gregory Glacier in Mackinder Valley. The postglacial pollen sequence (Coetzee, 1967) is essentially the same as at Sacred Lake.

Pollen recovered from a mire on Kilimanjaro at 2650 m (Plate 7) provided paleoclimatic and stratigraphic information for the last ~4600 yr (Coetzee, 1967). The sequence began with an Ilex Forest-fern maximum zone that is considered coeval with the climatic optimum on Mount Kenya. This pollen zone grades into a cooler and drier period beginning ~2200 yr BP, lasting to the present.

Correlation between East Africa and the Western European chronostratigraphy remains tentative, but is strengthened by the increased number of new radiocarbon dates. The Kalambo interstadial maximum and minimum ages appear essentially contemporaneous with the Paudorf Interstade of Central Europe. The Lower Würm (Liki) moraines still are poorly dated; however, the upper Würm appears to correlate with the late Liki (Liki II) on Mount Kenya. The subdivision of the European late glacial into six zones cannot be directly compared with the two pollen zones (Ui and Uii) at Sacred Lake, nor even with four pollen units in the Cherangani Hills (Plate 7). For Mount Kenya there appears to be only one cold period during late glacial time. Perhaps the altitude of the site is not critical enough to record these temperature fluctuations.

The well-documented European postglacial chronology may be broadly correlated with Mount Kenya. Not much is known about the early postglacial interval on Mount Kenya (Preboreal, Boreal in Europe) but presumably, according to palynological evidence (Coetzee, 1967), paleoclimate was becoming warmer (and wetter?). This warming trend peaked during the Atlanticum (broadly equated with the climatic optimum on Mount Kenya). The Subboreal in Europe was also a time of "early" Neoglaciation in the Alps and in North America. However, there is no evidence for a resurgence of glacial activity on the East African mountains during this time. Renewed glacial activity (Neoglaciation) in East Africa (on the Ruwenzori, on Kilimanjaro, and on Mount Kenya) took place largely during the sub-Atlanticum in Europe.

13.5 Correlation with the Marine Record

Long distance correlation with ocean core records from the Angola and Canary basins is made possible by multidisciplinary studies of piston cores that show paleoenvironmental changes throughout the Middle and Upper Quaternary (Jansen, Kuijpers and Troelstra, 1986). Sediments in the Zaire deep-sea fan show distinct breaks in accumulation rates and mineral composition. Sediment accumulation rates were higher during glacial ages and terrigeneous clays (mainly smectite) were transported into the Angola Basin by the Zaire River. Smectite clay minerals in these cores, supported by fossil pollen, reflect increasing aridity in tropical Africa and increasing desertification in the Congo Basin of Zaire and Angola. During interglacial ages, the expansion of the tropical rain forest stabilized land surfaces and cut off the supply of sediment to the Zaire River. During these interglacial stages weathering produced kaolinite and low-crystalline smectites, which are typical of weathering products produced today in the Zaire rain forest (van der Gaast and Jansen, 1984).

The data shown in Figure 13.5 show a decrease in sediment accumulation rates and in the amount of high-crystalline smectite. These data, from core T78-38, are similar to sediment logs from six piston cores recovered from the Zaire deep-sea fan (Jansen *et al.*, 1986). The decrease in sediment accumulation is accompanied by a decrease in high-crystalline smectite reflecting a change from glacial to interglacial climate in equatorial Africa (Jansen *et al.*, 1986).

In other areas of the equatorial Atlantic and Pacific, both oceanic and continental marine records document a shift to warmer conditions between 3.0 x 10^5 and 4.5 x 10^5 yr age. In core V26-41, at 19°N, the *Globorotalia menardii* curve reflects warmer conditions at 4.5 x 10^5 yr (Figures 13.6A and B). Documented increases in mean sea surface temperature occurred in the equatorial Atlantic at approximately the same time – 4.2 x 10^5 to 3.7 x 10^5 yr (van Donk, 1976). The carbonate content in core RC11-209 from the equatorial Pacific indicates a warmer (interglacial) climate between 4.3 x 10^5 and 4.0 x 10^5 yr (Figure 13.6C). These data correlate closely with SST data for the equatorial Pacific (Valencia, 1977). Moreover, the Radiolaria of core RC10-65 recovered from the eastern equatorial Pacific Basin shows an increasing upwelling assemblage (Jansen *et al.*, 1986). This correlates closely with palynological data from a long borehole near Bogota, Colombia, that show higher timberline between 3.7 x 10^5 and 3.9 x 10^5 yr ago (Hooghiemstra, 1984).

Cores from the Canary Basin (30°-33°N) show changes in foraminifera between 3.1 and 3.7 x 10^5 yr ago that support the hypothesis of a mid-Brunhes warmer climate. Increases in faunal assemblages, such as tropical-subtropical species including *Globorotalia menardii/tumida* and *Pulleniatina* spp. occurred during the height of the mid-Brunhes warm event (Jansen *et al.*, 1986). The most northern range of these species at present is 28°N. Because the southward movement of cool Northern Hemispheric water is typical of oceanic circulation during glacial stages (enhanced meridional circulation), the decrease of foraminifera like *G. menardii* is considered to signal a glacial age.

A recovered core from the Atlantic Ocean shows variations in sea surface temperatures (SST) (Figure 13.7A) that clearly define the mid-Brunhes warming shift of Jansen *et al.* (1986) (dashed area), as well as other warm signals that may correlate with younger and older interglaciations on Mount Kenya. Incidentally, the warmest signal coincides with the Teleki-Liki Interglaciation (Figure 2.27) on Mount Kenya that is considered the warmest and wettest of the Brunhes interglacial paleoclimates. The three older interglacial paleoclimates depicted in Figure 13.7A may equate with the Lake Ellis-Naro Moru, Gorges-Lake Ellis and pre-Gorges interglaciations described in Chapters 4 and 5.

The Indian Ocean core (Figure 13.7B) shows the relative strengths of the last four warmer intervals, with the longest and warmest cycles tentatively correlated with the Lake Ellis-Naro Moru and Teleki-Liki interglaciations. Although this core is admittedly incomplete, it tends to

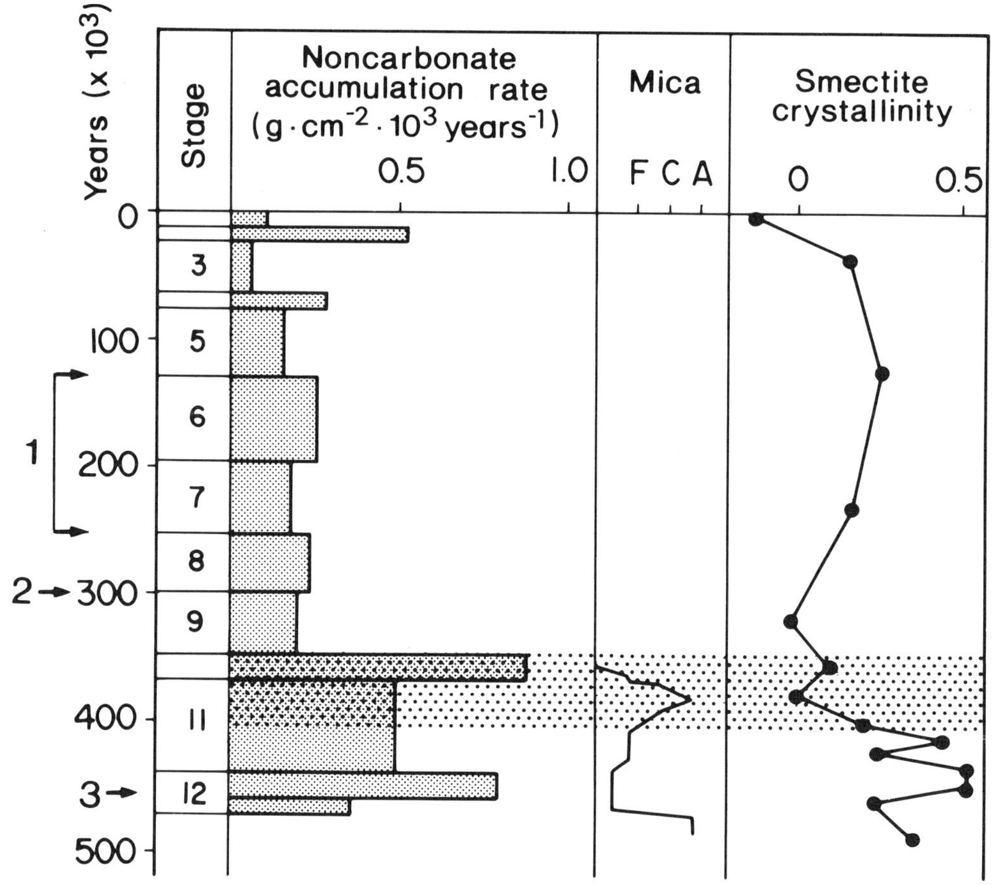

Figure 13.5: Rates of noncarbonate accumulation, abundance of mica and smectite crystallinity of pelagic core T78-38 from the Zaire deep-sea fan. The stippled area represents the mid-Brunhes Interglaciation of Jansen *et al.* (1986).

show the mid Brunhes warming shift and its relationship to a second Indian Ocean core taken at 41°S (Jansen *et al.*, 1986).

Fluctuations of $CaCO_3$ in marine sediments during glacial/interglacial stages for the past 1.4 x 10^6 yr are indicative of fluctuations from glacial to interglacial climatic conditions. The shift to lower amounts of $CaCO_3$ during interglaciations (Figure 13.8) is typical of marine sediments in many basins (Jansen *et al.*, 1986). Although $CaCO_3$ fluctuations may have other causes, in the Atlantic Ocean this is considered to signal a glacial paleoclimate. The stippled zone in Figure (13.7) may well correlate with the Naro Moru-Teleki Interglaciation on Mount Kenya. If this correlation holds, then the low $CaCO_3$ values at ~ 1.5 m, ~ 5m and 6-7m may each correlate with younger and older interglaciations, respectively.

As pointed out by Jansen *et al.* (1986) climatic changes throughout the Tertiary are mainly the product of plate movements, which changed the configuration of continents and oceans. Over shorter time frames of 100,000 yr, many paleoclimatic changes are closely connected to glacial and interglacial cycles that may have been forced by orbital variations, with periodicities of 4.0 x

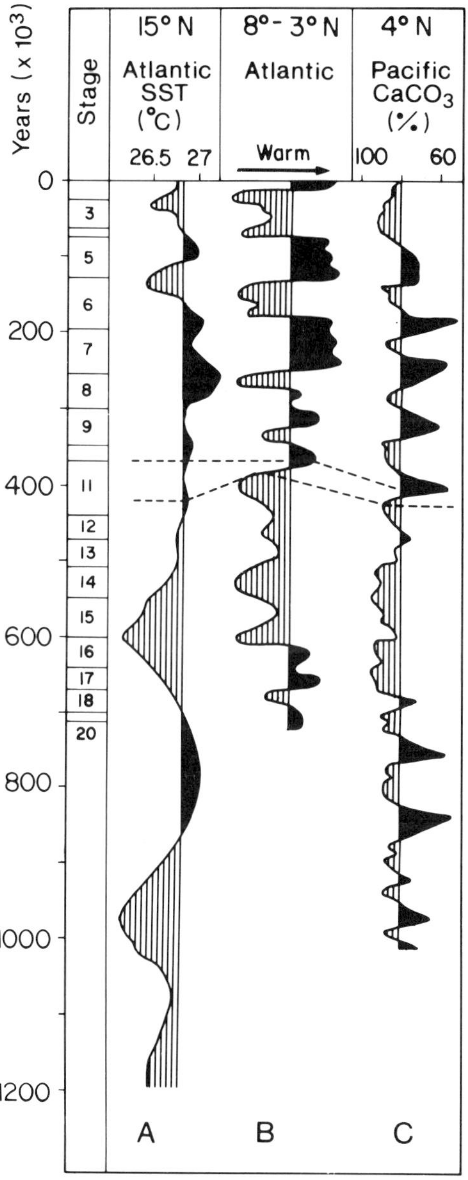

Figure 13.6: Quaternary paleoclimatic record from equatorial regions. Dashed area is the mid-Brunhes Interglaciation of Jansen *et al.* (1986). (SST)

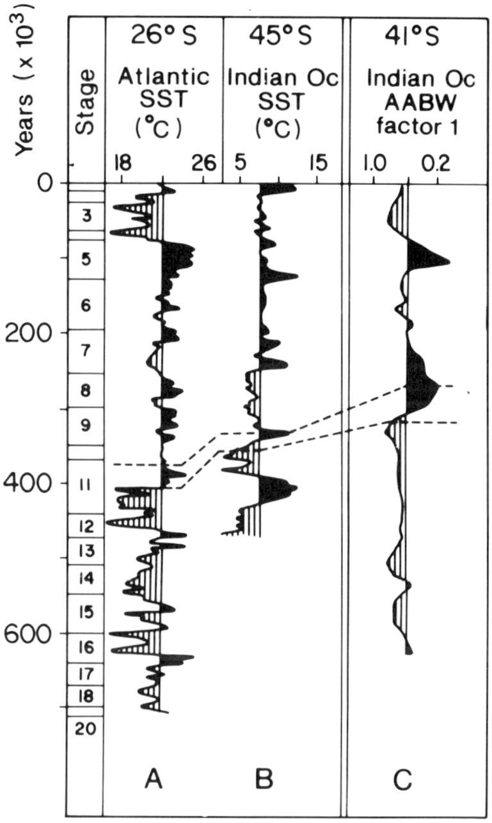

Figure 13.7: Long records of sea surface temperatures (SST) and AABW factors for the Atlantic and Indian Oceans.

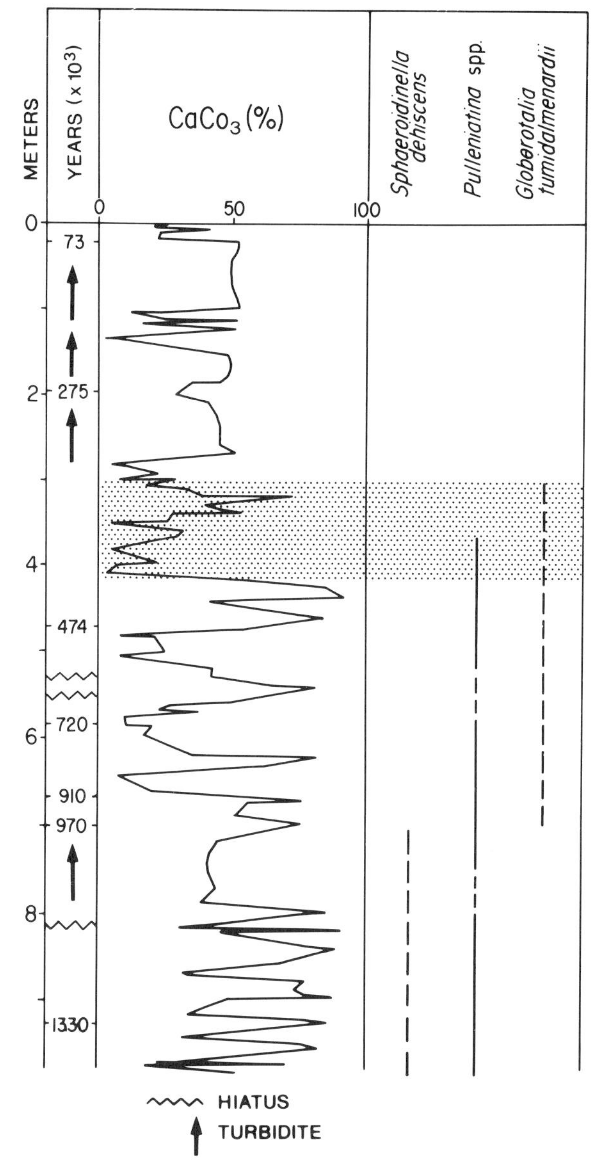

Figure 13.8: Fluctuations in the amount of $CaCO_3$ and distribution of tropical and subtropical foraminifera in core 82PCS18 from the Canary Basin. Stippled area shows postulated mid-Brunhes warming shift of Jansen *et al.* (1986).

10^5 yr and/or variations in solar radiation. If there is an asymmetric response to orbital eccentricities and/or solar radiation, then we might expect interglaciations on the order of every 3.0 x 10^5 to 4.0 x 10^5 yr or less. The precision with which we can date glacial and interglacial deposits in the tropical mountains will eventually allow more precise correlations with the marine record.

13.6 Correlation with the Rocky Mountain Glacial Record

There are some approximate similarities and differences between the glacial records of Mount Kenya and the Rocky Mountains that provide important information on the number of glaciations and paleoclimatic changes during the early and middle Quaternary. For the early and middle Quaternary in the Rocky mountains the established sequence of deposits, from oldest to youngest includes Washakie Point, Cedar Ridge, Sacagawea Ridge and Bull Lake tills (diamicts) (Richmond 1962, 1965; Karlstrom, 1987). The superposed sequence of these tills (diamicts) at exposures in the Wind River Mountains of western Wyoming, in Glacier Park (northern Rocky Mountains of the U.S.) and near Waterton Park (southern Alberta) leaves little doubt as to their relative ages and great antiquity. While at present it is impossible to provide absolute ages for these tills, it is certain that they predate the last glaciation and that the youngest (Bull Lake) is most likely correlated with isotope stage 6. Paleosols interbedded with these tills are generally Aridisols suggesting that interglacial paleoclimates were dry or, at least, subhumid supporting sagebrush ecosystems. Similar sequences of tills are known on Mount Kenya, but these are restricted to recognizing ice advances within the Gorges Glaciation (early Quaternary), or to differentiating Teleki and Naro Moru drifts (middle Quaternary). There are no known exposures of complete sequences of early through middle Quaternary tills on Mount Kenya.

The broadest possible correlation of pre-Liki and pre-Pinedale (= Würm/Wisconsinan) tills is given in Table 13.1:

Table 13.1
Tentative correlation of tills (diamicts) pre-dating the last glaciation on Mount Kenya (Liki) and in the Rocky Mountains (Pinedale).

Mount Kenya	Rocky Mountains
Teleki	Bull Lake
Naro Moru	Sacagewea Ridge[a] Cedar Ridge
Lake Ellis Gorges pre-Gorges	Washakie Point ? ?

a. U-dated at 440 ± 120 ka (Richmond, 1986)

Absolute maximum and minimum ages are not available for most of the Rocky Mountain deposits, but it is likely the two older tills belong to the early middle Quaternary ($< .73 \times 10^6$ yr), and the two younger ones to the late middle Quaternary. Only detailed paleomagnetic dating of the Rocky Mountain tills will provide a firm correlation to the Mount Kenya record.

The Pinedale and Liki sequences bear some striking similarities (Figure 13.9) even considering that the time of the onset of the last glaciation is unknown, and that the Liki/Pinedale interstades are not dated with a great deal of precision. Thus, the early Pinedale and early Liki stades are presumed to have occurred between ~100,000 and ~50,000 yr (Figure 13.9). There is no evidence for a middle Liki glacial event on Mount Kenya. However, in the Rocky Mountains there is some evidence from Yellowstone Park (Richmond, 1976) for interstadial deposits, containing wood or plant detrital material, dating between 38,000 and >45,000 yr BP.

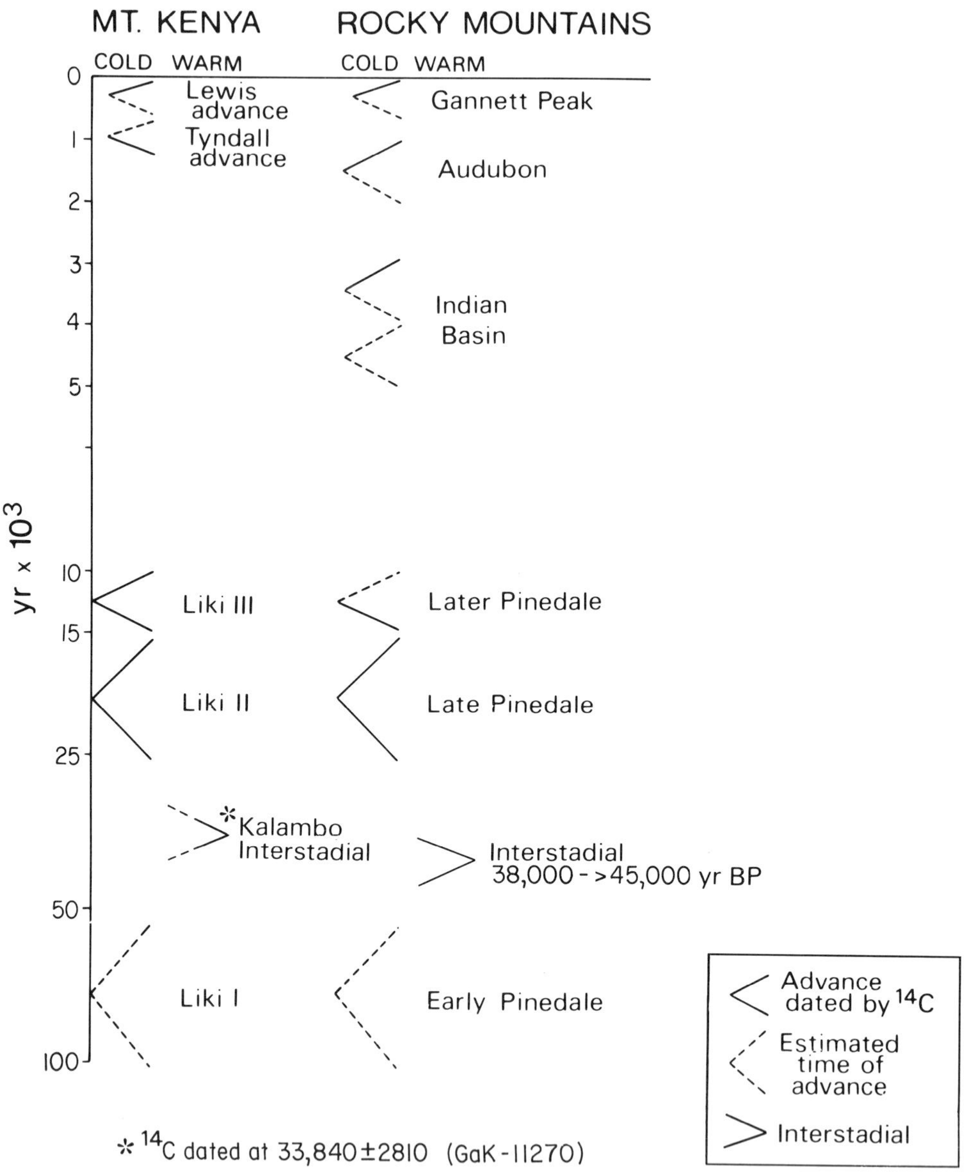

Figure 13.9: Correlation of late Pleistocene and Holocene glacial and interstadial deposits in the Rocky Mountains and in East Africa.

The late Pinedale in the Rocky Mountains probably began as early as ~25,000 yr BP making it equivalent to the Liki II stade on Mount Kenya. The later Pinedale in the Rocky Mountains is poorly dated, but probably equates to the late glacial, between ~13,000 and 12,000 yr BP. One of the problems in dating later Pinedale moraines, is that ^{14}C dates come *not* from the moraines, but from bogs or mires on the moraine surfaces or from lakes dammed by end moraines (Mahaney, 1987a, 1988c). Because of the uncertainty in the amount of time involved in infilling of bog and lacustrine basins, such bog- or lake-bottom dates provide only equivocal minimum ages for later Pleistocene deposits. This would amount to several centuries or even a few millennia (Mahaney, 1987a). From the landform/weathering evidence, both in the Rocky Mountains and in East Africa, it would appear most or all of the older moraines in the high valleys are of late glacial age. Indeed, recent evidence indicates it was too *dry* and too warm to support early Holocene glacial expansion (at least in the Afroalpine) (Hastenrath, 1984; Zinderen Bakker and Coetzee, 1987).

The early Holocene in both areas appears to have been too warm to support the growth of glaciers. Towards the end of the middle Holocene, Neoglaciation in the Rocky Mountains began at *ca.* ~5000 yr BP. Although there are few dates for the onset of renewed cirque glacier activity (Mahaney, 1984b); Mahaney and Fahey, 1988), the end of the first advance (Indian Basin) probably occurred ~3000 yr BP (Mahaney, 1981). Indeed, ^{14}C dates from organic detritus in outwash in the Wind River Range (western Wyoming) places the minimum age of Indian Basin moraines at just under 3000 yr BP (Mahaney 1987d). There are no equivalent early Neoglacial deposits on Mount Kenya.

The Audubon advance (Figure 13.9) in the Rocky Mountains may be coeval with the Tyndall deposits on Mount Kenya. Once again, ^{14}C dates for the Tyndall advance were derived from outwash and debris flow deposits, not directly from the moraines. While the Tyndall advance presumably reached its culmination *ca.* 1000 yr BP, it is likely that it began even earlier. Indeed, emplacement of outwash in upper Teleki Valley at 4050 m.a.s.l. is ^{14}C dated at *ca.* 2,000 yr BP. It is likely that early emplacement of outwash (site TV4; Chapter 7) is tied directly to the earliest advance of Tyndall ice. If this is the case, the Tyndall advance would be equivalent to the Audubon advance in the Rocky Mountains.

The Gannett Peak advance of the American Rocky Mountains is considered to represent the classical Little Ice Age (Mahaney, 1978) and is equivalent to the Lewis advance on Mount Kenya. Owing to the absence of wood and other suitable organic materials for ^{14}C dating, it is generally impossible to date these young Neoglacial deposits. Generally, these later Holocene deposits are too far above timberline to allow the use of dendrochronological methods for absolute dating of substrates. Dating is usually achieved using historical data, photography of moraine positions over the last century, lichen size and cover, and weathering rind data (compared with older stratigraphic units; Chapter 8).

13.6 Conclusions

The correlation of glacial deposits in the tropical mountains shows a broadly synchronous sequence of events among the major summits for which there is some radiometric age control. The "classical" Würm/Wisconsinan record is fairly well documented and dated in a number of areas and shows similar maximum and minimum dates. Only the Mexican volcanoes appear to have had a different sequence of events with major advances during middle Wisconsinan/Würm time (~34,000 yr BP). However, as more dates become available this scenario may change considerably.

Late glacial moraines in the high valleys of many tropical mountains, although poorly dated, have enough ^{14}C controls to suggest that ice advance or recessional stillstand occurred between 12,000 and 13,000 yr BP. In the Ecuadorian Andes, Clapperton (1987) has published ^{14}C dates from material collected within late glacial moraines (a rather rare find) placing them within the Younger Dryas. Only on Mount Kenya has there been any attempt to prove that these moraines result from recessional stillstand (see Chapter 7). In other areas, some workers believe that local ice advances occurred during the late glacial or from the early to middle Holocene.

The Neoglacial record on most of the tropical mountains is limited to about the last ~1000 yr. Only on Kilimanjaro, Irian Jaya, the Mexican volcanoes, and the Andes of Ecuador, Peru, Chile and Colombia is there evidence of somewhat older Neoglacial landforms. The oldest of these is on Irian Jaya, where four advances have been recognized, and on Iztaccihuatl and Ajusco volcanoes in Mexico where several Holocene advances are recognized. On Mount Wilhelm, in New Guinea, only snowbank formation is evident (two phases of late Holocene accumulation). Presumably, precipitation patterns and fluctuations are as important to glacier growth and development, as are temperature variations. Recession of ice on Mount Kenya during the late 19th and early twentieth century may have resulted from both reduced precipitation (during 1880-1900 A.D.) and a gradual temperature increase (S. Hastenrath, personal communication, 1989).

CHAPTER 14

SUMMARY

Moraines of Gorges age on Mount Kenya represent the basal till and occur on the eastern flank as low as *circa* 2850 m. Together with glacially-crushed clasts in lahars (pre-Gorges) deposited during pre-Olduvai time, they represent the oldest glacial sediments on the mountain, and the oldest documented till in the tropical mountains. Because deposits of Gorges age are composed of basaltic clasts, without any nepheline syenite and/or porphyritic phonolite, they are considered to represent the earliest stage of glacial deposition following erosion of surface lavas on Mount Kenya. The plug rocks of the volcano are dated by K/Ar at 2.64 my (Everden and Curtis, 1965), which indicates the lavas eroded by the earliest glaciation probably date from the early Pleistocene. Indeed, beyond the limit of Gorges drift, bedrock composed principally of tuff displays well weathered paleosols 2.5 m thick with rather extensive "moorland-like" surface A horizons. The parent materials and lower subsoil C horizons in these paleosols yield a strong normal remanent magnetism with a reversed overprint indicating the tuff may have been emplaced some time ago, either in the upper Gauss Chron (>2.4 m y) or during the Olduvai subchron.

The magnitude of the Gorges Glaciation was considerable given that the mountain was probably several hundred meters higher than today—similar to the present-day Mount Kilimanjaro—when Gorges Glaciers began to cut major valleys on the eastern flank. On the western, southern and northern flanks of the mountain it appears that Gorges drift was removed by younger glacial episodes that emplaced Naro Moru and Teleki tills, which were in turn, removed by younger Liki ice. Deeply weathered paleosols, formed in Gorges drift, attest to the amount of weathering since deposition in the early Quaternary. As with other pre-Liki tills, Gorges drifts have high percentages of ferromagnetic minerals, especially magnetite (Vortisch, Mahaney and Fecher, 1987) which make them ideally suited for paleomagnetic measurements. All the sites sampled thus far have yielded strong normal remanent magnetism, sometimes with reversed overprints, suggesting that these tills were emplaced largely during the Olduvai subchron.

Loess caps on paleosols of Gorges age have either reversed remanent magnetism suggesting emplacement prior to 0.73-0.79 my, or they have normal magnetism without overprinting suggesting they were emplaced during the Brunhes Chron. Because the upper loess root zone is often affected by bioturbation it proved impossible to assess the degree to which younger airfall sediments might have been added to these paleosols during the Brunhes Chron. While it is not the intent here to discuss the paleosols at length, it is important to note that these older paleosols often have thick (30-70 cm) A horizons, with strong (10YR 2/1 and 1/1) colors, and high (20-30%) organic matter contents. Because it is possible to demonstrate some movement of clay, organic matter, and minor amounts of organically complexed Fe and Al in these paleosols, it is possible to assign them to the Alfisol order (Soil Survey Staff, 1975). The thick "moorland-like" Ah horizons probably reflect depression of vegetation belts during the last glaciation, as well as low microbial numbers in the ericaceous and Hagenia belt paleosols (Mahaney and Boyer, 1986a). These thick "moorland-like" A horizons are relict features, retained in place due to the inability of microorganisms to rapidly decompose organic matter (Mahaney and Boyer, 1986a).

The second glaciation on Mount Kenya is known mainly from the Lake Ellis and Hobley Valley areas (Barendregt and Mahaney, 1988). This glaciation emplaced drift composed of clasts of basalt, kenyte and phonolite giving rise to surface and buried paleosols belonging to the Alfisol order, and displaying relatively thick profiles consisting largely of hematite, gibbsite and halloysite/metahalloysite in about the same or lesser quantity as in the older post-Gorges paleosols.

Paleomagnetic analyses of tills near Lake Ellis and in Hobley Valley yielded strong reversed remanent magnetism indicating an age older than 0.73-0.79 my (Barendregt & Mahaney, 1988). The relatively thin till bodies (2-3 m), and closeness to the steeper slopes at ~3150 m on the eastern flank, suggest that while a significant volume of rock was removed during the Lake Ellis Glaciation, relatively little was used to form moraines. The presence of clasts of different composition (porphyritic phonolite mainly) indicate that valley glaciers excavated lava as well as rock from the volcanic plug in the Central Peaks area.

Loess overlying Lake Ellis till has about the same thickness as that overlying Gorges Drift (*circa* 0.5 m) and gives either a strongly reversed or normal remanent magnetism. Because these sites are generally higher than those on Gorges drift they are at or beyond the range of timberline fluctuations, and occur on flat-lying plateau areas that are imperfectly drained. As a result they have cemented subsoils (C horizons) (Fragiudalfs, Soil Survey Staff, 1975) with clasts and matrices alike coated with thick skins of MnO_2 (Derbyshire and Mahaney, 1989). In some cases the upper parts of loess caps in these paleosols give a strong normal remanent magnetism indicating that some airfall deposition of fine-grained material occurred during the Brunhes Chron, and also that signficant hiatuses can be expected between loess and till in some profiles.

In lower Teleki Valley (western flank) and in Gorges Valley (eastern flank) (Mahaney, 1989a), tills of Naro Moru age occur in juxtaposition to older and younger glacial deposits. In some cases they lie buried beneath extensive slope wash deposits and overlie well-weathered soils in residual regolith. As with the Lake Ellis units, these deposits contain clasts of nepheline syenite and porphyritic phonolite indicating that this was the second glaciation to reach the volcanic plug (presently at an elevation of 5199 m). Because these moraines are found as low as 2950 m in Teleki Valley and somewhat higher (3150 m) on the eastern flank, it is possible to postulate a certain asymmetry in the form of the ice cap covering the Central Peaks during Naro Moru time.

Paleosols in Naro Moru drifts have variable degrees of development depending on whether they occur on the wetter western or drier eastern flanks. On the wetter western flank they are Tropudalfs, relatively thick and well developed, with a higher degree of feldspar dissolution (Vortisch, Mahaney and Fecher, 1987). On the eastern flank they may be Fragiudalfs (Soil Survey Staff, 1975); that is, soils with cemented pans in the subsurface, or Tropudalfs. Loess caps in these paleosols vary in thickness from 0.3 to 0.6 m, and in some cases, discrete layers of tephra alternate with loess (Mahaney, 1989a; Vortisch, Mahaney and Fecher, 1987).

Paleomagnetic measurements in Naro Moru till and loess bodies show always normal remanent magnetism indicating that both the till and loess were emplaced during the Brunhes Chron ($<730,000$ yr). Since the tephras associated with the Naro Moru drifts are similar in age with tuff described by Charsley (1989) in the Nanyuki Formation, it is likely that Naro Moru Till was emplaced prior to $320,000 \pm 20,000$ yr or during the early Brunhes Chron ($<730,000$ yr).

The youngest of the pre-Liki tills (Teleki Till) is known from sites in Teleki, Gorges and Ruguti drainages on Mount Kenya, and on Ithanguni (Derbyshire and Mahaney, 1989). These tills usually contain a mix of clasts consisting of phonolite, porphyritic phonolite and nepheline syenite. In Teleki Valley at the type section TV23 (Vortisch, Mahaney and Fecher, 1987) the post-Teleki paleosol is a well weathered Alfisol that bears the full extent of weathering during the last interglacial period on Mount Kenya, as well as the deposition of loess followed by weathering during the Liki Glaciation and postglacial period. All paleosols of post-Teleki age and older contain some halloysite/metahalloysite, kaolinite, and gibbsite as the main weathering products (Vortisch, Mahaney and Fecher, 1987); in general, the amounts of these clay minerals increase in older paleosols provided they are exposed at the surface to continual subaerial weathering. For the formation of gibbsite, it is important that the paleosols are relatively permeable and capable of aggressive leaching (Vortisch, Mahaney and Fecher, 1987). Buried paleosols record, for the most part, weathering that occurred while they were exposed to subaerial processes, and in some cases they provide the only information on weathering during interglacial periods.

Because the Teleki Glaciation appears to have been only slightly stronger than the last glacial maximum (Liki) advance, tills of Teleki age are prevalent only in a few drainages. An increase in percentages of clasts from the central volcanic plug and a decrease in basaltic clasts in Teleki deposits, indicate that erosion of the main volcanic roof rocks was nearly complete by Teleki time (isotope stage 6).

The onset of the Liki Glaciation is probably equivalent with isotope stage 4, beginning *circa* < 100,000 yr BP. Most major valleys on Mount Kenya have end and lateral moraine complexes formed during Liki I and II substages at *circa* 3200 m and higher. Moraines are divided into I and II on the basis of topographic position, weathering parameters, and paleosol cover (Mahaney, 1979a, 1987a; Vortisch, Mahaney and Fecher, 1987). Liki II is probably equivalent to isotope stage 2, and in some valleys (*e.g.* Teleki, Hausberg) it is the only vestige of the Liki Glaciation.

Recovered bog-bottom sediments from Liki II moraine surfaces gave minimum ages of ~ 15,000 yr BP (Mahaney 1988a). Attempts at dating these moraines directly have been fruitless owing to the scarcity of organic materials in them. However, the beginning of the Liki II Glaciation appears to have occurred *circa* 25,000 yr BP as documented from buried paleosols in lower Teleki Valley. At site TV61 (Barendregt and Mahaney, 1988) a series of buried paleosols provides evidence for burial of a thick "moorland-like" A horizon sequence (also seen in the epipedons of older ground paleosols) *circa* 0.5 m thick at *circa* 25,000 yr BP. The pollen from this buried paleosol indicates first a cooler climate than today (Zone B), and later a colder climate (Zone C).

Following the end of the Liki II substage, ice receded upvalley at a rapid rate, so that by 12,500 yr BP most valley glaciers had their termini at *circa* 4000 m.a.s.l. (Mahaney 1987a, 1987b, 1987c). The available pollen sequence for the period 15,000-12,000 yr BP (Coetzee, 1967; Mahaney, 1985) shows a progression from colder to warmer to cold temperatures during the late glacial (Liki III) substage. At the end of the Liki III recessional stillstand, ice in most valleys either disappeared completely, or receded into major cirques with a spatial distribution similar to the present. There is no evidence for early to middle Holocene ice advances despite the interpretations of four cores from Naro Moru Tarn (Johansson and Holmgren, 1985), Thompson Tarn (Perrott, 1982), and Hausberg and Oblong tarns (Karlén, 1985). Data gathered from 33 cores recovered from major bogs and lakes in many valleys shows conclusively that the ~ 4000 m end moraines have later Pleistocene ages (Mahaney, 1987a,b,).

Following the climatic optimum on Mount Kenya—*circa* 2000 yr BP—pollen evidence indicates colder and drier climate (pollen zone Z). Over the last thousand years the record shows two ice advances (Mahaney, 1985); first, the Tyndall advance at approximately 1000 yr BP, and second, the Lewis advance which is undated but considered equivalent to the classical Little Ice Age (300-100 yr BP).

The Tyndall advance, known from moraines, outwash, and debris flows in front of the Lewis and Tyndall glaciers (Mahaney, 1987a, b, c), radiocarbon dates at *circa* 900-950 yr BP. Because these deposits are not found in other cirques on Mount Kenya it is likely that ice resurgence was confined to the southwestern flank of the Central Peaks area. Soils in the Tyndall deposits are thin Entisols with A/Cox profiles, showing significant differences from the paleosols on nearby Liki III moraines, which are Inceptisols with cambic (color) B horizons and greater depth. Lichens also provide a means of separating Liki III deposits from Tyndall deposits, as lichen cover in general, and maximum diameters of *Rhizocarpon* section *Rhizocarpon are* radically different. Lichen cover (including the genera *Usnea, Umbilicaria, Buellia*, and *Rhizocarpon*) ranges from 50 to ~ 80 percent on the older Liki III deposits, while on Tyndall-age surfaces it is usually 30-50 per cent. These deposits are probably coeval with similar moraines described in the Ruwenzori (Lac Gris) and on Kilimanjaro (Recent Glaciation). While no absolute dates are available it is probable that deposits were emplaced prior to the classical Little Ice Age.

The Lewis advance, which was still evident when Gregory (1894b) visited the mountain in 1893, emplaced about 90 percent of Neoglacial surficial materials presently exposed in the Central Peaks area. Moraines are bouldery, often with an open network of stones lacking matrix materials, and practically devoid of soil cover. Where present, soils form only 10 percent of the surface cover, and to a depth of *circa* 10-15 cm – C/Cu horizon sequences (Mahaney, 1979a, 1987a).

With the present climatic warming, induced by rising concentrations of CO_2 and other gases, it is probable that mean annual temperature and precipitation will increase on many tropical mountains. Because all of these high tropical peaks and plateau areas are marginal for the existence of glaciers, it is likely the annual snowline will rise and many or most of the glaciers will disappear. For Mount Kenya only the glacial record will remain as a testimony to the ice caps covered with "kirira" (snow) that once covered a large part of the Mount Kenya massif.

There is no doubt whatsoever that interpretations of the Quaternary record on Mount Kenya have undergone revolutionary changes over the last two decades. Instead of two Pleistocene glaciations and one Neoglacial advance described nearly a century ago, five Pleistocene glaciations are documented from the moraine record, and a sixth is postulated from glacially-crushed grains in the lahar deposits. In addition, there are four documented interglacial periods and two Neoglacial advances. Limited evidence is also available for a pre-Gorges interglacial that is + 2.0 my old. The old model of a "younger" and "older" glaciation (Baker, 1967) has been refined. It is by no means certain as to how these glacial periods correlate with other continental areas or with the marine record. However, the last glaciation and preceding interglacial period correlate with the Würm/Wisconsinan and Eemian/Sangamonian (in Europe and N. America) and isotope stages 2 through 5, respectively. This is the rough outline of the Upper Pleistocene record on Mount Kenya as it exists today.

Of the older Middle and Lower Pleistocene records, paleomagnetism allows at least a separation of deposits at ~.73-.79 my, into older and younger suites of landforms. The oldest group is considered to have been emplaced just prior to and following the Olduvai subchron (1.67-1.87 my). Because glacially-crushed grains are present in the lahars of early Quaternary age, it seems the mountain was glaciated soon after emplacement of the Mount Kenya volcanic rock suite or during the early Quaternary. Thus, episodic glaciation has left a wealth of deposits almost directly on the equator in East Africa. The unravelling of this record provides an important scenario of paleoclimatic events that have been important in shaping not only the mountain itself, but also the ecosystems which are still developing below and above the timberline.

REFERENCES

Axelrod, D.I. and Raven, P.H., 1978, Late Cretaceous and Tertiary vegetation history of Africa, In: *Biogeography and Ecology of Southern Africa,* M.J.A. Werger ed., The Hague, Dr. W. Junk 77-130.

Alexander, E.B., 1974, Extractable iron in relation to soil age on terraces along the Truckee River, Nevada., *Soil Sci. Soc. Amer. Proc.* 38, 121-4.

Amer. Comm. Strat. Nomen., 1970, *Code of stratigraphic nomenclature, Amer. Assoc. Petrol. Geol.,* 21p.

Andrews, J.T. and Webber, P.J., 1964, A lichenometrical study of the northwestern margin of the Barnes Ice Cap: a geomorphological technique. *Geographical Bulletin* (Ottawa), No. 22, 80-104.

Andrews, J.T., Bowen, D.Q. and Kidson, C., 1979, Amino acid ratios and the correlation of raised beach deposits in southwest England and Wales. *Nature,* 281, 556-8.

Arthur, J.W., 1921, Mount Kenya, *Geogr. Jour.,* 58, 9-25.

____, 1923, A Sixth Attempt on Mount Kenya, *Geogr. Jour.,* 62, 205-9.

Bada, J.L.,, 1972. The dating of fossil bones using the racemization of isoleucine, *Earth and Planetary Science Letters,* 15, 223-31.

Bada, J.L., Luvendyk, B. P., and Maynard, J. B., 1970, Marine Sediments: Dating by the racemization of amino acids, *Science,* 170, 730-2.

Baker, B.H., 1967, Geology of the Mount Kenya Area, Geological Report No. 79, *Kenya Geological Survey,* Nairobi, 78p.

Bambauer, H.U., Taborszky, F., Trochim, H.D. and Troger, W.E., 1971, *Optische Bestimmung der Gesteinsbildenden Minerale-Teil 1,* Bestimmungstabellen, Stuttgart, 188p.

Barendregt, R.W., 1984, Using paleomagnetic remanence and magnetic susceptibility data for the differentiation, relative correlation and absolute dating of Quaternary sediments. In Mahaney, W.C., ed. *Quaternary Dating Methods*, Amsterdam, Elsevier, 101-122.

Barendregt, R.W., and Mahaney, W.C., 1988, Paleomagnetism of selected Quaternary sediments on Mount Kenya, East Africa: a reconnaissance study, *Journal of African Earth Sciences*, 7(1), 219-225.

Belknap, D.F. and Wehmiller, J.F., 1980, Amino Acid racemization in Quaternary mollusks: examples from Delaware, Maryland and Virginia. in Hare, P. E., Hoering, T. C. and King Jr., K., (eds.), *Biogeochemistry of Amino Acids*, New York, John Wiley and Sons, 401-14.

Bender, M.L., 1974, Reliability of amino acid racemization dating and paleotemperature analysis on bones, *Nature,* 252, 378-81.

Benedict, J.B., 1967, Recent glacial history of an alpine area in the Colorado Front Range, U.S.A. I, Establishing a lichen-growth curve. *Journal of Glaciology*, 6, 817-32.

____, 1968, Recent glacial history of an alpine area in the Colorado Front Range, U.S.A. II, Dating the glacial deposits. *Journal of Glaciology,* 7, No 49, 77-87.

____, 1985, Arapaho Pass. Glacial geology and archeology at the crest of the Colorado Front Range, Center For Mountain Archeology, Research Report 3, Ward, Colordo, 197p.

Berger, G.W., 1985, Thermoluminescence dating applied to a thin winter varve of the late glacial South Thompson silt, south-central British Columbia, *Canadian Journal of Earth Sciences*, 22, 1736-1739.

Berger, G. W. and Mahaney, W. C., 1989, Test of thermoluminescence dating of buried paleosols from Mount Kenya, Kenya, *Sedimentary Geology*, in press.

Beschel, R.E., 1950, Flechten als Altermasstab rezenter Moränen. Zeitschrift für Gletscherkunde und Glazialgeologie, Geografiska Annaler, 48A, No. 1, 24-31.

____, 1955, Individuum und Alter bei Flechten. *Phyton: Annales Rei Botanicae* (Horn), 6, Nos. 1-2, 60-8.

____, 1957, A project to use lichens as indicators of climate and time. *Arctic,* 10, 60.

____, 1958, Lichenometrical studies in West Greenland. *Arctic*, 11, No. 4, 254.

____, 1961, Dating rock surfaces by lichen growth and its application to glaciology and physiography (lichenometry) in *Geology of the Arctic,* II, G.O. Raasch, ed., Univ. of Toronto Press, 1044-62.

____, 1965, Epipetric succession and lichen growth rates in the eastern Nearctic. In *International Association for Quaternary Research, Seventh International Congress,* Boulder, Colorado, 1965. Abstracts, general sessions, 25-6.

Birkeland, P.W., 1969, Quaternary paleoclimatic implications of soil clay mineral distribution in a Sierra Nevada-Great Basin transect, *Jour. Geol.,* 77, 289-302.

____, 1973. Use of relative age dating methods in a stratigraphic study of rock glacier deposits, Mt. Sopris, Colorado. *Arctic and Alpine Research,* 5, No. 4, 401-16.

____, 1984, *Soils and Geomorphology,* Oxford, N.Y., 372p.

Birkeland, P.W. and Janda, R.J., 1971, Clay mineralogy of soils developed from Quaternary deposits of the eastern Sierra Nevada, California, *Geol. Soc. Amer. Bull.,* 82, 2495-2514.

Birkeland, P.W., Rodbell, D.T., and Short, S.K., 1989, Radiocarbon dates on deglaciation, Cordillera Central, Northern Peruvian Andes, *Quaternary Research*, in press.

Bishop, W.W., 1962a, Pleistocene Correlation in the Uganda Section of the Albert-Edward Rift Valley. *Actes 4e Pan Afr. Cong., Préhist. Kon. Mus. Midden-Afrika, Tervuren,* Sér. 8 (40) 245-54.

____, 1962b, World-wide dating by Correlation of Climatic Changes. *Adv. Science,* 491-4.

____, 1965, Quaternary Geology and Geomorphology in the Albertine Rift Valley, Uganda. *Abstracts VII INQUA Congress,* Denver, 36.

Blackwelder, E., 1915, Post Cretaceous history of the mountains of central western Wyoming. *Journal Geol.* 23, 97-117; 192-217; 307-40.

Blume, H.P. and Schwertmann, U., 1969, Genetic evaluation of profile distribution of aluminum, iron and manganese oxides, *Soil Science Soc. Amer. Proc.,* 33, 438-44.

Bouma, J., 1969, *Microstructure and stability of two sandy loam soils with different soil management,* Wageningen, Center for Agric. Publ. and Doc., 100 p.

Bouyoucos, G.J., 1962, Hyrdometer method improved for making particle size analyses of soils, *Agron. Jour.,* 54, 464-5.

Bower, C.A. and Wilcox, L.V., 1965, Soluble salts, in Black, C.A., ed., *Methods of Soil Analysis: part 2,* Madison, Wisconsin, Amer. Soc. Agron. 933-51.

Boyer, M.G. and Mahaney, W.C., 1985, Some microbiological aspects of surface and buried soils on Mount Kenya, East Africa, *Plant and Soil*, 86:309-319.

Bremner, J.R., 1965, Total nitrogen, in Black, C.A., ed., *Methods of Soil Analysis,* Madison, Wisc., Amer. Soc. Agronomy, 149-76.

Burgess, A., 1963, *Soil Organisms*, North-Holland Publishing Company, Amsterdam: 151-157.

Calkin, P.E. and Ellis, J.M., 1984, Development and application of a lichenometric dating curve, Brooks Range, Alaska, in Mahaney, W. C., ed., *Quaternary Dating Methods,* Amsterdam, Elsevier, 227-46.

Caviedes, C.N., and Paskoff, R., 1975, Quaternary glaciations in the Andes of North-Central Chile, *Journal of Glaciology*, 14: 155-170.

Charanjitsingh, C. and Gangwar, P.C., 1977, Zinc, iron, and citric acid in buffalo semen, *Indian Jour. Animal Science*, 47, 250-256..

Charsley, T.J., 1989, Composition and agc of older outwash deposits along the northwestern flank of Mount Kenya. *In Quaternary and Environmental Research on the East African Mountains,* Mahaney, W.C. (ed.) Balkema, Rotterdam, 165-174.

Christensen, M., 1981, Species diversity and dominance in fungal communities, Wicklow, D. and Carrol, G. eds., *The Fungal Community, Marcel Dekker, N. Y.:201-232.*

Clapperton, C., 1972, The Pleistocene moraine stages of west-central Peru, *Journal of Glaciology*, 11: 255-263.

____, 1986, Glacial geomorphology, Quaternary glacial sequence and paleoclimatic inferences in the Ecuadorian Andes, In V. Gardiner, ed., *International Geomorphology*, N.Y., Wiley, 843-869.

Clapperton, C. and McEwan, C., 1985, Late Quaternary moraines in the Chimborazo Area, Ecuador, *Arctic and Alpine Research*, 17(2): 135-142.

Clapperton, C. and Vera, R., 1986, The Quaternary glacial sequence in Ecuador: a reinterpretion of the work of Walter Sauer, *Jour. Quaternary Science 1(1): 45-56.*

Coe, M.J., 1967, The Ecology of the Alpine Zone of Mount Kenya. The Hague, Junk, 136p.

____, 1989. Biogeographic affinities of the high mountains of tropical Africa. In: ed. W.C. Mahaney, *Quaternary and Environmental Research on East African Mountains*, Balkema, Rotterdam, 257-278.

Coetzee, J.A., 1964, Evidence for a considerable depression of the vegetation belts during the upper Pleistocene on the East African mountains, *Nature,* 204, No. 4958, 564-6.

____, 1967, Pollen analytical studies in East and Southern Africa, *Palaeoecology of Africa,* 3, 146p.

Coffin, D.E., 1963, A method for the determination of free iron in soils and clays, *Canadian Jour. Soil Sci.,* No. 43, 7-17.

Colman, S.M., 1986, Levels of time information in weathering measurements, with examples from weathering rinds on volcanic clasts in the Western U.S. In S.M. Colman and D.P. Dethier, eds., *Rates of Chemical Weathering of Rocks and Minerals,* Academic Press, N.Y., 379-94.

Cooke, H.B.S., 1958, Observations relating to Quaternary environments in East and Southern Africa. Alex, L. du Toit Memorial Lecture No. 5, *Geol. Soc. S. Africa,* Amex 60, 73p.

Cooley, W.D., 1852, *Inner Africa Laid Open,* London.

Cooremans, B. and Mahaney, W.C., 1989, Palynology of the TV61 section on Mount Kenya, East Africa, *In* Heine, K., *Palaeoecology of Africa*, in press.

Cotton, A.D., 1930, A visit to Kilimanjaro, *Kew. Bull.,* 3, 97-121.

Crocker, R.L. and Major, J., 1985, Soil development in relation to vegetation and surface age, Glacier Bay, Alaska, *Jour. Ecol.,* 43, 427-48.

Cunningham, R.K., 1963, The effect of clearing a tropical forest soil, *Jour. of Soil Science, 14: 334-345.*

Curry, R.R., 1969, Holocene climatic and glacial history of the central Sierra Nevada, Calif., In: Schumm, S.A. and Bradley, W.C., eds., *U.S. Contributions to Quaternary Research.,* Geol. Soc. Amer. Spec. Paper, 123, 1-47.

Curtis, C.D. and Spears, D.A., 1971, Diagenetic development of kaolinite, *Clays and Clay Minerals*, 79, 219-227.

Davis, P.T., 1985, Neoglacial moraines on Baffin Island, In: Andrews, J. T. ed., *Quaternary Environments – Eastern Canadian Arctic, Baffin Bay, and Western Greenland,* Allen and Unwin, Boston, 682-718B.

Davis, P.T. and Osborn, G., 1987, Age of pre-Neoglacial cirque moraines in the central North American cordillera, *Géographie physique et Quaternaire*, 41(3), 365-375.

Day, P., 1965, Particle fractionation and particle size analysis, in Black, C. A., ed., *Methods of Soil Analysis:* Madison, Wisc., Amer. Soc. Agron., 545-67.

Dejou, J., Guyot, J., and Robert, M., 1977, Evolution superficielle des roches cristallines et cristallophylliennes dans les régions témpereés, Paris, Inst. Rech. Agron., 464p.

Denny, C.S., 1940, Stone rings on New Hampshire Mountains, Amer. J. Sci., 238: 432-438.

Derbyshire, E., 1978, A pilot study of till microfabrics using the scanning electron microscope, In Whalley, W.B., ed., *Scanning Electron Microscopy In The Study of Sediment: A Symposium, Norwich, Geoabstracts: 41-59.*

Derbyshire, E., and Mahaney, W.C., 1985, Anciennes glaciations de Mont Kenya: problèmes et methodologiques, sur les Glaciations de Montagnes: Europe, Ameriques, Afriqué, Asie, Conference de la Commission de Géographie des Montagues du Comité National Françaises de Géographie, June, 1985.

____, 1989, A preliminary scanning electron microscope study of five diamictons from Mount Kenya. In W.C. Mahaney, ed., *Quaternary Environmental Research of East African Mountains,* Balkema, Rotterdam, 155-164.

DeSchlippe, P., 1956, *Shifting Cultivation In Africa*, Routledge and Kegan Paul, London, 304 p.

Deverall, B.J., 1968, Psychrophiles, in Ainsworth, G.C. and Sussman, A.S. *The Fungi*, Academic Press, N.Y. Vol. 3, p. 129-135.

Dickson, B.A., and Crocker, R.L., 1954. A chronosequence of soil and vegetation near Mount Shasta, California III, *Jour. Soil Sci.*, 5, 73-91.

Dodge, C.W., 1953, Some lichens of tropical Africa, *Annals Missouri Botanical Garden*, 40, 271-401.

____, 1957, Some lichens of tropical Africa, III, *Annals Missouri Botanical Garden*, 46, 39-193.

Dormarr, J.F. and Lutwick, L.E., 1983, Extractable Fe and Al as an indicator for buried soil horizons, *Catena*, 10, 67-73.

Downie, C. and Wilkinson, P., 1972, *The Geology of Kilimanjaro*, Geology Dept., U. of Sheffield, 253 p.

Dreimanis, A., Hütt, G., Raukas, A., Whippey, P.W., 1985, Dating methods of Pleistocene deposits and their problems: I, Thermoluminescence dating, ed., N.W. Rutter, *Geoscience Canada*, March, 1985.

Dutton, E.A.T., 1929, *Kenya Mountain*, London.

Embleton, G. and King, C.A.M., 1971, *Glacial and Periglacial Gemorphology*, Macmillan. Toronto, 608p.

Eslinger, E.V., Mayer, L.M., Durst, T.L., Hower J., and Savin, S.M., 1973, An X-ray technique for distinguishing between detrital and secondary quartz in the fine grained fraction of sedimentary rocks, *Journal of Sedimentary Petrology*, 43, 540-543.

Evenson, E.B., Dreimanis, A., and Newsome, H., 1977, The subaquatic flow tills: a new interpretation for the genesis of some laminated till deposits, *Boreas*, 6, 115-134.

Everden, J.F. and Curtis G.H., 1965. The potassium-argon dating of late Cenozoic rocks in East Africa and Italy, *Current Anthropology*, 6(4), 343-385.

Fisher, R.A., 1963, Dispersion on a sphere, Proc. Royal Soc., London, A217, 295-305.

Fiskell, J.G.A. and Calvert, D.V., 1975, Effects of deep tillage, lime incorporation and drainage on chemical properties of Spodosol profiles, *Soil Science*, 120(2): 132-139.

Folk, R.L., 1968, *Petrology of Sedimentary Rocks,* Hemphill, Drawer M., University Station, Austin, Texas.

Furon, R., 1963, *The Geology of Africa*, Oliver and Boyd, London, 377p.

Gonzalez, E., van der Hammen, Th., Flint, R.F., 1965, Late Quaternary glacial and vegetational sequence in Valle de Langunillas, Sierra Nevada del Cocuy, Colombia, *Leidse Geologische Mededlingen*, 32: 157-182.

Gradwell, M.W., 1957, Patterned ground data at a high country station, *N.Z.J. Sci. Tech.*, 38: 793-806.

Gregory, J.W., 1894a, Contributions to the physical geography of British East Africa. *Geogr. Jour.* 4: 289-315, 408-25, 505-24.

____, 1894b, The glacial geology of Mount Kenya, *Quart. J. Geol. Soc.*, 50, 515-30.

____, 1896, *The Great Rift Valley*, London.

____, 1900, The Geology of Mount Kenya *Quart. Journ. Geol. Soc.*, 56, 205-22.

____, 1921, *The Rift Valleys and Geology of East Africa*, London.

Haldorsen, S., 1983, The enrichment of quartz in tills, In Evenson, E.B., Schlüchter C., and Rabassa, J., eds., *Tills and Related Deposits*, Rotterdam, Balkema: 141-150.

Hamilton, A.C., 1982, *Environmental History of East Africa*, N.Y., Academic Press, 328 p.

Hamilton, A.C. and Perrott, A., 1978, Date of deglacierisation of Mount Elgon, *Nature*, 273:49.

Hancock, R.G.V., 1984, On the source of clay used for Cologne Roman pottery,*Archaeometry*, 26, 210-217.

Hancock, R.G.V., Mahaney, W.C. and Stalker, A. MacS., 1988, Neutron activation analysis of tills in the North Cliff Section, Wellsch Valley, Saskatchewan, *Sedimentary Geology*, 55, 185-196.

Hansen, B.C.S., Wright, H.E. Jr., and Bradbury, J.P., 1984, Pollen studies in the Junin area, central Peruvian Andes, Geol. Soc. Amer. Bull., 95:1454-1465.

Hare, P.E., 1974, Amino acid dating of bone – the influence of water, *Carnegie Institution of Washington Yearbook,* Vol. 73, 576-81.

Harmsen, R, 1968, Taxonomic and ecological studies on the Afroalpine flora of Mount Kenya, *Hochgebirgsforschung*, 1:171-194.

____, 1983, Abundance distribution and the evolution of community structure, *Evolutionary Theory*, v. 6, p. 283-292.

____, 1989, Recent evolution and dispersal of insects on the East African mountains, In Mahaney, W.C., ed., *Quaternary and Environmental Research on East African Mountains*, Balkema, Rotterdam, 245-256.

Harmsen, R., Spence, J.R. and Mahaney, W.C., 1990, Glacial-interglacial cycles and development of the Afroalpine ecosystem on East African mountains. I and II., *Journal of African Earth Sciences*, in press.

Hastenrath, 1973, On the Pleistocene glaciation of the Cordillera de Talamanca, Costa Rica, *Zeitschrift für Gletcherkunde und Glazialgeologie*, 9(2): 105-121.

____, 1977, Pleistocene mountain glaciation in Ethiopia, *Jour. of Glaciology, 18(2):309-313.*

____, 1981, *The Glaciation of the Ecuadorian Andes*, Balkema, Rotterdam, 160p.

____, 1984, *The Glaciers of Equatorial East Africa*, Dordrecht, Reidel, 353p.

Hedberg, O., 1951, Vegetation belts of the East African Mountains, *Svensk Bot. Tidsk.*, 45, 140-202.

____, 1957, Afroalpine vascular plants: A taxonomic revision. *Symbot, Bot Uppsala*, 15, 411p.

____, 1964, *Features of Afroalpine Plant Ecology*, Almquist and Wiksells, Uppsala, 144p.

____, 1968, Taxonomic and ecological studies on the Afroalpine flora of Mount Kenya, *Hochgebirgsforschung*, 1:171-194.

____,1969, Evolution in a tropical high mountain. *Biol Jour. Lim. Soc.*, 1:135-148

____,1970, Evolution of the Afroalpine flora, *Biotropica*, 2(1):16-23.

Heiken, G., 1972, Morphology and petrography of volcanic ashes, *Geological Society of American Bull.*, 83, 1971-1988.

Heine, K., 1973, Zür glazialmorphologie und präkeramischen archäologil des Mexikanischen hochlandes während des Spätglazials (Wisconsin) und Holozäns, *Erdkunde*, 27: 161-180.

____, 1978, Neue beobachtungen zur chronostratigraphie der mittelwisconsin zeitlichen vergletscherungen und böden mexikanischer vulkane, *Eiszeitalter und Geigenwart*, 28: 139-147.

____, 1984, Comment on "Pleistocene glaciation of volcano Ajusco, Central Mexico and comparison with the standard Mexican glacial sequence" by S.E. White and S. Valastro Jr., *Quaternary Research*, 22:242-246.

____, Late Quaternary glacial chronology of the Mexican volcanoes, *Der Geowissenschaften*, 7: 197-205.

Heinzelin, J. deB., and Mollaret, H., 1956, Biotopes de haute altitude Ruwenzori, in *Expl. Parc. Natl. Albert*, ser. 2, fasc. 3.

Hildebrant, 1878, *On his travels in East Africa Proceedings*, Roy. Geogr. Soc., 22, 446-53.

Hodgson, J.M., 1976, Editor, *Soil Survey Field Handbook, Soil Survey Tech. Monog.* No. 5, Rothamsted Experimental Station, Harpenden, Herts, U. K., 99p.

Höhnel, L. von, 1894, Discovery of Lakes Rudolph and Stefanie, 1887-1888. 2 Vols., London.

Holmgren, G.S., 1967, A rapid citrate dithionite extractable iron procedure, *Soil Sci. Soc. Amer. Proc.*, 31, 210-11.

Hooghiemstra, H., 1984, Vegetational and climatic history of the high plains of Bogotá, Colombia: a continuous record of the last 3.5 million years, Gantner Verlag, 368 p.

Hope, G.S., 1986, Development of present day biotic distributions in the New Guinea mountains. in: *Flora and Fauna of Alpine Australasia. Ages and Origins*, B.A. Barlow, ed., CSIRO, Melborne, pp. 129-145

Hope, G.S., Peterson, J.A., Radok, U., Allison, I., eds., 1976, *The Equatorial Glaciers of New Guinea: Results of the Australian Universities Expeditions to Irian Jaya*, 1971-73, Rotterdam, Balkema, 244 p.

Hurni, H., 1989, Late Quaternary of Simen and other mountains in Ethopia, In Mahaney, W.C., ed., *Quaternary and Environmental Research on East African Mountains*, Rotterdam, Balkema, 105-120.

Hutchins, D.E., 1909, *Report on the Forests of British East Africa*, London.

Imbrie, J. *et al*, 1984, The orbital theory of Pleistocene climate: support from a revised chronology of the marine ^{18}O record, In Berger, A., *et al*, *Milankovitch and Climate*, Part l, NATO ASI, Series C, Mathematical and Physical Sciences, Dordrecht, Reidel, 126: 296-305.

Iniss, W.E. and Ingraham, J.L., 1978, Microbial life at low temperatures: mechanisms and molecular aspects, *in* Kushner, D.J., ed., *Microbial Life in Extreme Environments*. Academic Press, London, 73-104.

Innes, J.L., 1985a, Moisture availability and lichen growth: the effects of snow cover and streams on lichenometric measurements, *Arctic and Alpine Research*, 17, 417-24.

____, 1985b, An examination of some factors affecting the largest lichens on a substrate, *Arctic and Alpine Research*, 17: 98-106.

____, 1985c, A standard *Rhizocarpon* nomenclature for lichenometry, *Boreas*, 14: 83-85.

____, 1986, Dating exposed rock surfaces in the Arctic by lichenometry: the problem of thallus circularity and its effect on measurement errors, *Arctic*, 39: 253-259.

Irving, E., 1964, *Paleomagnetism and its application to geological and geophysical problems*, N.Y., Wiley.

Isaac, G.L., 1967a, The Stratigraphy of the Peninj Beds – an Early Middle Pleistocene formation West of Lake Natron, Tanzania. In: *Background to African Evolution*, Clark, J. D. and Bishop, W. W. (eds.), Chicago Univ. Press.

Jackson, M.L., 1956, *Soil Chemical Analysis - Advanced Course*, published by the author, Dept. Soil Science, Univ. of Wisconsin, Madison, Wisconsin.

Jansen, J.H.F., Kuijpers, A., and Troelstra, S.R., 1986, A mid-Brunhes climatic event: long term changes in global atmosphere and ocean circulation, *Science*, 232:619-622.

Jochimsen, M., 1973, Does the size of lichen thalli really constitute a valid measure for dating glacial deposits, *Arctic and Alpine Research*, 5, 414-7.

Johansson, L. and Holmgren, K., 1985, Dating of a moraine on Mount Kenya, *Geografiska Annaler*, 67A, 123-8.

Johnson, R.G., 1982, Brunhes-Matuyama magnetic reversal dated at 790,000 yr BP by marine-astronomical correlations, *Quaternary Research*, 17: 135-147.

King, K. Jr., 1980, Applications of amino acid biogeochemistry for marine sediments, In: Hare, P.E., Hoening, I.C. and King, Jr. K., (eds.), *Biogeochemistry of Amino Acids*, N.Y., John Wiley and Sons, 377-91.

Kiver, E.P., 1974, Holocene glaciation in the Wallowa Mountains, Oregon, in Mahaney, W.C., ed., *Quaternary Environments: Proceedings of a Symposium, Geog. Monogr. No. 5,* York Univ., 69-95.

Kondrat'yev, A.A. and Gritsenko, V.V., 1971, Changes produced by deep tillage in the agronomical properties of sod-podzolic soils, *Soviet Soil Science*, 3:286-294.

Koopmans, B.N. and Stauffer, P.H., 1968, Glacial phenomena on Mount Kinabalu, Sabah, Malaysian Geol. Survey Bull., 8, 25-35.

Krapf, J.L., 1860, *Travels, Researches and Missionary Labours in East Africa*, London.

Krinsley, D.H. and Donahue, J., 1968, Environmental interpretation of sand grain surface textures by electron microscopy, Geol. Soc. Amer. Bull., 79: 743-748.

Krog, H. and Swinscow, T.D.V., 1986, The lichen genera *Lecidea* and *Umbilicaria* in East Africa, *Nordic Journal Botany*, 6, 75-88.

Kvenvolden, K.A., Peterson, E., and Brown, P.S., 1970, Racemization of amino acids in sediments from Saanich Inlet, *Science*, 169, 1079-1082.

Lawson, D.E., 1981, Sedimentological characteristics and classification of depositional processes and deposits in the glacial environment. U.S. Army Corps of Engineers, CRREL Report 81-27, Hanover, N.H.

Levin, F.I., 1969, *Improvement of sod-podzolic soils by cultivation*, Jerusalem, Israel Prog. Scientific Translation, 109 p.

Li, J., Derbyshire, E., and Xu, S., 1984, Glacial and paraglacial sediments in the Hunza Valley, northwest Karakoram, Pakistan, a preliminary analysis, In Miller, K.J., ed., Proc. Internatl. Karakoram Proj., 1980, Cambridge Univ. Press, 2: 496-535.

Liddicoat, J.C. and Coe, R.S., 1979, Mono Lake geomagnetic excursion, *Jour. Geophysical Research*, 84, 261-71.

Limmer, A.W. and Wilson, A.T., 1980, Amino Acids in buried paleosols, *Journal of Soil Science*, 31 (1), 147-53.

Livingstone, D.A., 1962, Age of deglaciation in the Ruwenzori Range, Uganda, *Nature*, No. 4831, 859-60.

Lockwood, S.L., 1981, *The Fungal Community*, Marcel Dekker, New York, 319-349.

Löffler, E., 1972, Pleistocene glaciation in Papua and New Guinea, *Zeitschrift für Geomorphologie*, 13:32-58.

____, 1982, Pleistocene and present-day glaciations *Monographiae Biologicae*, 42:39-54.

Low, A.J., 1972, The effect of cultivation on the structure and other physical characteristics of grassland and arable soils (1945-1970), *Jour. of Soil Science*, 23(4):363-380.

____, 1973, Soil structure and crop yield, *Jour. of Soil Science*, 24(2):249-259.

Lutwick, L.E. and Dormarr, J.F., 1973, Fe status of Brunisolic and related soil profiles, *Canadian Journal of Soil Science*, 53, 85-97.

Mabberly, D.J., 1973, Evolution in the giant groundsels, *Kew Bulletin*, 28:61-96.

____, 1975a The pachycaul lobelias of Africa and St. Helena. *Kew Bulletin*, 29, 535-584.

____, 1975b, The giant *Lobelias*: pachycauly, biogeography, ornithophily and continental drift, *New Phytol*, *74: 365-374.*

____, 1976, The origin of the Afroalpine pachycaul flora and its implications, *Gardens Bull. (Singapore), 29: 41-55.*

Mackay, J.R. and Mathews, W.H., 1974, Needle ice striped ground. *Arctic and Alpine Research*, 6:79-84.

Mackinder, H.J., 1900, A Journey to the Summit of Mount Kenya, British East Africa, *Geogr. Jour.* 15(5): 453-86.

Mahaney, W.C., 1972, Late Quaternary history of the Mount Kenya Afroalpine area, East Africa, in E.M. van Zinderen Bakker, ed., *Palaeoecology of Africa*, 6, 39-42.

____, 1973, Neoglacial chronology in the Fourth of July Cirque, central Colorado Front Range, *Bull. Geol. Soc. Amer.*, 83, No. 1, 61-70.

____, 1974, Soil stratigraphy and genesis in deposits of Neoglacial age in the Arapaho and Henderson cirques, central Colorado Front Range, in Mahaney, W.C., (ed.), *Quaternary Environments; proc. of a Symposium, Geographical Monographs*, No. 5, 197-240.

____, 1975, Soils of post-Audubon age, Teton Glacier area, Wyoming: *Arctic and Alpine Research*, 7, (2), p. 141-53.

____, 1978, Late Quaternary stratigraphy and soils in the Wind River Mountains, western Wyoming, in *Quaternary Soils*, Mahaney, W.C., ed., Geoabstracts, Norwich, U. K., 223-64.

____, 1979a, Reconnaissance Quaternary Stratigraphy of Mount Kenya, *Palaeoecology of Africa*, 11, 63-70.

____, 1979b, Effects of agriculture on physical and chemical characteristics of forest soils in Naro Moru area in Kenya, *East African Agriculture and Forestry Journal*, 45, No. 2, 67-75.

____, 1979c, Soil erosion along the Naro Moru track – A western approach to Mount Kenya, *East African Agricultural and Forestry Journal*, 45, No. 2, 58-66.

____, 1980, Late Quaternary Rock Glaciers, Mount Kenya, East Africa, *Journal of Glaciology*, 25: 492-7.

____, 1981, Paleoclimate reconstructed from paleosols: evidence from East Africa and The Rocky Mountains, in *Quaternary Paleoclimate*, Mahaney, W.C., ed., Geoabstracts Ltd., Norwich, U.K., 227-247.

____, 1982a, Correlation of Quaternary glacial and periglacial deposits on Mount Kenya, East Africa, with the Rocky Mountains of western U. S., XI INQUA Congress, *Abstracts*, 1, U.S.S.R., August 1-9, 1982, 208.

____, 1982b, Chronology of glacial deposits on Mount Kenya, East Africa, van Zinderen Bakker, E. M., ed., *Palaeoecology of Africa*, 14, 25-43.

____, 1983a, Chronology of Quaternary glacial deposits on Mount Kenya, East Africa, *Geol. Assoc. Canada, Abstracts,* Volume, 256.

____, 1983b, Quaternary paleoclimate reconstructed from glacial deposits and paleosols, Mount Kenya, East Africa, in *International Symposium on Late Cainozoic Paleoclimates of the Southern Hemisphere in Swaziland, South Africa.*

____, 1984a, Soils of the Mount Kenya area, their formation, ecological and agricultural significance (with soil map, scale 1:125000): Discussion, *Mountain Research and Development*, 4, No. 3, 284-5.

____, 1984b, Late glacial and postglacial chronology of Mount Kenya, East Africa, in E.M. van Zinderen Bakker and J.A. Coetzee, eds., *Palaeoecology of Africa*, 16, 327-41.

____, 1984c, Loess in Pleistocene soils on Mount Kenya, East Africa, *Loess Letter Supplement No. 6*, Univ. of Waterloo, Dept. Earth Science, 20p.

____, 1984d, Radiometric dating of Quaternary glacial and nonglacial deposits, Mount Kenya, East Africa, Report on Grant 2306.81 to National Geographic Society, Research Report Series.

____,1984e, Indian Basin advance: new name for early Neoglacial till in western and north-central Wyoming, *Northwest Science,* 58:94-102.

____, 1985, Late glacial and Holocene paleoclimate of Mount Kenya, East Africa, *Zeitschrift für Gletscherkunde und Glazialgeologie*, Band 21, 203-11.

____, 1986, Environmental impact in the Afroalpine and subalpine belts of Mount Kenya, East Africa, *Mountain Research and Development*, 6 (3) 247-56.

____, 1987a, Reinterpretation of dated moraines at 4000m in the Mount Kenya Afroalpine area, *Palaeogeography Palaeoclimatology, Palaeoecology*, 60, 47-57.

____, 1987b, Dating of a moraine on Mount Kenya, Geografiska Annaler, 69A(2): 359-363.

____,1987c Notes on the behavior of the African buffalo (*Syncerus caffer caffer*) on Mount Kenya, *African Journal of Ecology*, 25: 199-202.

____, 1987d, Tentative growth curve for *Rhizocarpon geographicum* s.l. in Stroud Basin, Wind River Range, western Wyoming, *Northwest Science*, 61(1): 13-19.

____, 1988a, Late Pleistocene deglaciation of Mount Kenya, *Zeitschrift für Geomorphologie*, 32(2): 227-230.

____, 1988b, Amino acids in buried soils on Mount Kenya, *Journal of African Earth Sciences*, in press.

____, 1988c, Discussion of late Pleistocene age of the Type Temple Lake Moraine, Wind River Range, Wyoming, U.S.A., *Géographie physique et Quaternaire*, 42, 337-339.

____, 1989a, Quaternary glacial geology of Mount Kenya, in Mahaney, W.C., ed., *Quaternary and Environmental Research on East African Mountains*, Balkema, Rotterdam, 121-140.

____, 1989b, *Quaternary and Environmental Research on East African Mountains*, Balkema, Rotterdam, 483 p.

Mahaney, W.C. and Boyer, M.G., 1986a, Microflora distributions in paleosols: a method for calculating the validity of radiocarbon-dated surfaces *Soil Science*, 142(2), 100-107.

Mahaney, W.C. and Boyer, M.G., 1986b, Microflora distributions in Quaternary paleosols on Mount Kenya, East Africa, *Catena,* 13, 155-167.

Mahaney, W.C. and Boyer, M.G., 1986c, Needle ice on Mount Kenya and its effect on soil microfungi, *Acta Geocriogenica*, No. 4, 79-92.

Mahaney, W.C. and Boyer, M.G., 1987, Late glacial soil catena in upper Teleki Valley, Mount Kenya Afroalpine area, *Journal of African Earth Sciences*, 6(5): 731-740.

Mahaney, W.C. and Boyer, M.G., 1989, Notes on the morphology and genesis of mud polygons on Mount Kenya, East Africa, *Géographie physique et Quaternaire*, 42(1): 89-96.

Mahaney, W.C. and Charsley, T.J., 1986, Evidence for prehistoric man on Mount Kenya, *Nyame Akuna*, No. 27, 4-8.

Mahaney, W.C. and Ermuth, H.F., 1974, *The Effects of Agriculture and Urbanization on the Natural Environment*, Geographical Monograph No. 7, York U., Atkinson College, 152 p.

Mahaney, W.C. and Fahey, B.D., 1976, Quaternary soil stratigraphy of the Front Range Colorado, in Mahaney, W.C., ed., *Quaternary Stratigraphy of North America*: Stroudsburg, Pa., Dowden, Hutchinson and Ross, 319-52.

____, 1980, Morphology, composition and age of a buried paleosol on Niwot Ridge, Front Range, Colorado, U.S.A., *Geoderma*, 23, 209-18.

Mahaney, W.C. and Halvorson, D.L., 1986, Rates of mineral weathering in Rocky Mountain glacial sequences: evidence from western Wyoming, in *Rates of Chemical Weathering of Rocks and Minerals*, S. M. Colman and D. Dethier, eds., Academic Press, N.Y., 47-67.

Mahaney, W.C. and Rutter, N.W., 1989, Amino acid distributions in late glacial and Neoglacial soils in the Afroalpine area of Mount Kenya, East Africa, *Catena,* 16, 205-214.

Mahaney, W.C. and Sanmugadas, K., 1983, Early Holocene soil catena in Titcomb Basin, Wind River Mountains, western Wyoming, *Zeitschrift für Geomorphologie*, 27, No. 3, 265-81.

Mahaney, W.C. and Sanmugadas, K., 1985, Notes on the use of extractable iron and clay minerals for determination of soil age, *Geografisk Tidsskrift*, 85, 14-19.

Mahaney, W.C. and Spence, J.R., 1984, Glacial and periglacial sequence and floristics in Jaw Cirque, central Teton Range, western Wyoming, *Amer. Jour. Sci.*, 284, 56-81.

Mahaney, W.C. and Spence, J.R.,, 1985, Size variation of *Rhizocarpon* on moraine slopes in southern Norway: A Discussion, *Arctic and Alpine Research*, 17 (2), 211-212.

Mahaney, W.C. and Spence, J.R., 1986 Origin, evolution and paleoclimatic-ecological significance of sand dunes in the Mutonga Drainage, Mount Kenya, *Journal African Earth Sciences* 5(5):491-499.

Mahaney, W.C. and Spence, J.R., 1989, Lichenometry of Neoglacial moraines in Lewis and Tyndall cirques on Mount Kenya, East Africa, *Zeitschrift für Gletcherkunde und Glazialgeologie,* in press.

Mahaney, W.C., Fahey, B.D., and Lloyd, D.T., 1981, Late Quaternary glacial deposits, soils, and chronology, Hell Roaring Valley, Cascade Range, Washington, *Arctic and Alpine Research*, 13, No. 3, 339-56.,

Mahaney, W.C., Halvorson, D., Piegat, J., and Sanmugadas, K., 1984a, Evaluation of dating methods used to differentiate Quaternary deposits in the Wind River and Teton Ranges, Wyoming. In Mahaney, W.C., ed., *Quaternary Dating Methods*, Elsevier, Amsterdam, 355-374.

Mahaney, W.C., Boyer, M.G., Halvorson, D.L., and Sanmugadas, K., 1984b, Glacial chronology of the Rocky Mountains: some problems from observations in western Wyoming, in Mahaney, W.C., ed., *Correlation of Quaternary Chronologies*, Geo- Books, Norwich, U. K., 225-41.

Mahaney, W.C., Barendredt, R.W., Churcher, C.S. and Spence, J., 1986, Gorges Valley rock shelters, Mount Kenya Afroalpine area, East Africa, *Journal African Earth Sciences*, 5 (4), 321-327.

Mahaney, W.C., Boyer, M.G. and Rutter, N.W., 1986, Evaluation of amino acid composition as a geochronometer in buried soils on Mount Kenya, East Africa, *Géographie physique et Quaternaire*, 40 (2) 71-83.

Mahaney, W.C., Barendregt, R.W. and Vortisch, W.B., 1989, Quaternary glaciations and paleoclimate of Mount Kenya, East Africa, in Oerlemans, J., ed., *Glacier Fluctuations and Climatic Change,* Reidel, Dordrecht, Holland, 13-36.

Mahaney, W.C., Vortisch, W., and Spence, J.R., 1988, Pollen study and scanning electron microscopy of aeolian grains in a compound paleosol in the Mutonga Drainage, Mount Kenya, *Journal of African Earth Sciences*, 7(7/8): 895-902.

Mahaney, W. C., Vortisch, W., Barendregt, R. W., Fecher, K., 1989, Differentiation of tills and lahars using examples from southwestern Alberta and Mount Kenya, East Africa, In *Palaeoecology of Africa*, in press.

Mahaney, W.C., Harmsen, R., and Spence, J.R., 1990, Glacial-interglacial cycles and development of the Afroalpine ecosystem on East African mountains: I and II glacial and postglacial geological record and paleoclimate, *Journal of African East Sciences*, in press.

Mahaney, W.C., Hancock, R.G.V. and Barendregt, R.W., 1989, Stratigraphy, geochemistry and paleomagnetism of late Quaternary bedrock and paleosols, Karisoke Area, Virunga Mountains, Northwestern Rwanda, in preparation.

Matthews, B, 1954, A discussion on the physiology of man at high altitude, Proc. Royal Soc. Ser. B, 143:1-4.

Matthews, B, (1954), A discussion of the physiology of man at high altitude, Proc. Roy. Soc., Ser B, 143, 1-4.

Mazur, P., 1968, Survival of fungi after freezing and desiccation, *In* Ainsworth, G.C. and Sussman, A.S. *The Fungi*, Academic Press, N.Y.:325-394.

McGown, A. and Derbyshire, E., 1974, Genetic influences of the properties of tills, *Quart. Jour. Engineering Geology*, 10: 389-410.

McKeague, A., and Day, J., 1966, Dithionite and oxalate extractable Fe and Al as aids in differentiating various classes of soils. *Canadian Jour. Soil Sci.*, 46, 13-22.

Mehra, O.P. and Jackson, M.L., 1960, Iron oxide removal from soils and clays by a dithionite-citrate system buffered with sodium bicarbonate, in Swineford, A., ed., *Natl. Conf. on Clays and Clay Minerals, 1958*, London Pergamon, 317-27.

Mercer, J.H. and Palacios, O., 1977, Radiocarbon dating of the last glaciation in Peru, *Geology*, 5:600-604.

Messerli, B, Hurni, H., Kienholz, H., and Winiger, M., 1977, Bale Mountains: largest Pleistocene mountain glacier system of Ethiopia, INQUA Abstracts, Birmingham.

Miller, C.D., 1973, Chronology of Neoglacial deposits in the Northern Sawatch Range, Colorado, *Arctic and Alpine Research,* 5 (4), 385-400.

Miller, G.H. and Hare, P.E., 1975, Use of amino acid reactions in some arctic marine fossils as stratigraphic and geochronologic indicators, *Carnegie Institution of Washington Yearbook*, 74, 612-7.

Miller, J.M., 1956, Geology of the Kitale-Cherangani Hills area. Degree sheet 26, S. W. Quarter Govt. Printer, Nairobi, 34p.

Mitterer, R.M., 1974, Pleistocene stratigraphy in Southern Florida based on amino acid diagenesis in fossil Mercenaria, *Geology*, 2, 425-8.

____, 1975, Ages and diagenetic temperatures in Pleistocene deposits of Florida, based on isoleucine epimerization, In: *Mercenaria: Earth and Planetary Science Letters*, 28, 274-82.

Mloszewski, M.J., 1983, *The Behavior and Ecology of the African Buffalo*, Cambridge, Cambridge University Press, 256 p.

Mloszewski, M.J. and Mahaney, W.C., 1989, A check-list of wild African *Bovidae* related to montane habitats, with an expanded note on geophagic behavior of buffalo on Mount Kenya, In Mahaney, W.C., ed., *Quaternary and Environmental Research on East African Mountains*, Balkema, Rotterdam, 309-326.

Morrison, R.B., 1978, Quaternary soil stratigraphy—concepts methods and problems In: *Mahaney, W. C. ed., Quaternary Soils,* Geo Abstracts, 77-108.

Nelson, R.L., 1954, Glacial geology of the Frying Pan River drainage, Colorado, *Jour. Geol.*, 62, 325-43.

Nilsson, E., 1931, Quaternary glaciation and pluvial lakes in British East Africa and Abyssinia, *Geografiska Annaler*, H.13, 241-8.

____, 1935, Traces of ancient changes of climate in East Africa, *Geog. Annaler*, H.1-2, 1-21.

____, 1940, Ancient changes of climate in British East Africa and Abyssinia, *Geografiska Annaler*, H. 22, 1-79.

Nye, P.H., 1961, Organic matter and nutrient cycles under moist tropical forest, *Plant and Soil*, 13:333-346.

Nye, P.H. and Greenland, D.J., 1960, *The Soil Under Shifting Cultivation*, Commonwealth Bur. Soil Sci. Tech. Commun. No. 51.

Olsen, S.R. and Dean, L.A., 1965, Phosphorus, in Black, C.A. ed., *Methods of Soil Analysis*, Amer. Soc. Agron., Madison, Wisc., 35-49.

Orde-Brown, G. St.J., 1918, The South East Face of Mount Kenya., *Geog. Jour.*, 51: 389-92.

Osmaston, H., 1965, *The past and present climate and vegetation of Ruwenzori and its Neighbourhood*, D. Phil. Thesis, Oxford Univ., 300p.

____, 1967, The sequence of glaciations in the Ruwenzori and their correlation with glaciations of other mountains in East Africa and Ethiopia. In: *Palaeoecology of Africa II*, E.M. v. Z. Bakker, (ed.), Cape Town, Balkema, 26-8.

____,1989a, Glaciers, glaciations, and equilibrium line altitudes on Kilimanjaro, In Mahaney, W.C., ed., *Quaternary and Environmental Research on East African Mountains*, Rotterdam, Balkema, 5-30.

____, 1989b, Glaciers, glaciations and equilibrium line altitudes on the Ruwenzori, In Mahaney, W.C., ed., *Quaternary and Environmental Research on East African Mountains*, Rotterdam, Balkema, 31-104.

Outcalt, S.I., 1971a, An algorithm for needle ice growth, Water Res., 7:394-400.

____, 1971b, The climatology of a needle ice event: an experiment in simulation climatology. *Arch. Meteorol. Geophys.* Block Limatol. Ser. B. 19:325-338.

Oyama, M., and Takehara, H., 1970, *Standard Soil Color Charts*, Japan Research Council for Agriculture, Forestry and Fisheries.

Pawluk, S., 1978, The pedogenic profile in the stratigraphic section, in W.C. Mahaney (ed.), *Quaternary Soils*, Geo Abstracts, Norwich, U. K., 61-75.

Pedgley, D.E., 1966, Rainfall of Mount Kenya, *Weather*, 21, 176-88.

Peech, M.L., Alexander, T., Dean, L.A. and Reed, J.F., 1947, *Methods of soil analyses for soil fertility investigations*: Washington, D.C., U.S.D.A., Circ., 757, 25.

Pérez, F., 1987a, Downslope stone transport by needle ice in an Andean area (Venezuela), *Rev. de Geomorph. Dynamique*, 36 (2): 33-51.

____, 1987b, Needle ice activity and the distribution of stem rosette species in a Venezuelan Páramo, *Arctic and Alpine Research*, 19, 135-153.

____, 1989, A talus rock glacier in the Venezuelan Andes, in preparation.

Perrott, R.A., 1982, A high altitude pollen diagram from Mount Kenya, its implications for the history of glaciation in J. A., Coetzee and E. M. van Zinderen Bakker, eds., *Palaeoecology of Africa*, 14, Balkema, Rotterdam, 77-83.

Peterson, J.A. and Hope, G.S., 1972, Lower limit and maximum age for the last major advance of the Carstensz glaciers, W. Irian, *Nature*, 240:36-37.

Porter, Elliott, 1974, *The African Experience*, Avon Books, N.Y., 247p.

Porter, S.C., 1981, Lichenometric studies in the Cascade Range of Washington: establishment of *Rhizocarpon geographicum* growth curves at Mount Ranier, *Arctic and Alpine Research*, 13: 11-23.

Porter, S.C., Stuiver, M., and Yang, I.C., 1977, Chronology of Hawaiian glaciations, *Science*, 195: 61-63.

Potter, E.C. 1976, Pleistocene glaciation in Ethopia, *Journal of Glaciology*, 17:148-150.

Prance, G.T., 1982, A review of the phytogeographic evidences for Pleistocene climate changes in the Neotropics, *Ann. Missouri Bot. Gard., 69: 594-624.*

Radziunas, L., 1984, Amino acids in surface and buried soils on Mount Kenya: their value as relative and absolute age indicators, B.A., Hons. Thesis, Dept. of Geography, York Univ., Atkinson College, 29p.

Richmond, G.M., 1962, Three Pre-Bull Lake tills in Wind River Mountains Wyoming, U.S.G.S. Prof. Paper 450 D, p. 132-136.

____, 1965, *Glaciation of the Rocky Mountains, The Quaternary of the United States* Ed. Wright, H.E., and Frey, D.G., Princeton Univ. Press.

____, 1974, Geologic map of the Fremont Lake South quadrangle, Sublette County, Wyoming, U. S., Geol. Survey, Geol. Quad. Map GQ-1138.

____, 1976, Pleistocene stratigraphy and chronology in the mountains of western Wyoming, In Mahaney, W. C., editor, *Quaternary Stratigraphy of North America*, Stroudsburg, Dowden Hutchinson and Ross, 353-380.

____, 1986, Stratigraphy and correlation of glacial deposits of the Rocky Mountains, the Colorado Plateau and the Ranges of the Great Basin, In Sibrava, V., Bowen, D.Q., and Richmond, G.M., editors, *Quaternary Glaciations in the Northern Hemisphere* (Report of the I.G.C.P. Proj. 24) Pergamon Press, Oxford, U.K., 99-128.

Ross, G.J., Wang, C., and Protz, R., 1984, Soil mineralogical evidence as an indicator of post and pre-Wisconsinan weathering in Canada, in Mahaney, W.C., ed., *Correlation of Quaternary Chronologies*, Norwich, Geo-Books, 191-202.

Ross, W., MacGregor, 1911, Two finds on Mount Kenya, *J.E.A.*, Nat. Hist. Soc., 11(3), 60.

Ruhe, R.V., 1956, Geomorphic surfaces and the nature of soils, *Soil Sci.*, 82, 441-55.

Rull, V., 1987, Evidence of a cold climatic oscillation contemporary with the Little Ice Age in the Venezuelan Andes, *Bull. Asoc. Venezolano de Arguelogia*, 4:13-27.

Rull, V., Salgado-Labouriau, M.L., Schubert, C., and Valastro, S., 1987, Late Holocene temperature depression in the Venezuelan Andes: palynological evidence, *Palaeo-3*, 60: 109-121.

Runemark, H., 1956, Studies in *Rhizocarpon*, I., Taxonomy of the yellow species in Europe, *Opera Botanica*, 2 (1) 1-152.

Rutter, N.W., Crawford, R.J. and Hamilton, R., 1980, Correlation and relative age dating of Quaternary strata in the continuous permafrost zone of northern Yukon with D/L ratios of wood, freshwater molluscs and bone, In: Hare, P.E., Hoering, T.C. and King Jr. K. (eds.), *Biogeochemistry of Amino Acids*, N.Y., John Wiley and Sons, 463-75.

Rutter, N.W. and Crawford, R.J., 1984, utilizing wood in amino acid dating, In: Mahaney, W.C. (ed.), *Quaternary Dating Methods*, N.Y., Elsevier Science Publishing Co. Inc., 195-209.

Salgado-Labouriau, M.L., Schubert, C., and Valastro, S., 1987, Paleoecologic analysis of a late Quaternary terrace from Mucubají, Venezuelan Andes, *Jour. Biogeography*, 4: 313-325.

Salt, G., 1954, A Contribution to the Ecology of Upper Kilimanjaro, *J. Ecol.*; 42, 2, 375-423.

Schollenberger, C.J. and R.H. Simon, 1945, Determination of exchange capacity and exchangeable bases in soils – ammonium acetate method: *Soil Sci.* 59, 13-24.

Schubert, C., 1972, Late glacial chronology in the Northeastern Venezuelan Andes, *24th Int. Geological Congress*, Section 12: 103-109.

____, 1974, Late Pleistocene Mérida Glaciation, Venezuelan Andes, *Boreas*, 3: 147-152.

Schubert, C. and Rinaldi, M., 1987, Nuevos datos sobre la chronologia del Estadio Tardio de la Glaciacion Mérida, Andes Venezolanos, *Acta Cientifica Venezolana*, 38: 135-136.

Shackleton, R.M., 1945, *Geology of the Nyeri Area*, Dept. 12, Geol. Survey, Kenya.

Sharp, R.P., 1969, Semiquantitative differentiation of glacial moraines near Convict Lake, Sierra Nevada, Calif., *Jour. Geol.*, 77, 68-91.

Simpson, B.B., 1975, Pleistocene changes in the flora of the high tropical Andes, *Paleobiology, 1:273-294.*

Sinclair, A.R.E., 1977, *The African Buffalo*, Chicago, Univ. of Chicago Press, 355 p.

Simpson, B.B., 1975, Pleistocene change in the flora of the high tropical Andes, *Paleobiology, 1:273-294.*

Smith, J.M.B., 1977, Origins and ecology of the tropicalpine flora of Mt. Wilhelm, New Guinea, *Biol. J. Linn.* Soc., *9:87-131.*

____, 1986, Origins of Australasian tropicalpine and alpine floras. In: B.A. Barlow, ed., *Flora and Fauna of Alpine Australasia. Ages and Origins, CSIRO, Melbourne: 109-128.*

Smith, J.M.B. and Cleef, A.M., 1988, Composition and origins of the world's tropicalpine floras, *Jour. Biogeography*, 15:631-645.

Smith, R.I.L., 1984, Colonization and recovery of cryptogams following recent volcanic activity on Deception Island, South Shetland Islands, *Brit. Antarctic Surv. Bull., 62: 25-51.*

Soane, B.D. and Pidgeon, J.D., 1975, Tillage requirements in relation to soil physical properties, *Soil Science*, 119(5):376-384.

Soil Survey Staff, 1951, *Soil Survey Manual*, Washington, U.S. Govt., Printing Office, 503p.

____, 1975, *Soil Taxonomy*, Agriculture Handbook 436: Washington, U.S.D.A., 754p.

Sømme, L. and Zachariassen, K.E., 1981, Adaptations to low temperature in high altitude insects from Mount Kenya, *Ecol. Ent., 6:199-204.*

Soons, J.M., 1967, Erosion by needle ice in the Southern Alps, New Zealand, In Wright, H. E. Jr., and Osburn, W. H., eds., *Arctic and Alpine Environments*, Indiana U. Press, 217-227.

Spence, J.R., 1989, Plant succession on glacial deposits of Mount Kenya, East Africa, in Mahaney, W.C. (ed.), *Quaternary and Environmental Research on East African Mountains*, Balkema, Rotterdam, 279-290.

Spence, J.R. and Mahaney, W.C., 1988a, Growth and ecology of *Rhizocarpon* section *Rhizocarpon* on Mount Kenya, East Africa, *Arctic and Alpine Research,* 20: 237-242.

Spence, J.R. and Mahaney, W.C., 1988b, *Calliergon stramineum* (Brid.) Kind. new to Africa from Mount Kenya, East Africa, *Lindbergia*, 14:67-68.

Spence, J.R. and Pócs, T., 1989. Distribution patterns in the Afroalpine moss flora of East Africa. In: *Quaternary and Environmental Research on East African Mountains*, ed. W.C. Mahaney, Balkema, Rotterdam, 291-308.

Spink, P.C., 1945, Further Notes on the Kibo inner crater and glaciers of Kilimanjaro and Mount Kenya, *Geogr. Jour.*, 106, 210-6.

____, 1949, The Equatorial Glaciers of East Africa, *Jour. Glaciol.*, 1, 277-83.

Stickney, J.C. and van Liere, E.J., 1953, Acclimatization to low oxygen tension, *Physiological Review*, 33:13-34.

Stork, A., 1963, Plant immigration in front of retreating glaciers with examples from the Kebnekaise area, Northern Sweden, *Geografiska Annaler*, 45, 1-21.

Tarling, D.H., 1983, *Paleomagnetism*, Chapman and Hall, N.Y.

Thompson, B.W., 1966, The mean annual rainfall of Mount Kenya, *Weather*, 21, 48-9.

Thomson, J., 1885, *Through Massai Land*, London.

Troll, C., 1958, Structure, Soils, Solifluction, and Frost Climates of the Earth. U.S. Army Snow Ice and Permafrost Establishment. Trans. 43, 121p.

____, 1960, The relationship between the climates, ecology and plant geography of the southern cold temperate zone and of the tropical high mountains, *Proc. Royal Soc. Lond. Ser B, 152:529-532*.

Troll, C. and Wien, K., 1949, Der Lewisgletcher am Mount Kenya, *Geografiska Annaler*, 31, 257-74.

Valencia, M. J., 1977, Pacific Pleistocene paleoclimatic stratigraphies: a comparative analysis of results, *Quaternary Research*, 8: 339-354.

Valentine, K.W.G. and Dalrymple, J.B., 1976, Quaternary buried paleosols: a critical review, *Quaternary Research*, 6, 209-222.

Van der Hammen, T., Barelds, J., de Jong, H., and de Veer, A.A., 1981, Glacial sequence and environmental history in the Sierra Nevada de Cocuy (Colombia), *Palaeogeography, Palaeoclimatology, Palaeoecology*, 32: 247-340.

van Donk, J., 1976, ^{18}O Record of the Atlantic Ocean for the entire Pleistocene Epoch, *Geol. Soc. Amer. Mem.* 145, 147-163.

van Zinderen Bakker, E.M. and Maley, J., 1979, Late Quaternary pelaeoenvironments of the Sahara Region, in *Palaeoecology of Africa*, 11, Balkema, Rotterdam, p.83-104.

van Zinderen Bakker, E.M. and Coetzee, J.A., 1988, A review of late Quaternary pollen studies in East, Central and Southern Africa, *Review of Paleobotany and Palynology*, 55, 155-174.

Verosub, K.L., 1984, Mono Lake geomagnetic excursion found at Summer Lake, Oregon, *Geology*, 12, 643-6.

____, 1986, Principles and applications of paleomagnetism in the dating of young sediments, *Dating Young Sediments*, A.J. Harford *et al*, ed., CCPT Tech. Secretariat, Bangkok.

Von Höhnel, L., 1894, *Discovery of Lakes Rudolf and Stefanie*, V.1, Reprinted by Frank Cass, London, 1968, 362-80.

Vortisch, W.B., 1982, Clay mineralogical studies of some tills in northern Germany, *Geologica et Palaeontologica*, 15, 67-91.

Vortisch, W.B., Mahaney, W.C., and Fecher, K., 1987, Lithology and weathering in a paleosol sequence on Mount Kenya, *Geologica et Paleontologica*, 21, 245-55.

Walker, T.W. and Syers, J.K., 1976, The fate of phosphorus during pedogenesis, *Geoderma*, 15, 1-19.

Walkley, A. and Black, I.A., 1934, An examination of the Degtjareff method for determining soil organic matter, and a proposed modification of the chromic acid titration method, *Soil Science*, 37, 29-38.

Walter, H., 1960, Einführung in die Phytologie, 3(1), Stardortslehre, Stuttgart, E. Ulmer, 65-9.

Warren, R.G., 1956, NPK residues from fertilizers and farmyard manure in long term experiments at Rothamsted, *Fert Soc. Proc.*, 37:233-242.

Washburn, A. L., 1973, *Periglacial processes and Environment*, Edward Arnold, London, 328 p.

Watson, J.P., 1970, Termites and the soil balance, *Proc. Intern. Erosion Symposium*, Prague, 2, 83-88.

Wehmiller, 1971, Racemization of amino acids in Marine Sediments, *Science*, V. 173, 907-11.

____, 1982, A review of amino acid racemization studies in Quaternary molluscs: stratigraphic and chronologic applications in coastal and interglacial sites, Pacific and Atlantic coast, United States, United Kingdom, Baffin Island and tropical islands, *Quaternary Science Reviews*, 1, 83-120.

Wehmiller, J.F. and Hare, P.E., 1970, Amino acid diagenesis in fossil carbonates, Geological society of America, Abstracts with program, 2, 718.

____, 1971, Racemization of amino acids in marine sediments, *Science, 173:907-911.*

Wehmiller, J.F., Hare, P.E. and Kugala, G.A., 1976, Amino acids in fossil corals: racemization (epimerization) reactions and their implication for diagenetic models and chronological studies, *Geochimica et Cosmochimica Acta*, 40, 736-76.

White, S.E., 1962, Late Pleistocene glacial sequence for the west side of Iztaccihuatl, Mexico, *Geological Soc. Amer. Bull.*, 73: 935-938.

____, 1987, Quaternary glacial stratigraphy and chronology of Mexico, *Quaternary Science Reviews*, 5(1-4): 201-205.

White, S. and Valastro, S., Jr, 1984, Pleistocene glaciation of volcano Ajusco, central Mexico, and comparison with standard Mexican glacial sequence, *Quaternary Research*, 21:21-35.

Widden, P., 1977, *Truelove Lowland, Devon Island, Canada. A High Arctic Ecosystem*, University of Alberta Press, Calgary: 505-530.

Williams, P.J., 1957, Some investigations into solifluction features in Norway, Geog. Jour: 123:42-58.

Whittig, L.D., 1965, X-ray diffraction techniques for numerical identification and mineralogical composition, in Black, C.A., ed., *Methods of Soil Analysis*: Madison, Wisc., Amer. Soc. Agronomy, 671-96.

Wintle, A.G. and Huntley, D.J., 1982, Thermoluminescence dating of sediments, *Quaternary Science Reviews*, 1, 31-53.

Wright, H. E. Jr., 1984, Late glacial and late Holocene moraines in the Cerros Cuchpanga, Central Peru, *Quaternary Research*, 21:275-285.

Zeuner, F.E., 1948, Climate and early man in Kenya, *Man*, V.48, 3-6.

____, 1949, Frost soils on Mount Kenya and the relation of frost soils to aeolian deposits, *Journal of Soil Science*, 1, 20-30.

Zielinski, G.A., and Davis, P.T., Late Pleistocene age of the Type Temple Lake Moraine, Wind River Range, Wyoming, U.S.A., *Géographie physique et Quaternaire*, 41, 397-401.

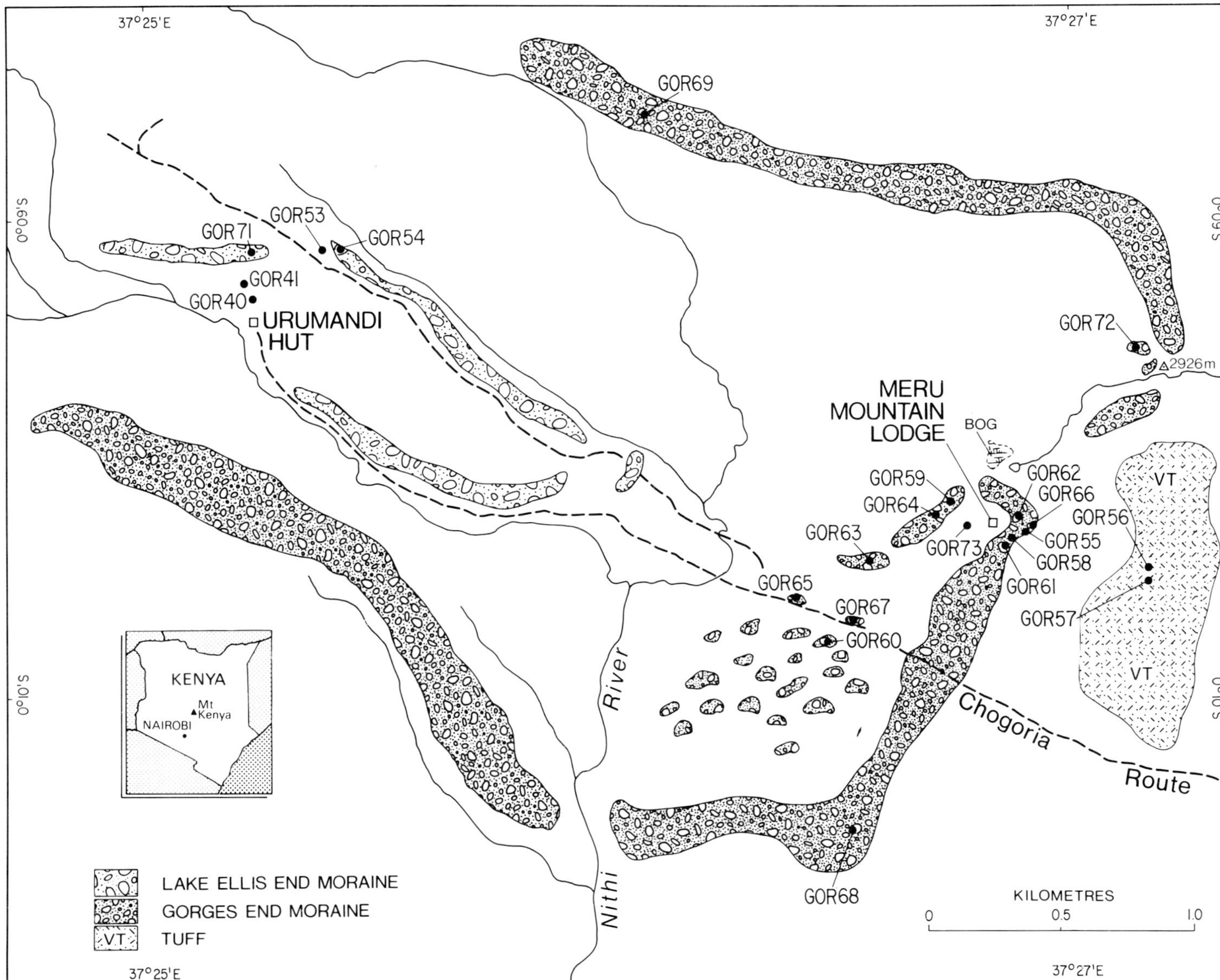

Lake Ellis and Gorges moraines on the eastern flanks of Mount Kenya

PLATE 2

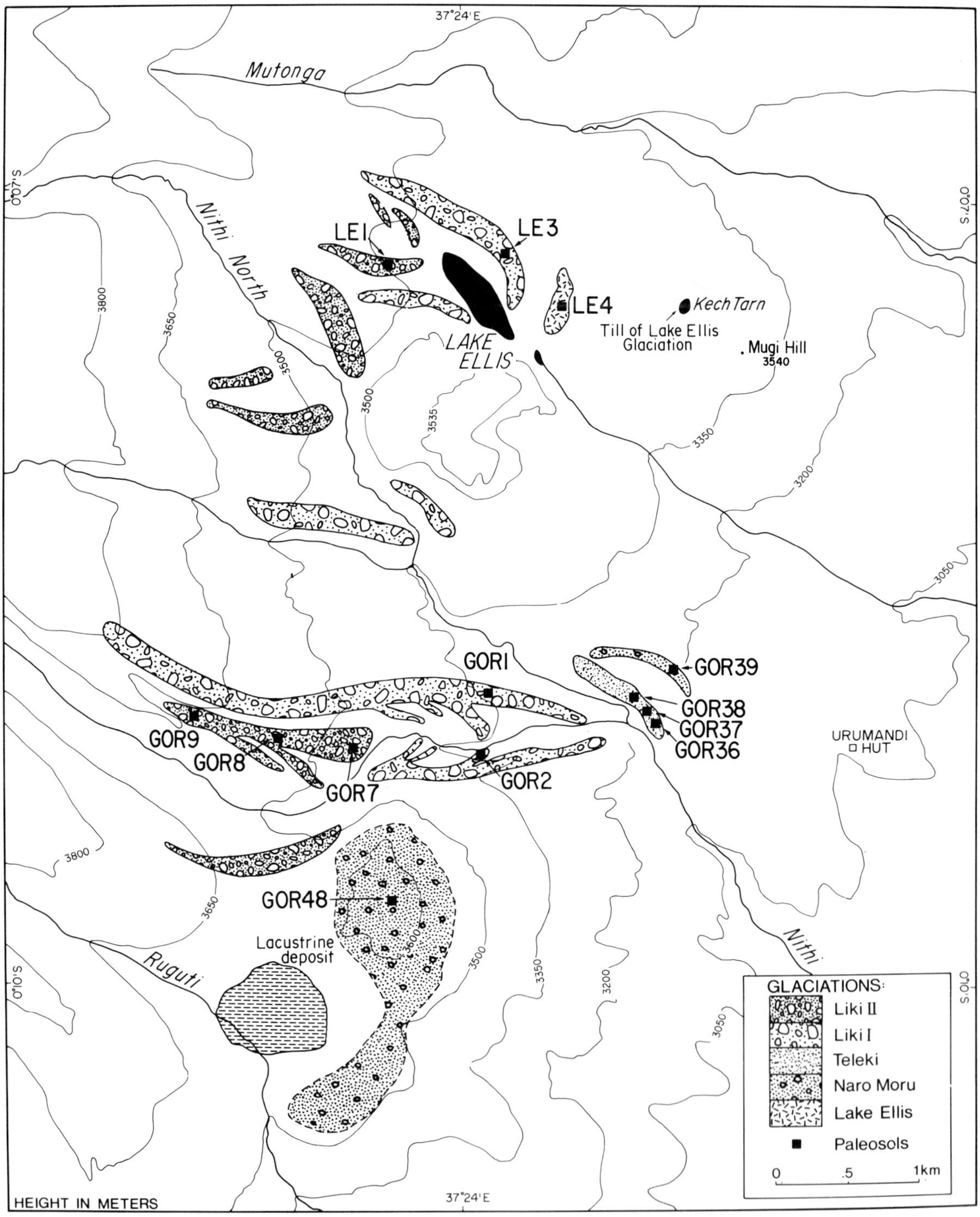

Teleki, Naro Moru and Liki moraines in Gorges Valley, Hinde Valley, and Lake Ellis areas, eastern flanks of Mount Kenya

PLATE 3

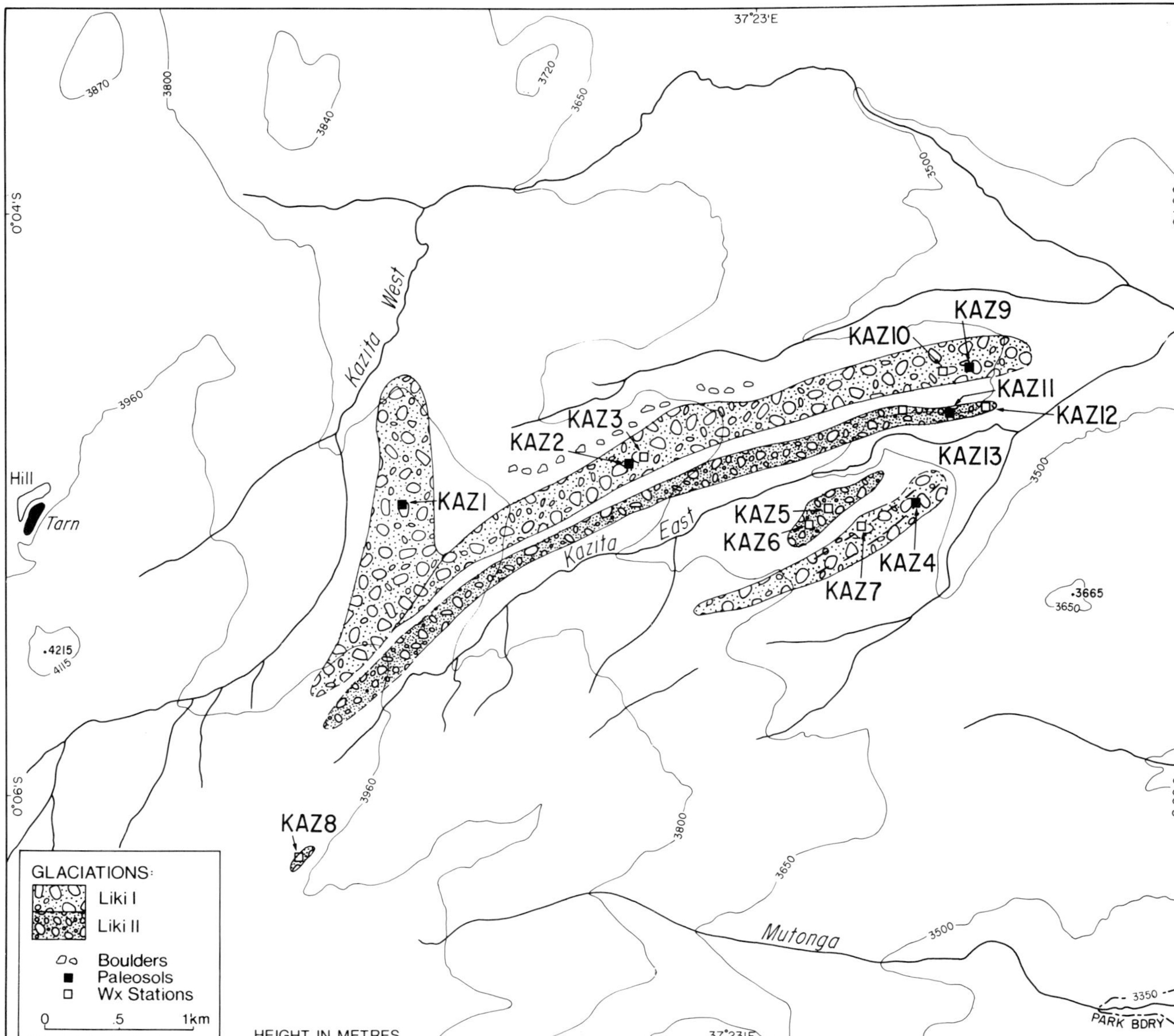

Liki Moraines in the Kazita East and West drainages of Mount Kenya

PLATE 4

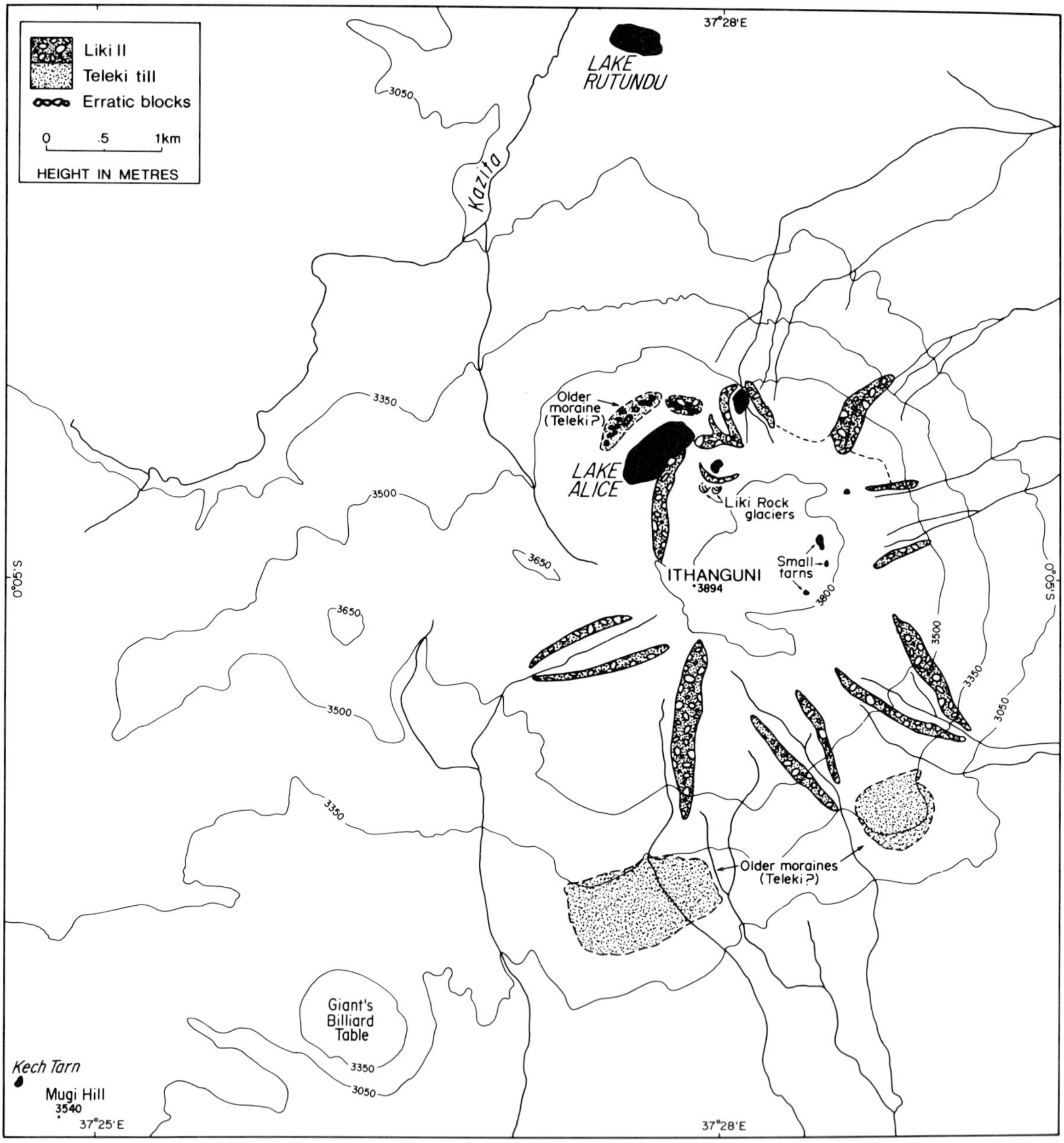

Liki moraines on Ithanguni, northeastern flank of Mount Kenya

PLATE 5

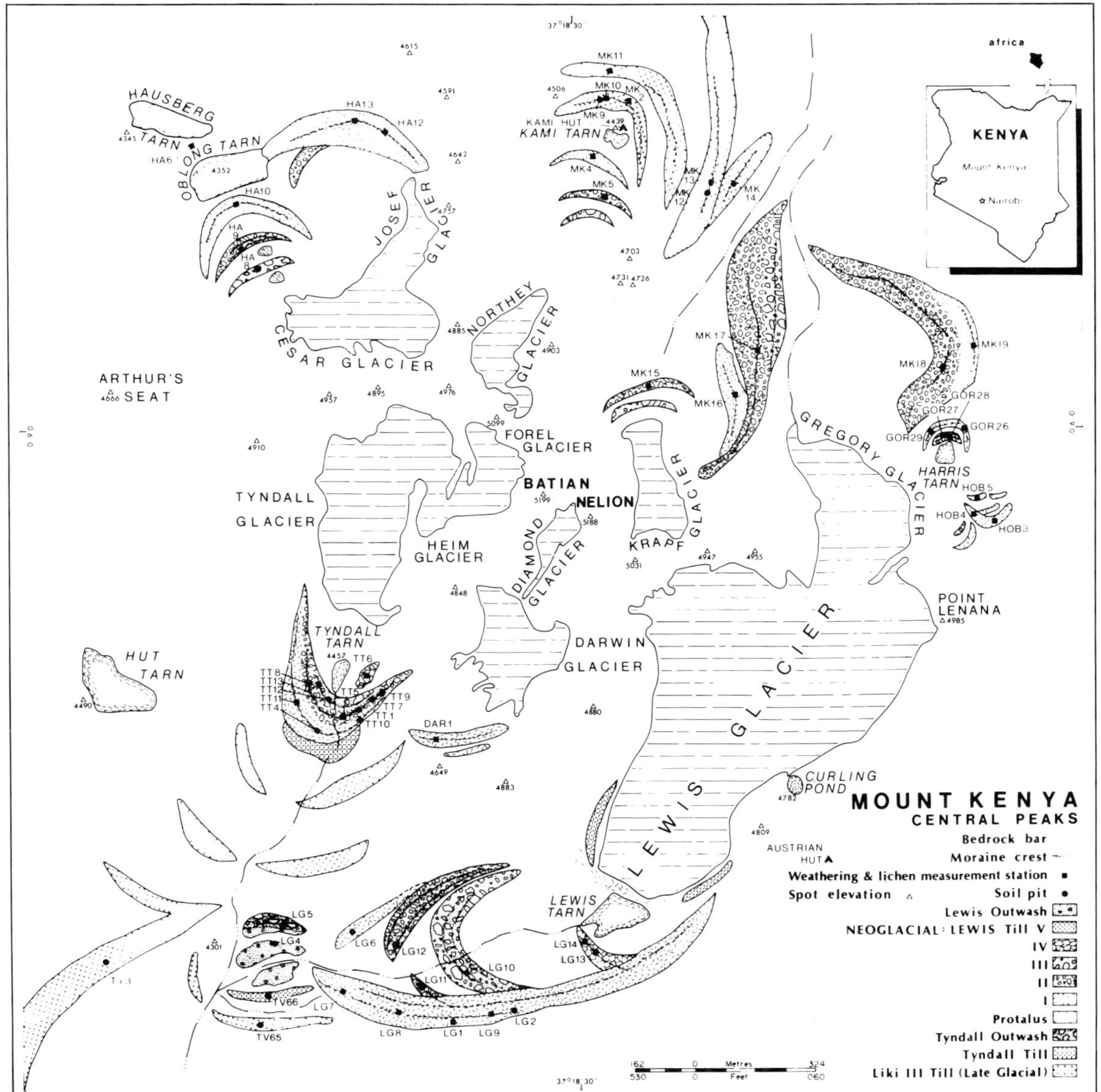

Neoglacial and Liki III moraines on Mount Kenya

PLATE 6

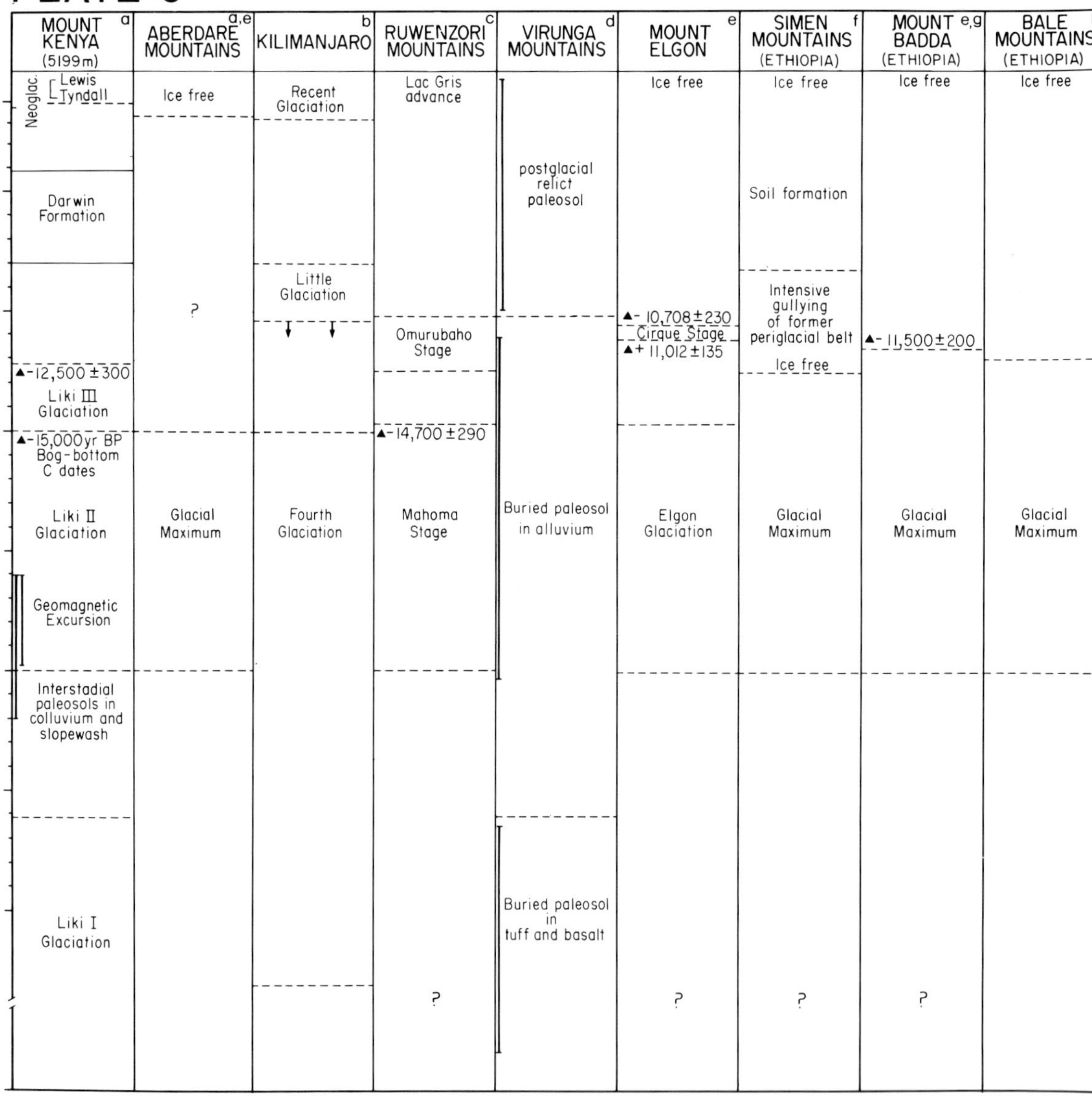

a. Mahaney, this volume

b. Osmaston, 1989a; Downie and Wilkinson, 1972

c. Osmaston, 1989b; Livingston, 1962

d. Mahaney, Barendregt and Hancock, 1989

e. Hamilton, 1982

f. Hurni, 1989

g. Potter, 1976; Messerli et al, 1977

h. Messerli et al, 1977

Correlation diagram of Quaternary glacial deposits of the Tropical Mountains (continued on facing page and overleaf)

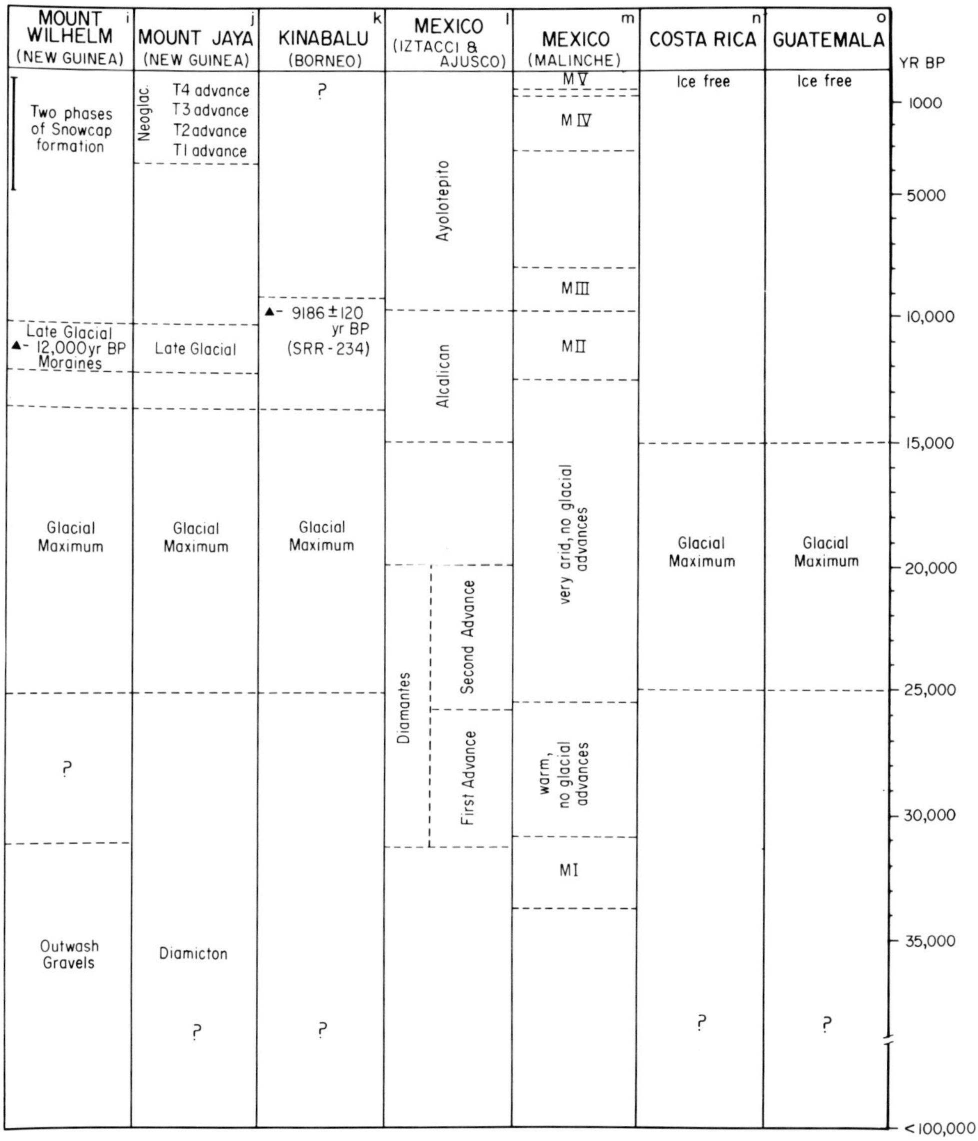

i. Löffler, 1972

j. Peterson and Hope, 1972; Löffler, 1982

k. Koopmans and Stauffer, 1968

l. White, 1962; White and Valastro, 1982

m. Heine, 1973, 1978, 1984, 1988

n. Hastenrath, 1973

o. Hastenrath, pers. communication, 1988

PLATE 6

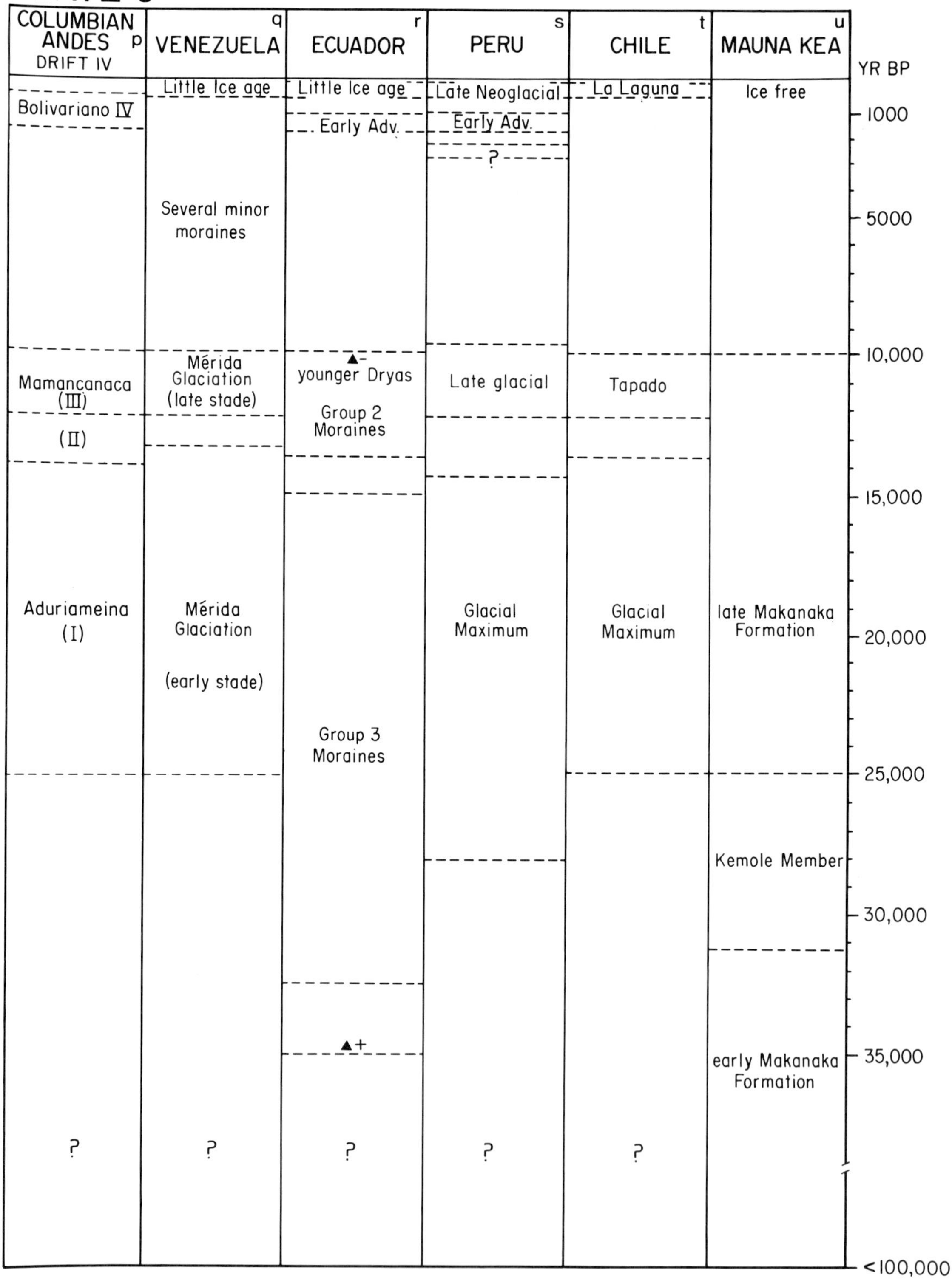

p. Gonzalez et al, 1965; van der Hammen, et al, 1981

q. Schubert, 1972; Pérez, 1989; Rull et al, 1987

r. Hastenrath, 1981; Clapperton, 1986

s. Clapperton, 1972; Mercer and Palacios, 1977; Birkeland, 1989

t. Caviedes and Paskoff, 1975

u. Porter et al, 1977

▲+ MAXIMUM ^{14}C DATE

▲− MINIMUM ^{14}C DATE

PLATE 7

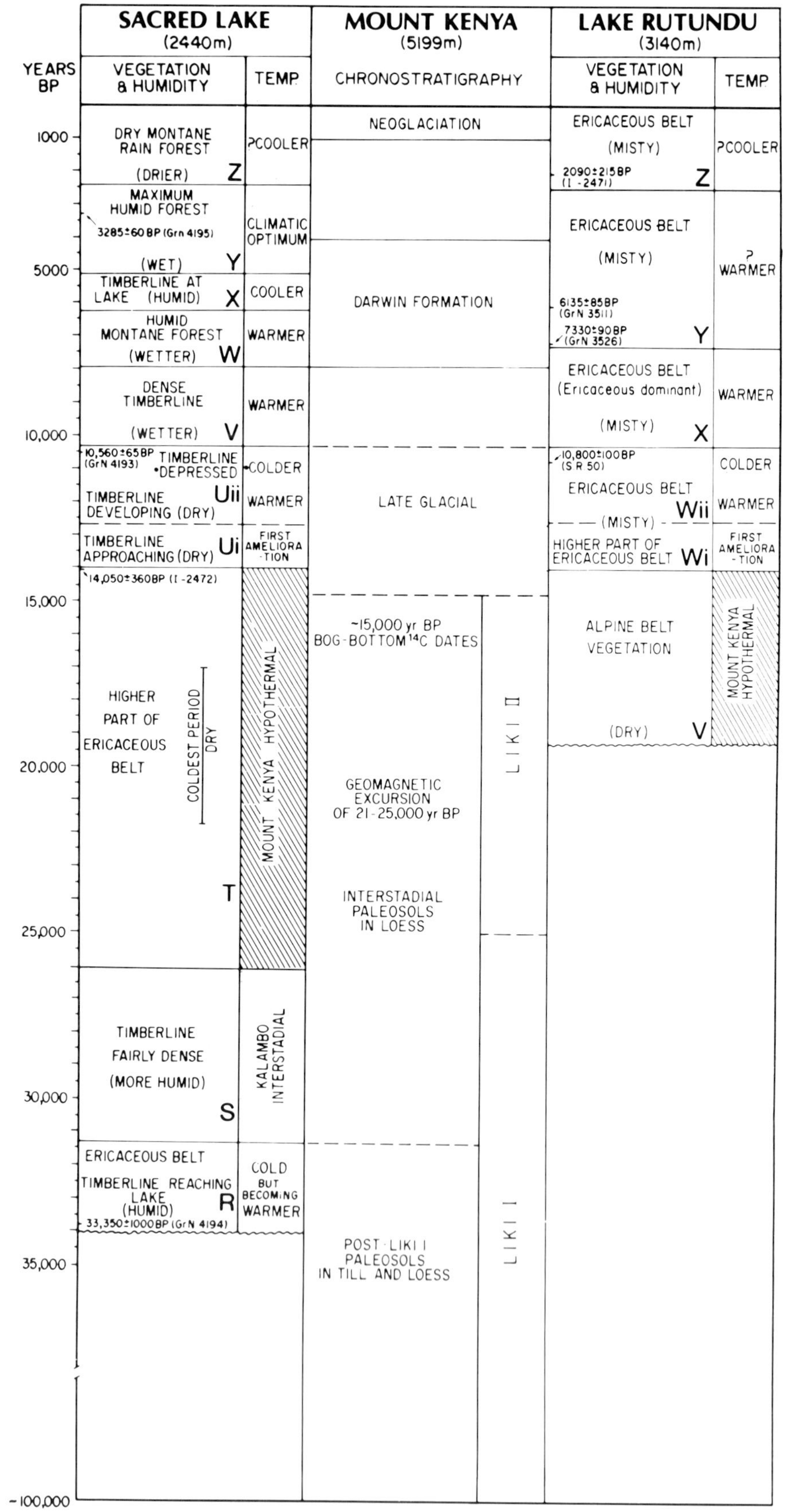

SACRED LAKE (2440m)
MOUNT KENYA (5199m)
LAKE RUTUNDU (3140m)
YEARS BP
VEGETATION & HUMIDITY
TEMP.
CHRONOSTRATIGRAPHY
VEGETATION & HUMIDITY
TEMP.
1000
5000
10,000
15,000
20,000
25,000
30,000
35,000
~100,000
DRY MONTANE RAIN FOREST (DRIER) Z
?COOLER
MAXIMUM HUMID FOREST
3285±60 BP (Grn 4195)
(WET) Y
CLIMATIC OPTIMUM
TIMBERLINE AT LAKE (HUMID) X
COOLER
HUMID MONTANE FOREST (WETTER) W
WARMER
DENSE TIMBERLINE (WETTER) V
WARMER
10,560±65 BP (GrN 4193)
TIMBERLINE DEPRESSED
COLDER
TIMBERLINE DEVELOPING (DRY) Uii
WARMER
TIMBERLINE APPROACHING (DRY) Ui
FIRST AMELIORA-TION
14,050±360 BP (I-2472)
HIGHER PART OF ERICACEOUS BELT
COLDEST PERIOD DRY
T
MOUNT KENYA HYPOTHERMAL
TIMBERLINE FAIRLY DENSE (MORE HUMID) S
KALAMBO INTERSTADIAL
ERICACEOUS BELT
TIMBERLINE REACHING LAKE (HUMID) R
33,350±1000 BP (GrN 4194)
COLD BUT BECOMING WARMER
NEOGLACIATION
DARWIN FORMATION
LATE GLACIAL
~15,000 yr BP BOG-BOTTOM ¹⁴C DATES
GEOMAGNETIC EXCURSION OF 21-25,000 yr BP
INTERSTADIAL PALEOSOLS IN LOESS
POST-LIKI I PALEOSOLS IN TILL AND LOESS
LIKI II
LIKI I
ERICACEOUS BELT (MISTY)
?COOLER
2090±215 BP (I-2471)
Z
ERICACEOUS BELT (MISTY)
? WARMER
6135±85 BP (GrN 3511)
7330±90 BP (GrN 3526)
Y
ERICACEOUS BELT (Ericaceous dominant) (MISTY) X
WARMER
10,800±100 BP (S R 50)
COLDER
ERICACEOUS BELT (MISTY) Wii
WARMER
HIGHER PART OF ERICACEOUS BELT Wi
FIRST AMELIORA-TION
ALPINE BELT VEGETATION
(DRY) V
MOUNT KENYA HYPOTHERMAL

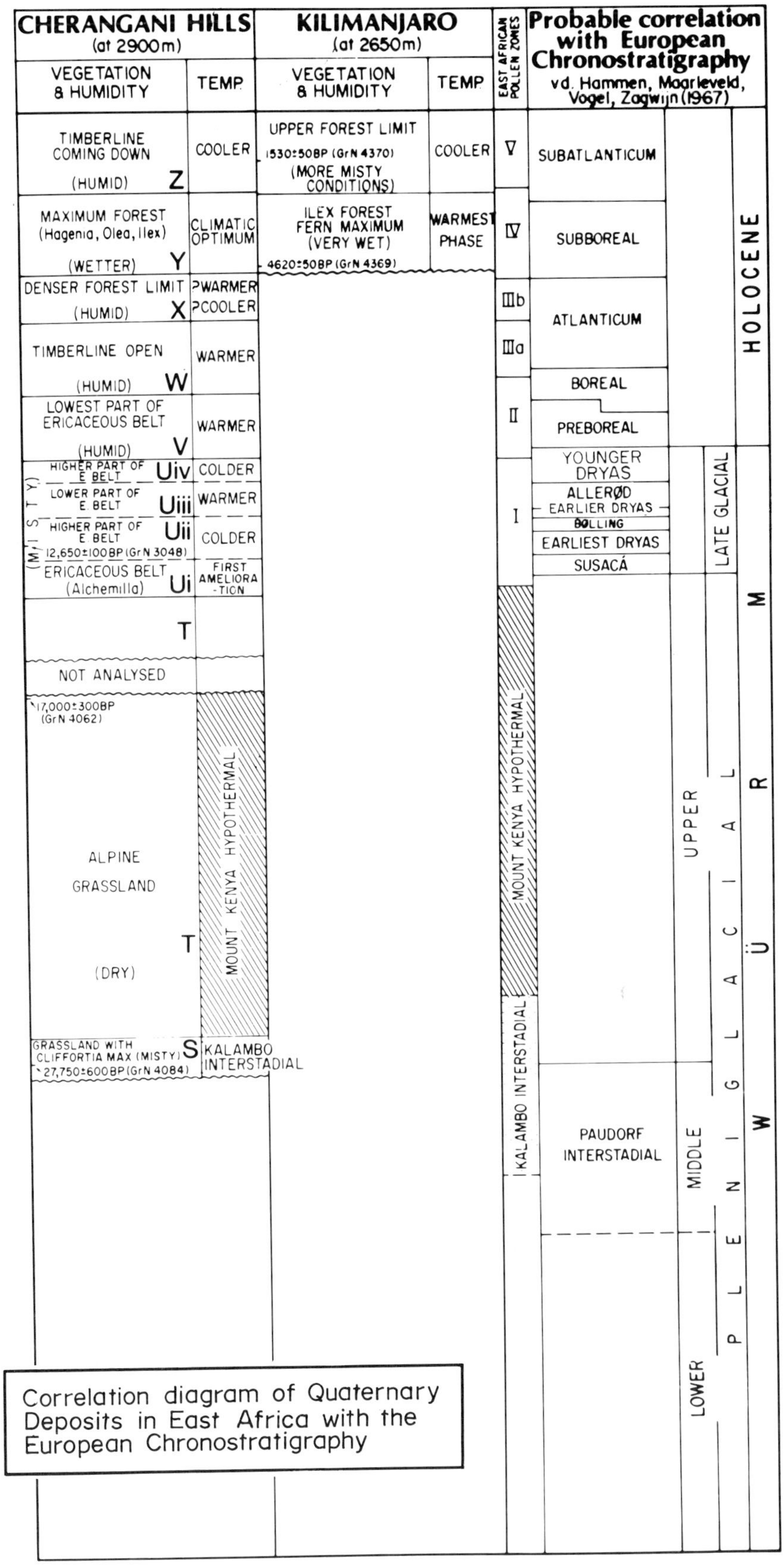

Correlation diagram of Quaternary Deposits in East Africa with the European Chronostratigraphy